Birefringent Thin Films and Polarizing Elements

Birefringent Thin Films and Polarizing Elements

Ian J Hodgkinson
Qi hong Wu

Department of Physics
University of Otago
New Zealand

World Scientific
Singapore • New Jersey • London • Hong Kong

Published by

World Scientific Publishing Co. Pte. Ltd.

5 Toh Tuck Link, Singapore 596224

USA office: 27 Warren Street, Suite 401-402, Hackensack, NJ 07601

UK office: 57 Shelton Street, Covent Garden, London WC2H 9HE

British Library Cataloguing-in-Publication Data
A catalogue record for this book is available from the British Library.

BIREFRINGENT THIN FILMS AND POLARIZING ELEMENTS

ISBN-13 978-981-02-2906-1
ISBN-10 981-02-2906-2

To Jenny and Qiong zhu Hu

Contents

List of Tables   xv

List of Figures   xvii

Glossary   xxiii

1 Introduction   1
   1.1   Structural Classification of Crystals   2
   1.2   Optical Classification of Crystals   2
   1.3   Structure of Birefringent Films   4
   1.4   Optical Classification of Birefringent Films   5
   1.5   Layout of the Book   7

I Propagation in Biaxial Media   9

2 Propagation Equations   11
   2.1   Maxwell's Equations   12
   2.2   Propagation in Free Space. Mathematical Methods   12
         2.2.1   SI units   16
   2.3   Propagation in Isotropic Media   17
   2.4   Propagation in Anisotropic Media   18
   2.5   Energy Flow   21
   2.6   Notation for Biaxial Media   22
         2.6.1   Material Axes   22
         2.6.2   Propagation Axes   24
         2.6.3   Rotations   24
         2.6.4   Computations   26
   2.7   Propagation in a Common Direction in a Biaxial Medium   26
         2.7.1   Maxwell's Equations   26
         2.7.2   Fresnel's Equation   27

      2.7.3   Eigenequations for Normalized Fields   28

3 Basis Vectors   31
  3.1   Partially Coherent States   32
      3.1.1   Coherence   32
      3.1.2   Stokes Parameters   33
      3.1.3   Stokes Vectors   34
      3.1.4   Degree of Polarization   35
      3.1.5   Unpolarized Light   35
      3.1.6   Partially Polarized Light   35
      3.1.7   Polarized Light   36
      3.1.8   Basis Vectors   36
  3.2   Coherent States   38
      3.2.1   Jones Vectors   38
      3.2.2   Elliptical Polarization   38
      3.2.3   Circular Polarization   39
      3.2.4   Linear Polarization   40
      3.2.5   Basis Vectors   40
      3.2.6   Photons   41
      3.2.7   Ellipsometric Parameters   41
  3.3   Propagation in Layered Biaxial Media   43
      3.3.1   Fresnel's Quartic Equation   43
      3.3.2   Propagation in the Deposition Plane   44
      3.3.3   Uniaxial Media   45
      3.3.4   Isotropic Media   45
      3.3.5   Basis Travelling Wave Fields   46
      3.3.6   Power   48

4 Transfer Matrices   49
  4.1   Mueller Calculus   50
      4.1.1   Rotated Elements   50
      4.1.2   Elements in Series   51
      4.1.3   Mueller Calculus Computations   52
  4.2   Jones Calculus   52
      4.2.1   Linear Polarizer   54
      4.2.2   Retardation Plate   54
      4.2.3   Quarter-Wave Plate   54
      4.2.4   Rotated Elements   55
      4.2.5   Elements in Series   55
      4.2.6   Periodic Arrangements   56

	4.2.7	Jones Calculus Computations	56
4.3		Relationship of Mueller and Jones Calculus	56
4.4		Berreman Calculus	57
	4.4.1	Field Matrix $\hat{F}$	57
	4.4.2	Field Coefficients $\vec{a}$	58
	4.4.3	Total Field $\vec{m}$	59
	4.4.4	Phase Matrix $\hat{A}_d$	59
	4.4.5	Characteristic Matrix $\hat{M}$	60
	4.4.6	System Matrix $\hat{A}$	62
	4.4.7	Properties of $\hat{M}$	62
	4.4.8	Computation of Film Parameters from $\hat{M}$	63
4.5		Abelès and Heavens Calculus	65
	4.5.1	Isotropic Layer	65
	4.5.2	Deposition Plane	67
	4.5.3	Berreman Calculus Computations	69
4.6		Relationship of Jones and Berreman Calculus	74
	4.6.1	Jones Matrix with Interference	74
	4.6.2	Jones Matrix with Reflections but without Interference .	75

5   Reflection and Transmission   **77**
5.1		General Case – All Media Biaxial	78
	5.1.1	Crystal-Crystal Interface	81
5.2		Sorting Columns of $\hat{F}$	81
5.3		Isotropic Cover and Substrate	85
	5.3.1	Amplitude Reflection and Transmission Coefficients . . .	87
	5.3.2	Irradiance Reflectance Coefficients	88
5.4		All Media Isotropic	89
	5.4.1	Phase Changes on Reflection and Transmission	90
5.5		Computations Using the BTF Toolbox	90
	5.5.1	General Birefringent Coating	90
	5.5.2	PS Coatings	91

6   Guided Waves   **93**
6.1		Modal Condition	94
	6.1.1	General Case	94
	6.1.2	Isotropic Cover and Substrate	95
	6.1.3	Uncoupled Modes	97
	6.1.4	Poles of R	98
	6.1.5	Examples	98
6.2		Modal Cutoffs	100

6.3 Modal Contours 100
6.4 Modal Field Structure 104
6.5 Modal Polarization 106
6.6 Modal Overlap 108
6.7 Modal Order 111
6.8 Power Flow 111
6.9 Prism Couplers 112

II Characterization of Anisotropic Films   115

7 Deposition of Microstructures   117

7.1 Vacuum Deposition 118
      7.1.1   Apparatus 118
      7.1.2   Deposition Parameters 119
7.2 Columnar Structures and Effective Media 119
      7.2.1   Uniaxial Media 119
      7.2.2   Biaxial Media 119
      7.2.3   Effective Anisotropic Media 123
      7.2.4   Zig-Zag and Wavy Anisotropic Media 123
      7.2.5   Helical Microstructures 125
7.3 Computer Modelling of Deposition 127
      7.3.1   Serial Deposition of Hard Spheres 127
      7.3.2   Visual Analysis of Simulations 127
      7.3.3   Radial Distribution Function 128
      7.3.4   Two-Dimensional Angular Distribution 131
      7.3.5   Column Angle 131
      7.3.6   Birefringence 131
      7.3.7   Conclusions from Simulations of Deposition 131

8 Form Birefringence   135

8.1 Perpendicular Incidence Ellipsometry 136
      8.1.1   Computation of Ellipsometric Parameters 136
      8.1.2   Characteristic Ellipsometric Curves 138
      8.1.3   Experimental Values 144
8.2 Measurement of Principal Refractive Indices 144
      8.2.1   In Situ Measurements 145
      8.2.2   Use of Narrowband Filters 146
      8.2.3   Photometric Method 147
      8.2.4   Waveguide Method 147

8.3 Modelling Form Birefringence 149
   8.3.1 Bragg-Pippard Equations 149
   8.3.2 Inversion of the Bragg-Pippard Equations 151

9 Effective Media   **153**
9.1 Herpin Indices for Isotropic Layers 154
9.2 Biaxial Layers with a Common Deposition Plane 155
   9.2.1 A and B Normal Columnar 157
   9.2.2 A and B Parallel, Tilted Columnar 159
   9.2.3 A and B Coplanar, Tilted Columnar with $\psi_A = -\psi_B$ . . 162
9.3 Biaxial Layers Deposited in Different Planes 164

10 Anisotropic Scatter   **167**
10.1 Scatter into the Air 168
10.2 Scatter From Stress-Related Cracks 169
10.3 Scatter Patterns Formed on the Film 174
10.4 Scatter into the Substrate 175
10.5 In Situ Measurement of Scatter 175
   10.5.1 Dependence of Haze on Δ 177
   10.5.2 Haze from Herring-Bone Stacks 177
10.6 Simple Theory of Scatter 180

11 Fluid Transport   **183**
11.1 Fluid Patches 184
   11.1.1 Recording Fluid Patches 185
   11.1.2 MDM Narrowband Filters 185
11.2 Scatter from Fluid Patches 188
   11.2.1 Scatter Anisotropy 188
   11.2.2 Theory of Scatter 191
   11.2.3 General AR Coating 194
   11.2.4 High Reflectance Coating 194
   11.2.5 Narrowband Interference Filter 194
11.3 Influence on Birefringence 197
   11.3.1 Change of Birefringence in Fluid Patches 197
   11.3.2 Principal Refractive Indices 201
   11.3.3 Cooling and Venting 202

12 Metal Films   **203**
12.1 Growth and Post-Deposition Sputter Etching 204
12.2 Direct Recording of Optical Anisotropies 206
   12.2.1 Silver and Gold 209

12.2.2 Aluminium 209
12.2.3 Aging 216
12.2.4 Argon Ion Sputter Etching 217
12.3 Computer Modelling of Anisotropy in Metals 217
12.3.1 Bulk Metals 221
12.3.2 Depolarization Factors 222
12.3.3 Isotropic Resonance 223
12.3.4 Anisotropic Resonance 225
12.4 Modelling Deposition and Etching 227
12.4.1 Simulated Deposition of Gold 228
12.4.2 Simulated Deposition of Silver 231
12.4.3 Simulated Deposition of Aluminium 231
12.4.4 Simulated Deposition / Etch Paths 231
12.5 Summary 236

III Applications of Birefringent Media 237

13 Linear polarizers   **239**
13.1 Real Polarizers 240
13.2 Dichroic Polarizers 241
13.3 Tilted Plate and Thin Film Polarizers 243
13.3.1 Plate Polarizers 243
13.3.2 Coated-Plate Polarizers 244
13.3.3 Embedded Thin Film Polarizers 246
13.3.4 Birefringent Fabry-Perot Polarizing Filter 247
13.4 Crystalline Prism Polarizers 251
13.4.1 Glan-Foucault Prism 252
13.4.2 Wollaston Prism 253
13.4.3 Rochon Prism 253

14 Phase Retarders   **255**
14.1 Crystalline Wave Plates 256
14.1.1 Quartz and Magnesium Fluoride 256
14.1.2 Multiple-Order Wave Plates 257
14.1.3 Zero-Order Wave Plate 259
14.1.4 Achromatic Wave Plates 259
14.1.5 Wide-Field Elements 264
14.1.6 Variable Phase Compensators 265
14.2 Birefringent Thin Film Analogues 267

14.2.1 Thin Film Wave Plates 267
14.2.2 Thin Film Babinet Compensator 270
14.2.3 Thin Film Soleil-Babinet Compensator 270
14.2.4 Thin Film Berek Compensator 271

15 Birefringent Filters   273
15.1 Polarization State Filters 274
        15.1.1 Linear Polarizer 274
        15.1.2 Circular Polarizer 274
        15.1.3 Rotator 274
        15.1.4 Depolarizer 275
15.2 Wavelength Filters 276
        15.2.1 Lyot-Ohman Filter 276
        15.2.2 Solc Filters 279
        15.2.3 Filters for Tuning Dye Lasers 282

16 Birefringent Coatings   289
16.1 Isotropic Coatings 290
16.2 General Birefringent Coating 290
16.3 PS Coatings 294
16.4 Design Considerations for PS Coatings 294
        16.4.1 Making an Anisotropic Version 296
        16.4.2 Replacing an Intermediate Index 299
        16.4.3 Identical Response Profiles Separated in Wavelength . . 299
        16.4.4 Spoiling the s-response 299
16.5 Normal and Hybrid Monitoring 300
16.6 PS Sampler 302
        16.6.1 Anisotropic Antireflection Coating 302
        16.6.2 Anisotropic Reflector 304
        16.6.3 Anisotropic-Phase Reflector 306
        16.6.4 Achromatic Antireflection Coating 308
        16.6.5 Achromatic Fifty Percent Reflector 310
        16.6.6 Single-Cavity Narrowband Filter 312
        16.6.7 Multi-Cavity Narrowband Filter 314
        16.6.8 Edge Filter 316
        16.6.9 Common-Index Thin Film Polarizer 318
        16.6.10 Multi-Cavity Linear Polarizer 320

A Birefringent Thin Films Toolbox   **323**
   A.1 Quick Reference 324
   A.2 Commands and Functions 327

Notes and References   **365**

Index   **371**

List of Tables

1.1   Classification of crystals   3

2.1   Linear relationships for optical media 13
2.2   Constants used in electromagnetism 17
2.3   Propagation surfaces for anisotropic optical materials 21
2.4   Maxwell's equations for plane, harmonic waves 27
2.5   Eigenequations for a common direction in a biaxial medium . . 29
2.6   Eigenequation for determining fields in a biaxial medium 30

3.1   Polarization states 42

4.1   Mueller and Jones matrices 53
4.2   Heavens and Abelès matrices for isotropic layers 67
4.3   Heavens and Abelès matrices for the deposition plane 70

6.1   Planar waveguides 94
6.2   Modal conditions on the elements of the system matrix 97
6.3   Bound modes of planar waveguides 99

8.1   Materials for birefringent coatings 144
8.2   Refractive indices of zirconium oxide 148
8.3   Refractive indices of titanium oxide layers 149

12.1 Optical constants of aluminium, gold and silver 222

14.1 Refractive indices of magnesium fluoride and quartz 258

16.1 Properties of PS coatings 294

List of Figures

1.1   Crystallographic axes and unit cell of a crystal   2
1.2   Columnar microstructure in a titanium oxide film   5
1.3   Model of a birefringent thin film   6

2.1   Fields of electromagnetic waves in vacuum   16
2.2   Fields of electromagnetic waves in an isotropic medium   18
2.3   Fields of electromagnetic wave in an anisotropic medium . . .   19
2.4   Material axes for a biaxial optical medium   23
2.5   Propagation axes for a biaxial optical medium   24

3.1   Stokes vector, Jones vector and Berreman vector   32
3.2   Ellipsometer for measuring state of polarization   33
3.3   Representation of unpolarized light   37
3.4   Elliptical polarization states   38
3.5   Circular polarization states   39
3.6   Linear polarization   40

4.1   Transfer matrices   50
4.2   Mueller matrices for a series of elements   51
4.3   Transformation property of the phase matrix $\hat{A}_d$ 60
4.4   Transformation performed by the characteristic matrix $\hat{M}$ . . .   61
4.5   Stack of biaxial layers   61
4.6   Transformation performed by the system matrix $\hat{A}$ 62
4.7   Typical PS coating   72
4.8   Transmission through a wave plate, with and without interference 74

5.1   Labelling scheme used for the amplitudes of the four basis vectors 78
5.2   Plots of $\alpha = n\cos\theta$ versus $\beta = n\sin\theta$ for a biaxial medium . .   82
5.3 R–T coefficients plotted as functions of β   84
5.4   Real and imaginary parts of α near a cusp   86

6.1   Evanescent fields in the bounding media of a biaxial waveguide   94

6.2   Log plot of modal test expression $|A_{11}A_{33} - A_{13}A_{31}|$ 99
6.3   Upper modal cutoff β   101
6.4   Modal contours for isotropic and anisotropic waveguides   102
6.5   Curves marking the onset of evanescence   103
6.6   Snapshot and excursions of fields in an isotropic waveguide . .   105
6.7   Snapshot and excursions of fields in an anisotropic waveguide .   106
6.8   Snapshot and excursions of fields for a hybrid mode   107
6.9   Polarization of the modes of the anisotropic waveguide   108
6.10   Polarization of modes in a region of strong coupling   109
6.11   Mechanism for modal overlap in the deposition plane   110
6.12   Input and output prism couplers   112
6.13   Prism acting as simultaneous input/output coupler   113
6.14   Geometry for maximum coupling   113

7.1   Vacuum chamber for depositing microstructures   118
7.2   Scanning electron micrograph of a cerium oxide film   120
7.3   Bunching of columns   121
7.4   Retardance of cerium oxide films   122
7.5   Platelets in a zirconium oxide bi-layer   122
7.6   Interface formed by two layers of zirconium oxide   123
7.7   Zig-zag columns in a magnesium fluoride film   124
7.8   Helical microstructure in a zirconium oxide film   125
7.9   Individual helical columns in magnesium fluoride   126
7.10   Periodic boundary conditions   127
7.11   Cross-section of sequential deposition at $+50°$ and $-50°$ 128
7.12   Small clusters of spherical particles   129
7.13   Radial distribution function $g(r)$ for simulated deposition at $0°$ 130
7.14   Two-dimensional angular distributions   132
7.15   Column angle versus deposition angle   133

8.1   Perpendicular incidence ellipsometry   137
8.2   Perpendicular incidence transmittance profiles T_p and T_s . . .   139
8.3   Calculated ellipsometric profiles for anisotropic dielectric film .   140
8.4   Experimental ellipsometric profiles for an anisotropic film . . .   141
8.5 Δ v. d profiles calculated for absorbing anisotropic films   . . .   142
8.6 Δ v. d profiles calculated for inhomogeneous anisotropic films   143
8.7   Transmittance spectra of birefringent MDM filter   146
8.8   Realization of principal indices from transmission peaks   147
8.9   Transmittance curves for a titanium oxide film   148
8.10   Polarization of an ellipsoidal crystallite   150

9.1   Effective indices and column angle for isotropic layers 156
9.2   Effective indices and column angle for normal columnar layers   158
9.3   Effective indices and column angle, parallel tilted columns   . . 160
9.4   Effective normal incidence refractive anisotropy 161
9.5   Phase retardation recorded for periodic stacks 162
9.6   Residual anisotropy in herring bone layers 163
9.7   Effective indices and dielectric axes, general case 165

10.1   Anisotropic scatter photographed on a plane white screen . . . 168
10.2   Scatter distributions recorded for a titanium oxide film 170
10.3   Apparatus for scanning scatter distributions 171
10.4   Direct acquisition of projected scatter distributions 172
10.5   Anisotropic scatter in front of a titania film 173
10.6   Stress-related cracks in a tilted columnar film 174
10.7   Bright patterns superposed on a titania film 176
10.8   Method used to measure light scattered into the substrate . . . 177
10.9   Anisotropic scatter flux trapped in substrates 178
10.10   Apparatus for *in situ* measurements of anisotropic scatter . . . 178
10.11   Anisotropic scatter recorded during the deposition of zirconia . 179
10.12   Anisotropic scatter from a zirconium oxide herring-bone stack   180
10.13   Cone of maximum light scatter 181
10.14   Elliptical section of the zero-order interference cone 182

11.1   Circular fluid patches in an interference filter 184
11.2   Displacement of peak transmittance by a circular fluid patch . 186
11.3   Elliptical moisture patches in silicon oxide deposited obliquely   187
11.4   Apparatus for recording scatter from fluid patches 188
11.5   Pattern of light scattered by circular fluid patches 189
11.6   Moisture patches and scatter from magnesium fluoride 190
11.7   Scatter anisotropy as moisture patches grow and merge 191
11.8   Random array of moisture patches, identical aperture functions 192
11.9   Thin slice of material in an optical coating 193
11.10   Sensitivity factors for scatter from thin slices 195
11.11   Calculated reflection scatter parameter for multilayer reflector   196
11.12   Calculated values of the transmission scatter parameter 198
11.13   Apparatus for measuring scatter from interference filters . . . 198
11.14   Measured values of transmittance and scatter 199
11.15   Fluid patches and fringes of equal chromatic order 200
11.16 A_n and A_f for Ag-MgF$_2$-Ag filters 201
11.17   Retardance of thin film wave plates, cooling and venting . . . 202

12.1   Anisotropy in thin metal films, shape of globular particles . . .   205
12.2   Changes in morphology during growth and etching   207
12.3   Overview of changes in particle shape and density of gold films   208
12.4   Apparatus for monitoring optical anisotropy in metal films   . .   210
12.5 R, T, and A recorded during the deposition of silver   211
12.6   Anisotropy profile $T_s - T_p$ v. mass thickness measured for gold   212
12.7   Intrinsic anisotropy of aluminium deposited without oxygen . .   213
12.8   Residual anisotropy of aluminium caused by oxygen   214
12.9   Dependence of reflection anisotropy on deposition parameters .   214
12.10 Dependence of reflection anisotropy on angle of incidence . . .   215
12.11 Arrangement of sources for coating a mirror with aluminium .   216
12.12 Reflection anisotropy depends mainly on the last film deposited   217
12.13 T measured during the deposition and aging of a silver film . .   218
12.14 Knee-shaped features in the transmittance of a gold film . . .   218
12.15 Dependence of anisotropy of gold films on initial mass thickness   219
12.16 Anisotropy recorded during deposition and etching of gold . .   220
12.17 Aggregated media, anisotropy defined by crystallites and voids   223
12.18 Refractive index resonance in crystallite-defined material . . .   224
12.19 Dependence of resonance on wavelength and polarization . . .   226
12.20 Profiles of T simulated by the Bragg-Pippard model   229
12.21 Ellipsometric profiles for gold   230
12.22 Simulations of R, T, and A for a silver film deposited at 45°   .   232
12.23 Simulated normal incidence anisotropy curves for aluminium .   233
12.24 Structural hysteresis loops for packing density and cd fraction   234
12.25 Simulations of anisotropy as gold is deposited and etched . . .   235

13.1   Light transmission by a real polarizer   241
13.2   Wire grid polarizer.   242
13.3   Stretched metallic island polarizer   243
13.4   Single-plate and multiple-plate polarizers.   244
13.5   Coated-plate polarizer.   245
13.6 T_p and T_s for a reflecting stack on a glass plate   245
13.7   Extinction coefficient for a reflecting stack on a glass plate . .   246
13.8   Polarizing cube beam-splitter   247
13.9   Transmittances for a polarizing beam splitting cube   248
13.10 Extinction coefficient for a polarizing beam splitting cube . . .   248
13.11 Fabry-Perot polarizing filter   249
13.12 Transmittance curves for a Fabry-Perot polarizing filter   250
13.13 Glan-Foucault prism polarizer   252
13.14 Wollaston prism polarizer   253

13.15 Rochon prism polarizer 254

14.1   R and T of zero-order quarter-wave plate 260
14.2   Defining the transmittance $T(\lambda)$ of a wave plate 260
14.3   Retardance and T of an achromatic quarter-wave plate 262
14.4   Retardance and T of an achromatic half-wave plate 263
14.5   Wide-field element 265
14.6   Babinet compensator 266
14.7   Soleil-Babinet compensator 267
14.8   Single film wave plate 268
14.9   Double film wave plate 268
14.10 Composite single film wave plate 269
14.11 Composite double film wave plate 269
14.12 Thin film Babinet compensator 270
14.13 Thin film Soleil-Babinet compensator 270

15.1   Circular polarizer 274
15.2   Half-wave plate rotator 275
15.3   Optically-active rotator 275
15.4   Depolarizer 275
15.5   Lyot-Ohman birefringent filter 276
15.6   Basic unit of Lyot-Ohman filter 277
15.7   Transmittance of Lyot-Ohman birefringent filter 278
15.8   Folded Solc filter 280
15.9   Transmittance of 10-plate folded Solc filter 281
15.10 Fan Solc filter 281
15.11 Birefringent filter for tuning a laser 283
15.12 Transmittance of birefringent plate in laser cavity 284
15.13 Transmittance of the birefringent filter 2 1 6 285
15.14 Transmittance of the filter 2g 2 7g 1 5g 6 g 286
15.15 Periodic birefringent filter. 286
15.16 Transmittance of stack of 10 identical birefringent plates . . . 287

16.1   Multilayered reflecting coating 291
16.2   Experimental principal refractive indices 292
16.3   Experimental column angles 293
16.4   Experimental values of the indices n_p and n_s 295
16.5   Zero-reflectance condition 297
16.6   Anisotropic antireflection coating 298
16.7   Index-matching condition 300
16.8   Monitor data for PS coatings 301

16.9   RI and reflectance of anisotropic antireflection coating   303
16.10 RI profile and reflectance of anisotropic reflector   305
16.11 RI, reflectance, and phase difference for anisotropic reflector .   307
16.12 RI and reflectance of achromatic antireflection coating   309
16.13 Refractive index profile and reflectance of achromatic reflector   311
16.14 RI profiles and transmittance of double-cavity filter   313
16.15 RI profiles and transmittance of double-cavity filter   315
16.16 RI profile and transmittance of edge filter   317
16.17 RI and transmittance of common-index thin film polarizer . .   319
16.18 RI profiles and transmittance of double-cavity linear polarizer   321
16.19 Extinction ratio of PS double-cavity linear polarizer   322

Glossary

a	air
$\vec{a}$	field coefficients (Berreman calculus)
a, b, c	lengths of crystallographic axes
$\mathbf{a}, \mathbf{b}, \mathbf{c}$	crystallographic axes
a_0, a_4	Fourier coefficients of ellipsometric signal
a_p	aperture function of fluid patches
ac	alternating current
alumin	refractive index of aluminium
$angles$	input parameter (Berreman calculus)
A	absorptance
$\hat{A}$	system matrix (Berreman calculus)
$\hat{A}_d$	phase matrix (Berreman calculus)
$\hat{A}_{dp}, \hat{A}_{ds}$	Heavens phase matrices
A_f	fluid transport anisotropy
A_n	refractive anisotropy
A_p, A_s	absorptances for p and s and polarizations
A_r, A_t	reflection and transmission anisotropy
A_s	scatter anisotropy
AR	antireflection coating
b_2, b_4	Fourier coefficients of ellipsometric signal
bpcd	Bragg-Pippard cd equations
bpcdi	inversion of Bragg-Pippard cd equations
bpvd	Bragg-Pippard vd equations
bpvdi	inversion of Bragg-Pippard vd equations
$\mathbf{B}, \vec{B}, \hat{B}$	magnetic induction
B_1, B_2, B_3	constants in Laurent and Sellmeier dispersion equations
BP	Bragg-Pippard
BTF	Birefringent Thin Film (Toolbox)

c	velocity of light in vacuum
cd	crystallite defined
clight	speed of light in vacuum
cmat	characteristic matrix (Berreman calculus)
cover	input parameter (Berreman calculus)
cw	continuous wave
C	cover
C, F, R	type of principal dielectric axis
C_1, C_2, C_3	constants in Laurent and Sellmeier dispersion equations
CRC	Chemical Rubber Company
d	physical thickness
d_C	optical coupling distance
dr	change in reflection due to fluid
dt	change in transmission due to fluid
$d\phi$	phase thickness of thin slab
$\mathbf{D}, \vec{D}, \hat{D}$	electric displacement
e	eccentricity of ellipsoidal particles
e	extraordinary (ray propagation)
epsilon	relativity permittivity matrix
epsilon0	permittivity of vacuum
$\mathbf{E}, \vec{E}, \hat{E}$	electric field
E_a	applied field
E_d	depolarization field
E-beam	electron beam
f	structure factor
f_A, f_B	fractions of material A and material B
fmat	field matrix (Berreman calculus)
fresnel	*BTF Toolbox* script file
fsr	free spectral range
$\mathcal{F}$	fringe finesse
$\vec{F}$	field vector (Berreman calculus)
$\hat{F}$	field matrix (Berreman calculus)
$\hat{F}_p, \hat{F}_s$	Heavens field matrices
FECO	fringes of equal chromatic order
FWHM	full width at half maximum
g	glass

g	relative wavenumber
$g(r)$	radial distribution function
gold	refractive index of gold
h	Planck's constant
herpin	*BTF Toolbox* script file
H	high index layer
$\mathbf{H}, \vec{H}, \hat{H}$	magnetic field
H_0	open transmittance of linear polarizer
H_{90}	closed transmittance of linear polarizer
HBC	Henderson-Brodsky-Chaudhari (deposition model)
indices	input parameter (Berreman calculus)
I	intermediate index layer
$\hat{I}$	identity matrix
IAD	ion-assisted deposition
ID	identification
$\mathbf{J}$	current density
jmat	Jones matrix
$\vec{J}, \hat{J}$	Jones vector, Jones matrix
$\vec{J_h}$	Jones vector for horizontal linear polarization
$\vec{J_{\mathcal{L}}}$	Jones vector for left circular polarization
$\vec{J_{\mathcal{R}}}$	Jones vector for right circular polarization
$\vec{J_v}$	Jones vector for vertical linear polarization
$\mathbf{k}$	wave vector
k_p, k_s	components of scatter wave vector
k_Y, k_Z	components of scatter wave vector
layer	input parameter (Berreman calculus)
L	length of phase compensator
L	depolarization factor
$\hat{L}$	auxiliary matrix (Berreman calculus)
$\mathcal{L}$	left circular polarization
L_1, L_2, L_3	depolarization factors
LDT	laser damage threshold
m	order of interference
m	order of wave plate
$\vec{m}$	total field (Berreman calculus)
material	input parameter (Berreman calculus)

mfp	mean free path
mgf2	refractive index of magnesium fluoride
mksa	metre-kilogramme-second-ampere (system of units)
mmat	Mueller matrix
mu0	permeability of vacuum
M	magnetization
$\hat{M}$	characteristic matrix (Berreman calculus)
$\hat{M}$	Mueller matrix
$\hat{M}_p, \hat{M}_s$	Abelès characteristic matrices
$\hat{M}_p'$	simplified Abelès characteristic matrix
MDM	metal-dielectric-metal
n	refractive index
n_f	refractive index of fluid
n_I	imaginary part of refractive index
n_p, n_s	refractive indices for normal incidence
n_R	real part of refractive index
n_y, n_z	refractive indices for normal incidence
n_1, n_2, n_3	principal refractive indices
N	number of layers, plates, periods or stages
N	number of peaks (ellipsometry)
NaN	not a number
o	ordinary (ray propagation)
p	packing density
p_0	packing density at resonance
p_x, p_y, p_z	components of Poynting vector
pmat	phase matrix (Berreman calculus)
poynting	Poynting flux vectors (Berreman calculus)
pscover	input parameter (Berreman calculus)
pslayer	input parameter (Berreman calculus)
psmaterial	input parameter (Berreman calculus)
psstack	input parameter (Berreman calculus)
pssubstrate	input parameter (Berreman calculus)
pssystem	input parameter (Berreman calculus)
P	layer with columns in x–y plane
P	power diffracted from fluid patches
$\mathcal{P}$	linear polarization

P	polarization (electrical)
PIE	perpendicular incidence ellipsometry
PMT	photomultiplier tube
PS	stack with P and/or S layers
q	refractive index dependent constant
quartz	refractive index of crystalline quartz
r	displacement vector
$\hat{r}$	intermediary matrix
r_1, r_2, r_3	radii of spheroid
reflect	reflectance and transmittance (Berreman calculus)
reverse	*BTF Toolbox* script file
rjmat	rotation matrix (Jones calculus)
rmmat	rotation matrix (Mueller calculus)
rxmat	rotation matrix (Berreman calculus)
rymat	rotation matrix (Berreman calculus)
rzmat	rotation matrix (Berreman calculus)
R	reflectance
R	retardance of wave plate
$\mathcal{R}$	right circular polarization
$\hat{R}$	reflectance and transmittance (Berreman calculus)
$\hat{R}_j$	rotation matrix for Jones calculus
$\hat{R}_m$	rotation matrix for Mueller calculus
R_p	reflectance for p polarization
R_{ps}	reflectance p from s
R_s	reflectance for s polarization
$\hat{R}_x$	rotation matrix (Berreman calculus)
$\hat{R}_y$	rotation matrix (Berreman calculus)
$\hat{R}_z$	rotation matrix (Berreman calculus)
s	dimensionless unit vector in direction of the wave vector
s_r	reflection sensitivity to fluid
s_t	transmission sensitivity to fluid
silver	refractive index of silver
smat	system matrix (Berreman calculus)
solc	transmittance of folded Solc filter
stack	input parameter (Berreman calculus)
substrate	input parameter (Berreman calculus)

S	layer with columns in x–z plane
S	substrate
$\mathbf{S}$	Poynting vector
$\vec{S}$	Stokes vector
S_0, S_1, S_2, S_3	Stokes parameters
SEM	scanning electron microscope
SI	Systèm International d'Unités
SPIE	Society of Photo-Optical Instrumentation Engineers
tao2	ri's and column angle of tantalum oxide
tio2	ri's and column angle of titanium oxide
T	transmittance
T_p	transmission efficiency of linear polarizer
T_p, T_s	transmittances for p and s polarizations
T_{ps}	transmittance p from s
T_T	total transmittance of linear polarizer
v	phase velocity
v_p, v_s	propagation speeds of fluid front
vd	void defined
vretard	*BTF Toolbox* script file
vscatter	*BTF Toolbox* script file
V	degree of polarization
w	uncompensated retardance
x, y, z	propagation axes
$\mathbf{x, y, z}$	unit vectors
zro2	ri's and column angle of zirconium oxide
z_0	impedance of vacuum
z0	impedance of vacuum
α	$n\cos\theta$
α, β, γ	angles between crystallographic axes
β	Snell's law quantity
γ	ratio of field components
Δ	ellipsometric parameter
Δn_{ij}	differences between principal refractive indices
Δn	birefringence at normal incidence
Δn	change in index caused by fluid

Δr	change in reflection caused by fluid
Δt	change in transmission caused by fluid
$\Delta \lambda$	change in wavelength caused by fluid
Δd_p	packing density parameter
Δd_f	structure factor parameter
Δp	packing density factor
$\Delta \kappa$	difference in wavenumber
$\Delta \lambda_{fsr}$	free spectral range (spectral range without overlap)
$\varepsilon, \boldsymbol{\varepsilon}$	relative permittivity (dielectric constant)
ε_R	real part of dielectric constant
ε_I	imaginary parts of dielectric constant
$\varepsilon_0 \varepsilon$	permittivity
ε_0	permittivity of vacuum
η	areal density of fluid patches
η	column twist (angle-1 for position of biaxial material)
θ	optical angle of incidence
θ_p, θ_s	angles of scatter
θ_v	vapour angle of incidence (deposition angle)
κ	wavenumber
λ	wavelength
λ_p, λ_s	wavelength of fringe of equal chromatic order
$\mu_0 \mu$	permeability
μ_0	permeability of vacuum
ξ	azimuthal angle (angle-3 for position of biaxial material)
ξ_r	rocking angle for folded Solc filters
ξ_r	rotation angle during deposition
ξ_r	rotation angle for fan Solc filters
ρ	charge density
σ	conductivity
ϕ	optical thickness
χ	electric susceptibility
χ_m	magnetic susceptibility
Ψ	ellipsometric parameter
ψ	column angle (angle-2 for position of biaxial material)
ω	angular frequency

Chapter 1

Introduction

Light and other electromagnetic waves are our most important carriers of information. In everyday life we make direct use of the properties of light in receiving and analyzing visual data. The focusing action of the lens of our eye depends on the speeds of light in air and in the biological materials of the eye, and the colour that we see for an object is determined by the wavelengths of the light waves that enter our eye to form an image on our retina.

Light can be modelled as a transverse wavemotion, with electric and magnetic field components vibrating at right angles to each other and at right angles to the direction of propagation. When the direction of vibration of the electric field is constant, light is said to be linearly polarized in that direction. Other polarization states can be distinguished. If the "tip" of the electric vector rotates around a circle, the light is said to be circularly polarized, and two senses of rotation are possible. Similarly, elliptically polarized states occur when the tip of the electric field vector traces an ellipse as time proceeds.

From a mathematical viewpoint both circular and elliptical light can be described as the superposition of two linearly polarized beams vibrating at right angles to each other. When a beam of light impinges on a surface, the plane of incidence is defined by the directions of the incident beam and the normal to the surface. In such a case it is convenient to choose the two linearly polarized states so that they vibrate parallel to the plane of incidence (designated p) and perpendicular to the plane of incidence (designated s).

Most lasers emit polarized light, and the polarization property provides a "tag" that can be "read" by a device used to control the propagation path of the laser beam. For example, an element known as a polarization beam splitter typically allows a p-polarized beam to pass unhindered but reflects an s-polarized beam by a substantial angle, often 90°. Polarization is an important property for optical switching and will be utilized in all-optical computers, imaging and

1

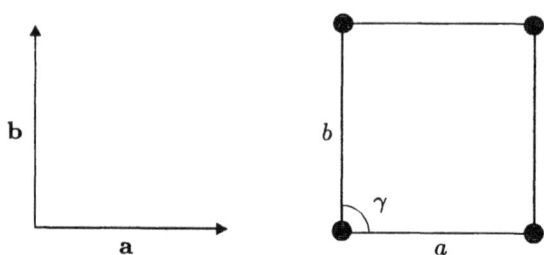

Fig. 1.1. Crystallographic axes and unit cell of a crystal.

data transfer systems that are currently under development.

Many optical elements are available for polarizing or changing the state of polarization of a beam of light. Most of these are macroscopic in size and constructed from a solid block of crystalline material such as quartz or magnesium fluoride.

1.1 Structural Classification of Crystals

The structural form of a crystal, as atoms on a 3–D lattice, was confirmed in early experiments involving the diffraction of x-rays.[1] The fundamental building block of a crystal is called a unit cell and, as cells are required to join together without leaving gaps, only a few shapes are possible for a unit cell. One face of a typical cell is illustrated in Fig. 1.1, and in Table 1.1[2,1] the allowed shapes are defined by the lengths a, b, c of the crystallographic axes **a, b, c** and the angles α, β, γ between the crystallographic axes.[a]

1.2 Optical Classification of Crystals

The wavelength of visible light ($\approx 500\,\text{nm}$) is much greater than the lattice constants of a crystal ($\approx 0.5\,\text{nm}$), and for this reason light that is travelling in a crystal "sees" a medium with macroscopically averaged properties. The speed of propagation in this "smoothed" medium depends on the direction of travel (and polarization) with respect to three mutually perpendicular axes called the principal dielectric axes. Relative to the speed of light in air, the speed in the crystal is lower by a factor called the refractive index and denoted by the symbol n. Thus the refractive index n depends on the direction of propagation in a

---

[a]The symbols α, β, γ represent different quantities in the remaining parts of the book.

Table 1.1. Classification of crystals.

Crystal System	Unit Cell Axes and Angles	Principal Dielectric Axes	Principal Refractive Indices	Optical Classification
Cubic	$a = b = c$ $\alpha = \beta = \gamma = 90°$	R,R,R	$n_1 = n_2 = n_3$	Isotropic
Hexagonal	$a = b \neq c$ $\alpha = \beta = 90°$ $\gamma = 120°$	R,R,F	$n_1 \neq n_2 = n_3$	Uniaxial
Tetragonal	$a = b \neq c$ $\alpha = \beta = \gamma = 90°$	"	"	"
Trigonal	$a = b = c$ $\alpha = \beta = \gamma$ $< 120°, \neq 90°$	"	"	"
Orthorhombic	$a \neq b \neq c$ $\alpha = \beta = \gamma = 90°$	F,F,F	$n_1 \neq n_2 \neq n_3$	Biaxial
Monoclinic	$a \neq b \neq c$ $\alpha = \gamma = 90° \neq \beta$	C,C,F	"	"
Triclinic	$a \neq b \neq c$ $\alpha \neq \beta \neq \gamma$	C,C,C	"	"

crystal, but all cases can be computed using three principal refractive indices, n_1, n_2, n_3, associated with three principal dielectric axes. These axes are defined so that the principal refractive index n_1 is the refractive index appropriate to light travelling with its electric field parallel to principal dielectric axis-1 etc.

Now we return to Table 1.1. In a *cubic* crystal the three refractive indices are equal, $n_1 = n_2 = n_3 = n$. Thus light travels at the same speed along any one of the principal axes. Further, by resolving fields for light travelling in a general direction into components along the principal axes we can deduce that the speed is independent of direction. Thus the refractive index is n for any direction of propagation and the optical classification of the medium is *isotropic*. The triad of axes can be rotated to any angular position, and for this reason the principal axes are labelled R in the table.

In the *hexagonal*, *tetragonal*, and *trigonal* systems one principal dielectric axis is fixed by the symmetry of the crystal and labelled F. These materials have two of the three refractive indices equal, and we write $n_1 \neq n_2 = n_3$ to conform with our notation for birefringent films. Light travels along principal dielectric axis-1, the *optic axis*, at the same speed irrespective of polarization direction. The optical classification of these materials is *uniaxial*.

The three refractive indices are different for *orthorhombic*, *monoclinic* and *triclinic* crystals, $n_1 \neq n_2 \neq n_3$. These are classified optically as *biaxial* because they have two directions along which the speed of propagation is independent of polarization. Another feature, shared by monoclinic and triclinic crystals, is non-alignment of one or more of the principal dielectric axes with the crystal axes. The angular position of such a dielectric axis depends on the colour of the light, and hence the symbol C in the table.

1.3 Structure of Birefringent Films

There are many potential applications of birefringent devices that are poorly matched to macroscopic technology. A new requirement is for very thin, parallel, birefringent layers that are compatible with emerging planar layer and nano-optical technologies.

Birefringent properties occur in thin films that are deposited obliquely in vacuum. The idea is not a new one, as reports of anisotropic effects in thin films go back more than a century,[3] and some observations on fluoride films were made in the late 1950's.[4] However, it is only during the past decade that the physical cause of the birefringence has been understood[5] and systematic efforts have been mounted to produce films of commercial quality. This book summarizes these recent developments.

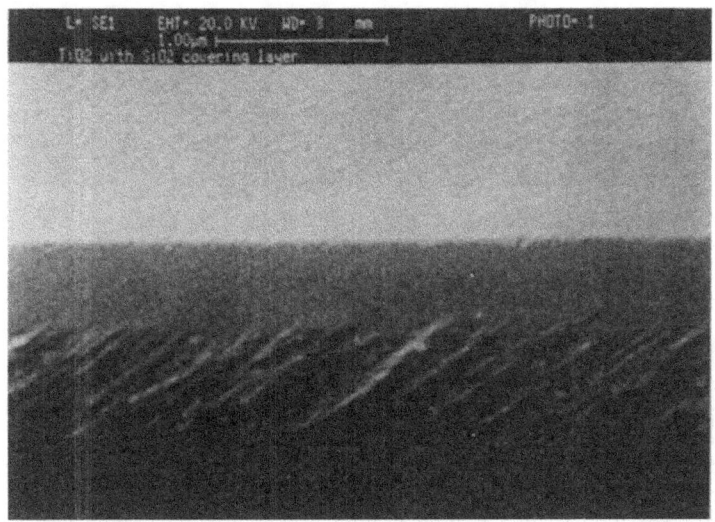

Fig. 1.2. Scanning electron micrograph of a fractured titanium oxide film overcoated with a layer of silicon oxide. The microstructural columns in the titania layer result from deposition at an oblique angle. The overcoat of silica was deposited at normal incidence and with ion beam assistance to produce a dense, non-porous protective layer. (Photograph from the authors' laboratory.)

Birefringence in thin films is caused by columnar microstructure.[b] An example of columnar microstructure is given in Fig. 1.2 for a titanium oxide film, and a model that incorporates essential features of the microstructure is shown in Fig. 1.3. The *form birefringence*, as it is called, depends on column shape to cause direction-related perturbations of refractive index by a mechanism known as depolarization.

1.4 Optical Classification of Birefringent Films

The locations of the three mutually orthogonal principal dielectric axes 1, 2, 3 of a columnar film are *fixed* by deposition geometry and symmetry. Referring to the model shown in Fig. 1.3, axis-1 is placed along the direction of the columns, axis-3 is perpendicular to the deposition plane, and axis-2 is in the deposition plane. Experiments show that the three principal refractive indices

---

[b]Other factors contribute to birefringence in thin films, but in a minor way. These include crystallinity and anisotropic stress.

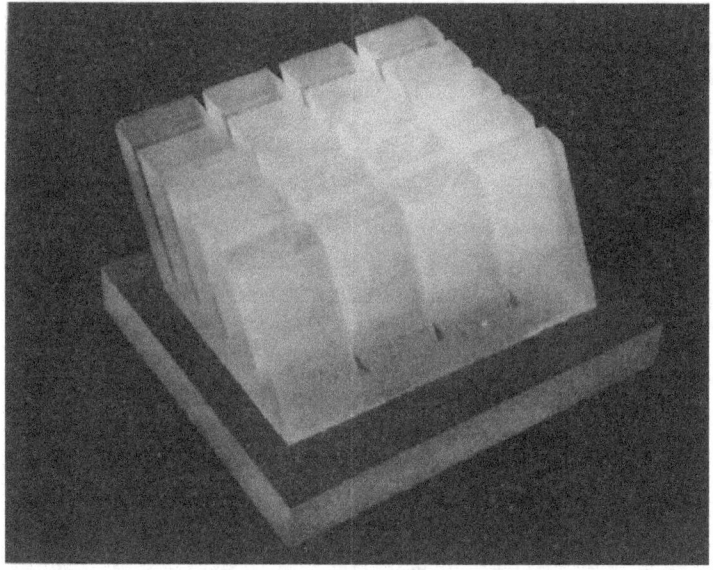

Fig. 1.3. Model of a birefringent thin film. The deposition plane is parallel to the left-hand face.

satisfy $n_1 > n_3 > n_2$. A comparison of these properties of a birefringent film with the properties of crystals listed in Table 1.1 shows that a match exists with the orthorhombic system. Thus we can state:-

A dielectric thin film deposited obliquely behaves like an orthorhombic crystal and exhibits biaxial optical properties.

1.5 Layout of the Book

As a consequence of the similarity of the optical properties of orthorhombic crystals and obliquely deposited films, a common mathematical framework can be used to describe the propagation of light in the materials. We have divided our book into three parts, and in the first part we develop the common mathematical framework. We start by considering Maxwell's equations in conventional vector form, and then rewrite the equations for the special case of plane wave solutions in matrix form.

Plane waves that can propagate in layers of the materials are represented by column vectors with four elements determined from electric and magnetic field components. Four-by-four matrices are used for transferring waves through interfaces and across layers, and relationships between matrix methods are explored in this section. We are pleased to acknowledge the significant contributions made by Samer Kassam to our development of the matrix method.

Our emphasis in the first part targets computation, rather than development of the last nuance of abstract theory. Basic lines of computer code are developed for computation of optical properties such as reflectance and transmittance. To make immediate use of this code the reader requires access to the MATLAB[6] mathematical software package. MATLAB offers several relevant and useful features for optical calculations – the basic data element is a matrix that does not require dimensioning, matrix elements are allowed to be real or complex without specifying a data type, and matrix functions are provided for standard operations such as finding eigenvalues and eigenvectors. Our software for bulk and layered anisotropic media is compiled as a *Birefringent Thin Films Toolbox* that can be downloaded electronically from the publisher. Documentation for the *BTF Toolbox* is provided in Appendix A of the book.

The second part of the book treats the deposition, characterization, and modelling of birefringent films, and includes discussion on the management of undesirable properties such as scatter and water transport. A chapter on the anisotropic optical properties of metal films is included. In general, in this second part, we make extensive use of examples from our laboratory and illustrations from work that we have published in collaboration with colleagues from around the world. Our acknowledgements to the authors and to the publishers are made in the figure captions using an amended format that is easier to read. These references are not repeated in the Notes and References section at the end of the book. We acknowledge, as well, contributions to the work made by research students, and the financial support that we have received from the University of Otago and from the New Zealand Foundation for Research, Science and Technology. Many staff of the University of Otago have provided technical support for the research described in this part of the book and we take this opportunity to record our appreciation of their expert skills.

In the third part of the book, applications of birefringent media are discussed in chapters with the headings Linear Polarizers, Phase Retarders, Birefringent Filters, and Birefringent Coatings. Many of the polarizing elements that are considered in these chapters are macroscopic in size, rather than thin film constructions. They are included both for completeness, and to provide hints as to how competitive birefringent thin film counterparts may be designed. The book is concluded with a "sampler" of ten anisotropic thin film coatings.

Part I

Propagation in Biaxial Media

Chapter 2

Propagation Equations

The interaction of light with matter has been modelled successfully at both microscopic and macroscopic levels. In the former case individual actions involving an atom and a photon can be considered and the quantum model, as it is called, is particularly useful for explaining phenomena such as the emission of characteristic line spectra by atoms and the photoelectric effect in which light releases charge carriers in an optical medium.

The second model is based on the work of Maxwell and uses a set of macroscopic parameters, mostly vectors, that are averaged over a volume large compared to the space occupied by an individual atom. Maxwell's model has proved to be successful in applications such as the propagation of light in optical media and is the better model for explaining effects such as interference and diffraction. As this book is concerned principally with the propagation of light in bulk and layered anisotropic optical media the Maxwellian model is used exclusively.

In this chapter we derive propagation equations for light in anisotropic media, and define the matrix notation that is used throughout the book. In turn, we

- list Maxwell's equations and the macroscopic parameters used in electromagnetism

- consider the propagation of electromagnetic waves in free space, in isotropic media, and in anisotropic media

- explain our matrix notation

- express Maxwell's equations in matrix form for the propagation of plane waves in anisotropic media.

2.1   Maxwell's Equations

In the Maxwellian model the electromagnetic state at a point in an optical medium is specified in terms of macroscopic parameters from the set:-

Electric field	**E**	Magnetic field	**H**
Polarization	**P**	Magnetization	**M**
Electric displacement	**D**	Magnetic induction	**B**
Charge density	ρ	Current density	**J**

Spatial and temporal derivatives of these parameters are connected in a group of four equations known as Maxwell's equations. In differential form Maxwell's equations can be written as

$$\nabla \times \mathbf{E} = -\mu_0 \frac{\partial \mathbf{H}}{\partial t} - \mu_0 \frac{\partial \mathbf{M}}{\partial t}$$

$$\nabla \times \mathbf{H} = \varepsilon_0 \frac{\partial \mathbf{E}}{\partial t} + \frac{\partial \mathbf{P}}{\partial t} \mathbf{J}$$

$$\nabla \cdot \mathbf{E} = -\frac{1}{\varepsilon_0} \nabla \cdot \mathbf{P} + \rho/\varepsilon_0$$

$$\nabla \cdot \mathbf{H} = -\nabla \cdot \mathbf{M}. \qquad (2.1)$$

Electric and magnetic fields provide the driving forces in optical media, and the consequences include electric current due to the motion of conduction electrons, polarization due to small relative displacements of bound charges, and magnetization due to induced magnetic moments. If an outcome such as electric current density is proportional to the strength of the driving force, $\mathbf{J} = \sigma\mathbf{E}$, then the medium is said to exhibit a *linear* response; alternatively the statement that Ohm's law is obeyed would convey the same meaning in this example. The constant of proportionality in the relation used to link $\mathbf{J}$ and $\mathbf{E}$ is called the conductivity. In Table 2.1 we have listed the most important linear relationships for optical media and the corresponding proportionality constants. This book deals exclusively with linear optical materials.[7]

2.2   Propagation in Free Space. Mathematical Methods

In free space we have $\rho = 0$, $\mathbf{P} = 0$, $\mathbf{M} = 0$, $\mathbf{J} = 0$, and Maxwell's equations simplify to

Table 2.1. Linear relationships for optical media.

Conductivity	σ	$\mathbf{J} = \sigma\mathbf{E}$
Electric susceptibility	χ	$\mathbf{P} = \chi\varepsilon_0\mathbf{E}$
Permittivity	$\varepsilon_0\varepsilon$	$\mathbf{D} = \varepsilon_0\varepsilon\mathbf{E} = \varepsilon_0\mathbf{E} + \mathbf{P}$
Magnetic susceptibility	χ_m	$\mathbf{M} = \chi_m\mathbf{H}$
Permeability	$\mu_0\mu$	$\mathbf{B} = \mu_0\mu\mathbf{H} = \mu_0(\mathbf{H} + \mathbf{M})$

$$\nabla \times \mathbf{E} = -\mu_0\frac{\partial \mathbf{H}}{\partial t} \tag{2.2}$$

$$\nabla \times \mathbf{H} = \varepsilon_0\frac{\partial \mathbf{E}}{\partial t} \tag{2.3}$$

$$\nabla \cdot \mathbf{E} = 0 \tag{2.4}$$

$$\nabla \cdot \mathbf{H} = 0. \tag{2.5}$$

Maxwell's first and second equations, Eq. (2.2) and Eq. (2.3), are particularly significant for electromagnetic waves because they connect spatial changes in one field to temporal changes in the other. Six equations are implied, and the form is illustrated by writing out one pair; the others can be obtained by cyclic permutations of the subscripts x, y and z:-

$$\frac{\partial E_z}{\partial y} - \frac{\partial E_y}{\partial z} = -\mu_0\frac{\partial H_x}{\partial t} \tag{2.6}$$

$$\frac{\partial H_z}{\partial y} - \frac{\partial H_y}{\partial z} = \varepsilon_0\frac{\partial E_x}{\partial t}. \tag{2.7}$$

In Cartesian coordinates ∇ has the form

$$\nabla \equiv \vec{i}\frac{\partial}{\partial x} + \vec{j}\frac{\partial}{\partial y} + \vec{k}\frac{\partial}{\partial z}, \tag{2.8}$$

and $\mathbf{E}$ and $\mathbf{H}$ are given by

$$\mathbf{E} = \mathbf{x}E_x + \mathbf{y}E_y + \mathbf{z}E_z \tag{2.9}$$

$$\mathbf{H} = \mathbf{x}H_x + \mathbf{y}H_y + \mathbf{z}H_z. \tag{2.10}$$

Here $\mathbf{x}$, $\mathbf{y}$ and $\mathbf{z}$ are unit vectors directed along the x, y and z axes respectively. Hence the third and fourth Maxwell equations, the divergence equations, can be written as

$$\frac{\partial E_x}{\partial x} + \frac{\partial E_y}{\partial y} + \frac{\partial E_z}{\partial z} = 0 \tag{2.11}$$

$$\frac{\partial H_x}{\partial x} + \frac{\partial H_y}{\partial y} + \frac{\partial H_z}{\partial z} = 0. \tag{2.12}$$

Both **E** and **H** can be made the subject of a differential wave equation by

- taking the curl of both sides of Eq.(2.2) and Eq.(2.3),

- making use of the operator identity

$$\nabla \times (\nabla \times \ ) = \nabla(\nabla \cdot \ ) - \nabla^2, \tag{2.13}$$

- using Eqs.(2.4) and (2.5) to substitute for the divergences.

The result is the pair of equations

$$\nabla^2 \mathbf{E} = \mu_0 \varepsilon_0 \frac{\partial^2 \mathbf{E}}{\partial t^2} \tag{2.14}$$

$$\nabla^2 \mathbf{H} = \mu_0 \varepsilon_0 \frac{\partial^2 \mathbf{H}}{\partial t^2}. \tag{2.15}$$

Similar expressions occur in other branches of physics, and as they are known to support harmonic wave solutions they are referred to as vector wave equations. In fact each of the field components E_x, E_y, E_z, H_x, H_y, H_z satisfies the vector wave equation, and hence simplification to a single scalar wave equation is possible; thus, for a scalar U,

$$\frac{\partial^2 U}{\partial x^2} + \frac{\partial^2 U}{\partial y^2} + \frac{\partial^2 U}{\partial z^2} = \mu_0 \varepsilon_0 \frac{\partial^2 U}{\partial t^2}. \tag{2.16}$$

The travelling wave fields, expressed in exponential notation as

$$\mathbf{E} = \mathbf{E}_0 e^{i(\mathbf{k} \cdot \mathbf{r} - \omega t)} \tag{2.17}$$

$$\mathbf{H} = \mathbf{H}_0 e^{i(\mathbf{k} \cdot \mathbf{r} - \omega t)}, \tag{2.18}$$

satisfy the differential wave equation. Here ω is the *angular frequency* of the waves, **r** is a displacement vector from the origin at (0,0,0) in 3–D space, and **k** is the *wave vector* of magnitude $2\pi/\lambda$ where λ is the *wavelength* in the medium (vacuum).

The constants in the wave equation indicate that plane wavefronts, surfaces of constant phase defined by

$$\mathbf{k}.\mathbf{r} - \omega t = \text{constant}, \tag{2.19}$$

travel in the direction of $\mathbf{k}$ at a *phase velocity* of

$$v = 1/\sqrt{\mu_0 \varepsilon_0}. \tag{2.20}$$

Maxwell predicted the phenomenon of electromagnetic waves, and calculated the speed that the waves would travel in vacuum. The result was so close to Fizeau's experimental value for the speed of light in air that Maxwell concluded light itself must be an electromagnetic wave disturbance.

Maxwell's equations simplify in a useful way for plane, harmonic wave solutions represented (for a wave of unit amplitude) by the complex exponential expression

$$\exp i(\mathbf{k}.\mathbf{r} - \omega t) = \exp i(k_x x + k_y y + k_z z - \omega t). \tag{2.21}$$

Taking the time differential of the complex expression gives

$$\frac{\partial}{\partial t} \exp i(\mathbf{k}.\mathbf{r} - \omega t) = -i\omega \exp i(\mathbf{k}.\mathbf{r} - \omega t), \tag{2.22}$$

and the spatial derivative is

$$\nabla \exp i(k_x x + k_y y + k_z z - \omega t) = i\mathbf{k} \exp i(k_x x + k_y y + k_z z - \omega t). \tag{2.23}$$

Hence the operator relations,

$$\frac{\partial}{\partial t} \rightarrow -i\omega \tag{2.24}$$

$$\nabla \rightarrow i\mathbf{k}, \tag{2.25}$$

are applicable to plane, harmonic waves, and Maxwell's equations for these waves are:-

$$\mathbf{k} \times \mathbf{E} = \mu_0 \omega \mathbf{H} \tag{2.26}$$

$$\mathbf{k} \times \mathbf{H} = -\varepsilon_0 \omega \mathbf{E} \tag{2.27}$$

$$\mathbf{k} \cdot \mathbf{E} = 0 \tag{2.28}$$

$$\mathbf{k} \cdot \mathbf{H} = 0. \tag{2.29}$$

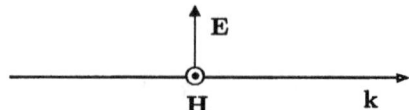

Fig. 2.1. Fields of an electromagnetic wave propagating in vacuum.

One advantage of this form is that the relative directions of the wave vector and the fields are displayed. Thus the four vector equations show that **k**, **E** and **H** form a mutually orthogonal right-handed triad, as illustrated in Fig. 2.1. As well, the equations show that the magnitudes of **H** and **E** are related by a constant,

$$H = \frac{E}{z_0}. \tag{2.30}$$

The constant z_0 is called the *impedance of vacuum* (free space) and is defined by

$$z_0 \equiv (\mu_0/\varepsilon_0)^{1/2}; \tag{2.31}$$

it has the approximate value of $377\,\Omega$.

2.2.1   SI units

Another property of light – related to the speed of propagation and pivotal to the development of Einstein's special theory of relativity – is that measurements of the speed in free space always yield the same value, independent of the relative motion of the source and observer. Thus the speed of light in vacuum is a fundamental constant and, in 1983, the value was defined to be the exact number

$$c = 2.99792458 \times 10^8 \, \text{m/s}. \tag{2.32}$$

The value of c now plays a primary role in the SI system of units (Systèm International d'Unités), which has evolved from the MKSA (metre-kilogramme-second-ampere) system. Since 1964 the unit of time, the *second* (s), has been defined as the duration of 9 192 631 770 periods of the radiation corresponding to the transition between the two hyperfine levels of the ground state of the cesium-133 atom. One *metre* (m), the unit of length, is now the distance travelled by light during a time interval of 1/299 792 458 of a second.

Table 2.2. Constants used in electromagnetism.

Quantity	Symbol	Value	SI Unit
Velocity of light in vacuum	c	2.997925	$10^8 \, \text{m/s}$
Permeability of vacuum	μ_0	4π	$10^{-7} \, \text{Vs/Am}$
Permittivity of vacuum	$\varepsilon_0 = 1/\mu_0 c^2$	8.854188	$10^{-12} \, \text{As/Vm}$
Impedance of vacuum	$z_0 = (\mu_0/\epsilon_0)^{1/2}$	3.767303	$10^2 \, \text{V/A}$

The quantity μ_0 is assigned the exact value of $4\pi \times 10^{-7} \, \text{Vs/Am}$ and the value of the constant ε_0 can be calculated as $\varepsilon_0 = 1/\mu_0 c^2 \approx 8.85 \times 10^{-12} \, \text{As/Vm}$. Table 2.2 provides a summary of the electromagnetic constants specified to six decimal places.

In the *BTF Toolbox* the SI constants c, μ_0, ϵ_0 and z_0 are available directly as **clight, mu0, epsilon0** and **z0**.

2.3 Propagation in Isotropic Media

In a medium that is isotropic, uncharged, nonmagnetic, and nonconducting we have $\rho = 0$, $\mathbf{M} = 0$, $\mu = 1$, and $\mathbf{J} = 0$. Maxwell's equations have the form

$$\nabla \times \mathbf{E} = -\mu_0 \frac{\partial \mathbf{H}}{\partial t} \tag{2.33}$$

$$\nabla \times \mathbf{H} = \varepsilon_0 \varepsilon \frac{\partial \mathbf{E}}{\partial t} \tag{2.34}$$

$$\nabla \cdot \mathbf{D} = 0 \tag{2.35}$$

$$\nabla \cdot \mathbf{H} = 0, \tag{2.36}$$

and the scalar wave equation is

$$\frac{\partial^2 U}{\partial x^2} + \frac{\partial^2 U}{\partial y^2} + \frac{\partial^2 U}{\partial z^2} = \mu_0 \varepsilon_0 \varepsilon \frac{\partial^2 U}{\partial t^2}. \tag{2.37}$$

Thus electromagnetic waves in the medium travel at phase speed

$$v = 1/\sqrt{\mu_0 \varepsilon_0 \varepsilon}, \tag{2.38}$$

and Maxwell's equations for plane, harmonic waves are

$$\mathbf{k} \times \mathbf{E} = \mu_0 \omega \mathbf{H} \tag{2.39}$$

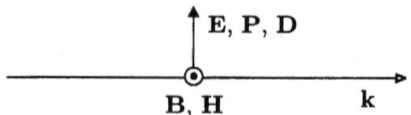

Fig. 2.2. Fields of an electromagnetic wave propagating in an isotropic medium

$$\mathbf{k} \times \mathbf{H} = -\varepsilon_0 \varepsilon \omega \mathbf{E} \tag{2.40}$$

$$\mathbf{k} \cdot \mathbf{D} = 0 \tag{2.41}$$

$$\mathbf{k} \cdot \mathbf{H} = 0. \tag{2.42}$$

As for electromagnetic waves propagating in vacuum, $\mathbf{k}$, $\mathbf{E}$ and $\mathbf{H}$ form a mutually orthogonal right-handed triad, the vectors $\mathbf{E}$, $\mathbf{P}$ and $\mathbf{D}$ are parallel, and so are $\mathbf{B}$ and $\mathbf{H}$ (Fig. 2.2). The relationship between the magnitudes of $\mathbf{H}$ and $\mathbf{E}$ is now

$$H = \frac{nE}{z_0}, \tag{2.43}$$

where

$$n \equiv c/v \tag{2.44}$$

is the refractive index.

2.4   Propagation in Anisotropic Media

In a nonmagnetic, nonconducting, electrically anisotropic medium $\rho = 0$, $\mathbf{M} = 0$, $\mathbf{J} = 0$, $\mu = 1$, and the relative permittivity is a symmetric tensor $\boldsymbol{\varepsilon}$. Maxwell's equations, in the standard form followed by the form for plane, harmonic wave solutions are

$$\nabla \times \mathbf{E} = -\mu_0 \frac{\partial \mathbf{H}}{\partial t} \tag{2.45}$$

$$\nabla \times \mathbf{H} = \varepsilon_0 \boldsymbol{\varepsilon} \frac{\partial \mathbf{E}}{\partial t} \tag{2.46}$$

$$\nabla \cdot \mathbf{D} = 0 \tag{2.47}$$

$$\nabla \cdot \mathbf{H} = 0, \tag{2.48}$$

and

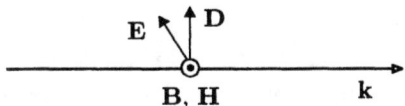

Fig. 2.3. Fields of an electromagnetic wave propagating in an anisotropic medium.

$$\mathbf{k} \times \mathbf{E} = \mu_0 \omega \mathbf{H} \tag{2.49}$$

$$\mathbf{k} \times \mathbf{H} = -\varepsilon_0 \varepsilon \omega \mathbf{E} \tag{2.50}$$

$$\mathbf{k} \cdot \mathbf{D} = 0 \tag{2.51}$$

$$\mathbf{k} \cdot \mathbf{H} = 0. \tag{2.52}$$

Recalling that $\varepsilon \mathbf{E} = \mathbf{D}$, we see from Eqs.(2.49)–(2.52) that $\mathbf{k}$, $\mathbf{D}$ and $\mathbf{H}$ form a mutually orthogonal right-handed triad, just as in an isotropic medium (Fig. 2.2). Similarly $\mathbf{B}$ and $\mathbf{H}$ are parallel, but in general $\mathbf{E}$ is neither parallel to $\mathbf{D}$ nor perpendicular to $\mathbf{k}$.

The vector wave equation is

$$\nabla \times (\nabla \times \mathbf{E}) + \varepsilon \frac{\partial^2 \mathbf{E}}{\partial t^2} = 0, \tag{2.53}$$

and the condition for harmonic, plane wave solutions is

$$\mathbf{k} \times (\mathbf{k} \times \mathbf{E}) + k_0^2 \varepsilon \mathbf{E} = 0. \tag{2.54}$$

At this stage we define a dimensionless unit vector $\mathbf{s}$ in the direction of the wave vector $\mathbf{k}$,

$$\mathbf{s} = \mathbf{k}/k, \tag{2.55}$$

and replace $\mathbf{k}$ in Eq.(2.54) with

$$\mathbf{k} = k_0 n \mathbf{s}. \tag{2.56}$$

This leaves a vector equation,

$$n^2 \mathbf{s} \times (\mathbf{s} \times \mathbf{E}) + \varepsilon \mathbf{E} = 0, \tag{2.57}$$

which we write out as three scalar equations:-

$$[n_1^2 - n^2(s_2^2 + s_3^2)]E_1 + n^2 s_1 s_2 E_2 + n^2 s_1 s_3 E_3 = 0 \qquad (2.58)$$
$$n^2 s_1 s_2 E_1 + [n_2^2 - n^2(s_1^2 + s_3^2)]E_2 + n^2 s_2 s_3 E_3 = 0 \qquad (2.59)$$
$$n^2 s_1 s_3 E_1 + n^2 s_2 s_3 E_2 + [n_3^2 - n^2(s_1^2 + s_2^2)]E_3 = 0. \qquad (2.60)$$

The subscripts 1, 2, 3 in these equations refer to the material frame in which $\boldsymbol{\varepsilon}$ is a diagonal tensor with non-zero elements $\varepsilon_{11} = n_1^2$, $\varepsilon_{22} = n_2^2$, $\varepsilon_{33} = n_3^2$.

The condition for a non-trivial solution (i.e. at least one electric field component not equal to zero) is that the determinant of the coefficients of the E's must be equal to zero,

$$\det \begin{bmatrix} n_1^2 - n^2(s_2^2 + s_3^2) & n^2 s_1 s_2 & n^2 s_1 s_3 \\ n^2 s_1 s_2 & n_2^2 - n^2(s_1^2 + s_3^2) & n^2 s_2 s_3 \\ n^2 s_1 s_3 & n^2 s_2 s_3 & n_3^2 - n^2(s_1^2 + s_2^2) \end{bmatrix} = 0. \qquad (2.61)$$

An alternative and more useful form is obtained when the determinant is expanded,

$$(s_1^2 n_1^2 + s_2^2 n_2^2 + s_3^2 n_3^2)n^4 - [(s_1^2 + s_2^2)n_1^2 n_2^2 + (s_2^2 + s_3^2)n_2^2 n_3^2 + (s_3^2 + s_1^2)n_3^2 n_1^2]n^2$$
$$+ n_1^2 n_2^2 n_3^2 = 0.$$
$$(2.62)$$

Both Eq.(2.61) and Eq.(2.62) are equivalent to Fresnel's equation, which is often written as

$$\frac{1}{n^2} = \frac{s_1^2}{n^2 - n_1^2} + \frac{s_2^2}{n^2 - n_2^2} + \frac{s_3^2}{n^2 - n_3^2}. \qquad (2.63)$$

Solving the quadratic in n^2, Eq.(2.62), for a particular direction of propagation specified by s_1, s_2, s_3 yields two positive values for n. When all propagation directions in 3–D space are considered, the allowed n's can be represented by a double-sheeted surface, $n(\mathbf{s})$, called the *refractive index surface*. An understanding of the general shape of the index surface is helpful for visualizing propagation in biaxial media.

Several other surfaces are used in crystal optics. The *wave vector surface* $k(\mathbf{s})$ which can be obtained by putting $s_j = k_j/k_0 n$ in Eq.(2.61) is just a scaled version of $n(\mathbf{s})$. Substitution of $s_j = nv_j/c$, also into Eq.(2.61), gives a reciprocal surface $v(\mathbf{s})$ called the *phase velocity surface*. Finally the *ray velocity surface*

Table 2.3. Propagation surfaces for anisotropic optical materials.[7]

Refractive index surface $n(s)$

$$\det \begin{bmatrix} n_1^2 - n^2(s_2^2 + s_3^2) & n^2 s_1 s_2 & n^2 s_1 s_3 \\ n^2 s_1 s_2 & n_2^2 - n^2(s_1^2 + s_3^2) & n^2 s_2 s_3 \\ n^2 s_1 s_3 x & n^2 s_2 s_3 & n_3^2 - n^2(s_1^2 + s_2^2) \end{bmatrix} = 0$$

Wave vector surface $k(s)$

$$\det \begin{bmatrix} n_1^2 k_0^2 - k_2^2 - k_3^2 & k_1 k_2 & k_1 k_3 \\ k_2 k_1 & n_2^2 k_0^2 - k_3^2 - k_1^2 & k_2 k_3 \\ k_3 k_1 & k_3 k_2 & n_3^2 k_0^2 - k_1^2 - k_2^2 \end{bmatrix} = 0$$

Phase velocity surface $v(s)$

$$\det \begin{bmatrix} n_1^2 v^4/c^2 - v_2^2 - v_3^2 & v_1 v_2 & v_1 v_3 \\ v_2 v_1 & n_2^2 v^4/c^2 - v_3^2 - v_1^2 & v_2 v_3 \\ v_3 v_1 & v_3 v_2 & n_3^2 v^4/c^2 - v_1^2 - v_2^2 \end{bmatrix} = 0$$

Ray velocity surface $u(S/S)$

$$\det \begin{bmatrix} c^2/n_1^2 - u_2^2 - u_3^2 & u_1 u_2 & u_1 u_3 \\ u_2 u_1 & c^2/n_2^2 - u_3^2 - u_1^2 & u_2 u_3 \\ u_3 u_1 & u_3 u_2 & c^2/n_3^2 - u_1^2 - u_2^2 \end{bmatrix} = 0$$

$u(S/S)$ gives the speed of propagation along the Poynting vector (energy flow vector, defined in the next section). Table 2.3 lists the determinantal form of the four surfaces.

2.5 Energy Flow

From the theory of electrostatics it can be shown that an energy density $\varepsilon\varepsilon_0 E^2/2$ is associated with an electric field in a dielectric medium, and similarly, a magnetic field acting alone produces an energy density of $\mu H^2/2$ in a non-magnetic medium. Note that for an electromagnetic wave, $\varepsilon\varepsilon_0 E^2/2 = \mu H^2/2$ because $E = vB$ and $v = 1/\sqrt{\mu\varepsilon\varepsilon_0}$.

Electromagnetic waves travelling in vacuum transport energy in the direction of propagation and the energy flow is shared between the electric and magnetic fields. Both the direction of flow, and the instantaneous total rate of flow through unit area normal to the flow direction, are given by the *Poynting vector* which has the general form

$$\mathbf{S} = \mathbf{E} \times \mathbf{H}. \tag{2.64}$$

The irradiance I is the time-averaged value of the magnitude of the Poynting vector,

$$I \equiv \langle S \rangle; \tag{2.65}$$

for waves travelling in an isotropic medium of refractive index n,

$$\langle S \rangle = \frac{n}{2z_0} |E_0|^2. \tag{2.66}$$

Note that in an anisotropic medium the direction of energy flow, given by $\mathbf{E} \times \mathbf{H}$, is not the same as the direction of wave propagation given by $\mathbf{D} \times \mathbf{H}$, because in general $\mathbf{E}$ is not parallel to $\mathbf{D}$.

2.6   Notation for Biaxial Media

2.6.1   Material Axes

For a biaxial medium the three mutually orthogonal principal axes are labelled 1, 2, 3, and quantities associated directly with the principal axes are identified by subscripts $_1$, $_2$, $_3$ (see Fig. 2.4). Thus the three equations,

$$\begin{aligned} \varepsilon_1 &= n_1^2 \\ \varepsilon_2 &= n_2^2 \\ \varepsilon_3 &= n_3^2, \end{aligned} \tag{2.67}$$

relate the principal dielectric constants and the principal refractive indices. For a uniaxial medium the label 1 is assigned to the optic axis, unless other factors make such an assignment inconvenient.

Column vectors, usually formed from the components of the fields associated with plane waves, are used extensively in the remaining parts of this book. As well, matrices are used to collate sets of column vectors associated with particular propagation rules and for use as transfer matrices. While it is true that a column vector is just a particular case of a matrix, it is also a fact that

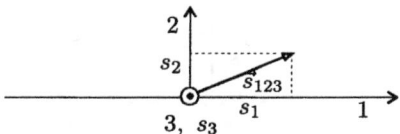

Fig. 2.4. Material axes 1, 2, 3 for a biaxial optical medium. In general the wavenormal $\vec{s}_{123}$ is not in the plane of the diagram.

equations become easier to read if the two can be distinguished readily. To achieve this the right-overarrow symbol $\vec{}$ is used as a type identifier for column vectors, and the hat symbol $\hat{}$ is used as an identifier for matrices.

As examples of column vectors,

$$\vec{s}_{123} = \begin{bmatrix} s_1 \\ s_2 \\ s_3 \end{bmatrix} \tag{2.68}$$

gives the direction of propagation of a plane wave in material frame coordinates, and

$$\vec{E}_{123} = \begin{bmatrix} E_1 \\ E_2 \\ E_3 \end{bmatrix} \tag{2.69}$$

represents the electric field of the wave in the same frame. The components E_1, E_2 and E_3 are amplitudes that may be signed or complex, but note that the complex exponential spatial and temporal phase terms of the wave are implied but not included explicitly.

As an example of the matrix notation,

$$\hat{\varepsilon}_{123} = \begin{bmatrix} \varepsilon_1 & 0 & 0 \\ 0 & \varepsilon_2 & 0 \\ 0 & 0 & \varepsilon_3 \end{bmatrix} \tag{2.70}$$

is the relative permittivity of the medium in material frame coordinates.

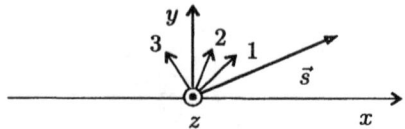

Fig. 2.5. Propagation axes x, y, z for a biaxial optical medium. The material axes 1, 2, 3 are not in the plane of the diagram.

2.6.2 Propagation Axes

Light propagation is specified with respect to a second coordinate frame, labelled x, y, z as shown in Fig. 2.5. At a plane interface (or for planar layered media) x is taken to be the normal to the surface and y and z are parallel to the surface. Whenever it is practical to do so, the light waves are assumed to propagate in the x-y plane, i.e. the x-y plane is the plane of incidence.

2.6.3 Rotations

The relative directions of the material and propagation axes are specified in the following way. We start with aligned axes (1 parallel to x, 2 parallel to y, 3 parallel to z) and rotate the material in turn by an angle η about x, by an angle ψ about z and by an angle ξ about x again. Transformations between axes are carried out using the rotation matrices

$$\hat{R}_x(\phi) = \begin{bmatrix} 1 & 0 & 0 \\ 0 & \cos\phi & -\sin\phi \\ 0 & \sin\phi & \cos\phi \end{bmatrix} \qquad (2.71)$$

and

$$\hat{R}_z(\phi) = \begin{bmatrix} \cos\phi & -\sin\phi & 0 \\ \sin\phi & \cos\phi & 0 \\ 0 & 0 & 1 \end{bmatrix}. \qquad (2.72)$$

For a tilted columnar material formed by vacuum deposition the angle η is equal to zero. Thus for a columnar film it is sufficient to rotate the material by an angle ψ about z to establish the column angle in the deposition plane, followed by a rotation of ξ about x to establish the position of the propagation plane.

As an example of the use of the rotation matrices, the electric field $\vec{E}$ in the propagation frame can be determined from the material frame field $\vec{E}_{123}$ by

$$\vec{E} = \hat{R}_x(\xi)\hat{R}_z(\psi)\hat{R}_x(\eta)\vec{E}_{123}. \tag{2.73}$$

Note that we have not used the subscripts x, y, z to identify $\vec{E}$ as a vector in the propagation frame. Such a default notation is useful because propagation frame values are used more frequently than material frame values.

The rotation matrices are unimodular, $\det \hat{R}_x = 1$, $\det \hat{R}_z = 1$, and the inverses correspond to negative angles of rotation, $[\hat{R}_x(\xi)]^{-1} = \hat{R}_x(-\xi)$ etc. Thus the inverse of the transformation given by Eq.(2.73) can be written as

$$\vec{E}_{123} = \hat{R}_x(-\eta)\hat{R}_z(-\psi)\hat{R}_x(-\xi)\vec{E}. \tag{2.74}$$

How are matrices transformed between the two coordinate systems? To answer this question we consider a particular case, transformation of the equation that connects the electric displacement and the electric field. We start in the material frame with

$$\vec{D}_{123} = \varepsilon_0\hat{\varepsilon}_{123}\vec{E}_{123}, \tag{2.75}$$

and use Eq.(2.74), together with a similar equation for the electric displacement, to obtain sequentially

$$\begin{aligned}
\hat{R}_x(-\eta)\hat{R}_z(-\psi)\hat{R}_x(-\xi)\vec{D} &= \varepsilon_0\hat{\varepsilon}_{123}\hat{R}_x(-\eta)\hat{R}_z(-\psi)\hat{R}_x(-\xi)\vec{E}, \\
\vec{D} &= \varepsilon_0\hat{R}_x(\xi)\hat{R}_z(\psi)\hat{R}_x(\eta)\hat{\varepsilon}_{123}\hat{R}_x(-\eta)\hat{R}_z(-\psi)\hat{R}_x(-\xi)\vec{E}, \\
\vec{D} &= \varepsilon_0\hat{\varepsilon}\vec{E}. \tag{2.76}
\end{aligned}$$

Hence the (symmetric) relative permittivity for the propagation plane, which we write as

$$\hat{\varepsilon} = \begin{bmatrix} \varepsilon_{xx} & \varepsilon_{xy} & \varepsilon_{xz} \\ \varepsilon_{xy} & \varepsilon_{yy} & \varepsilon_{yz} \\ \varepsilon_{xz} & \varepsilon_{yz} & \varepsilon_{zz} \end{bmatrix}, \tag{2.77}$$

can be computed using the equation

$$\hat{\varepsilon} = \hat{R}_x(\xi)\hat{R}_z(\psi)\hat{R}_x(\eta)\hat{\varepsilon}_{123}\hat{R}_x(-\eta)\hat{R}_z(-\psi)\hat{R}_x(-\xi). \tag{2.78}$$

2.6.4   Computations

In the *BTF Toolbox* the rotation matrices $\hat{R}_x$ and $\hat{R}_z$ are available as the functions **rxmat** and **rzmat**. In each case the argument is an angle in radian. The relative permittivity $\hat{\epsilon}$ for the propagation plane is obtained using the function **epsilon** with argument $[n_1\ n_2\ n_3\ \eta\ \psi\ \xi]$. For example, **epsilon**$([2.4\ 1.55\ 2.0\ 0\ 40\pi/180\ 30\pi/180])$ returns

$$
\begin{bmatrix}
4.3728 & 1.4318 & 0.8266 \\
1.4318 & 3.8423 & -0.0910 \\
0.8266 & -0.0910 & 3.9474
\end{bmatrix}.
$$

To avoid the tedium of writing out every detail of long input arguments, such as $[2.4\ 1.55\ 2.0\ 0\ 40\pi/180\ 30\pi/180]$, we have defined a hierarchy of arguments for the *BTF Toolbox*. Thus, the three principal refractive indices of a birefringent material are represented by

$$indices = [n_1\ n_2\ n_3],$$

and the three angles that define the angular position of the birefringent material are represented by

$$angles = [\eta\ \psi\ \xi].$$

Then the complete representation of the birefringent material can be obtained by combining *indices* and *angles*,

$$material = [indices\ angles],$$

and the call to the previous function would be **epsilon**$(material)$.

2.7   Propagation in a Common Direction in a Biaxial Medium

2.7.1   Maxwell's Equations

Plane, harmonic waves that propagate in a biaxial medium satisfy the vector form of Maxwell's equations listed in the left-hand side of Table 2.4.

The middle column of Table 2.4 shows a column vector form of Maxwell's equations for individual plane waves propagating in the x-y plane. These equations are derived from the equations in the right-hand column of the table by using the matrix

Table 2.4. Maxwell's equations for plane, harmonic waves. (Adapted from I.J. Hodgkinson, S. Kassam and Q.H. Wu, *Journal of Computational Physics* 133, 75, 1997. Copyright © 1997 Academic Press. Reprinted with permission.)

Vector Form	Column Vector Form	Matrix Form
$n_s \mathbf{s} \times \mathbf{E} = z_0 \mathbf{H}$	$n_s \hat{s} \vec{E} = z_0 \vec{H}$	$\hat{s} \hat{E} \hat{n}_s = z_0 \hat{H}$
$n_s \mathbf{s} \times \mathbf{H} = -\frac{1}{z_0} \epsilon \mathbf{E}$	$n_s \hat{s} \vec{H} = -\frac{1}{z_0} \hat{\epsilon} \vec{E}$	$\hat{s} \hat{H} \hat{n}_s = -\hat{\epsilon} \hat{E} / z_0$
$\mathbf{s}.\mathbf{D} = 0$	$\vec{s}' \vec{D} = 0$	$\vec{s}' \hat{D} = 0$
$\mathbf{s}.\mathbf{H} = 0$	$\vec{s}' \vec{H} = 0$	$\vec{s}' \hat{H} = 0$

$$\hat{s} = \begin{bmatrix} 0 & 0 & s_y \\ 0 & 0 & -s_x \\ -s_y & s_x & 0 \end{bmatrix} \qquad (2.79)$$

for the operation $\mathbf{s} \times$; the row vector

$$\vec{s}' = [s_x \ s_y \ 0] \qquad (2.80)$$

is the simple transpose of $\vec{s}$. Note that the matrix $\hat{s}$ is singular, $\det \hat{s} = 0$, and hence care is needed with matrix algebra involving $\hat{s}$.

The right-hand side of Table 2.4 lists Maxwell's equations in matrix form. Here all solutions to the problem of plane wave propagation in a common direction are combined together. Thus $\hat{E}$, for example, is formed from the $\vec{E}$'s and $\hat{n}_s$ is a diagonal matrix formed from the n_s's associated with the individual plane waves.

2.7.2 Fresnel's Equation

The n's for propagation in a common direction can be determined directly from Fresnel's equation, Eq.(2.62). The material frame components s_1, s_2 and s_3 of the wave vector are required and can be calculated from the propagation frame values using the rotation matrices,

$$\begin{bmatrix} s_1 \\ s_2 \\ s_3 \end{bmatrix} = \hat{R}_x(-\eta) \hat{R}_z(-\psi) \hat{R}_x(-\xi) \begin{bmatrix} s_x \\ s_y \\ 0 \end{bmatrix} . \qquad (2.81)$$

2.7.3   Eigenequations for Normalized Fields

Matrix solutions to the problem of propagation in a common direction in a biaxial medium can be obtained from the column vector form of Maxwell's equations (middle column of Table 2.4) in the following way. It is assumed that the dielectric matrix $\hat{\varepsilon}$ is known and the wave propagation direction $\vec{s}$ is specified for the x-y propagation plane. The refractive indices $n_s = \varepsilon^{1/2}$ associated with the two waves and the fields $\vec{E}$, $\vec{D}$, $\vec{H}$ that appear in the equations, together with the magnetic induction $\vec{B} = \mu_0 \vec{H}$, are the unknowns.

The first two equations in the middle column of Table 2.4 can be combined simultaneously, to eliminate $\vec{H}$. This leaves an equation for $\vec{E}$ which can be organized as an eigenequation. In the most simple form,

$$(-\hat{\varepsilon}^{-1}\hat{s}^2)\vec{E} = \eta_s \vec{E}, \qquad (2.82)$$

each electric field $\vec{E}$ that satisfies the equation is a *right eigenvector* and $\eta_s = 1/\varepsilon_s$ is an eigenvalue; in the form of a generalized eigenequation,

$$\hat{I}\vec{E} = \varepsilon_s(-\hat{\varepsilon}^{-1}\hat{s}^2)\vec{E}, \qquad (2.83)$$

$\hat{I}$ is an identity matrix (which is omitted later), the electric field $\vec{E}$ is again the eigenvector but now ε_s (rather than the reciprocal) is the eigenvalue.

The advantage of all this is that the electric field, or both the electric field and the refractive indices, can be obtained by making a single call to the MATLAB eig function. One important point though is that the returned eigenvectors are normalized ($E_x^2 + E_y^2 + E_z^2 = 1$) and uncertain in sign. Thus they represent nothing more than the vibration direction of the wave field.

Several ways exist for using the MATLAB eig function. The most simple form $\hat{E} = \text{eig}(-\hat{\varepsilon}^{-1}\hat{s}^2)$ yields a matrix $\hat{E}$ in which the solutions for the electric field are organized as the columns of a 3 × 3 matrix. The four fields can be found individually by appropriate calls to the eig function,

$$
\begin{aligned}
\hat{E} &= \text{eig}(-\hat{\varepsilon}^{-1}\hat{s}^2) \\
\hat{D} &= \text{eig}(-\hat{s}^2\hat{\varepsilon}^{-1}) \\
\hat{H} &= \text{eig}(-\hat{s}\hat{\varepsilon}^{-1}\hat{s}) \\
\hat{B} &= \text{eig}(-\hat{s}\hat{\varepsilon}^{-1}\hat{s}).
\end{aligned}
\qquad (2.84)
$$

Alternatively, the equation

$$[\hat{E}, \hat{\varepsilon}_s] = \text{eig}(\hat{I}, -\hat{\varepsilon}^{-1}\hat{s}^2) \qquad (2.85)$$

Table 2.5. Eigenequations for propagation in a common direction in a biaxial medium. (Adapted from I.J. Hodgkinson, S. Kassam and Q.H. Wu, *Journal of Computational Physics* 133, 75, 1997. Copyright © 1997 Academic Press. Reprinted with permission.)

Eigenequation	MATLAB solution	Equation satisfied by solution
$\vec{E} = \varepsilon_s(-\hat{\varepsilon}^{-1}\hat{s}^2)\vec{E}$	$[\hat{E}, \hat{\varepsilon}_s] = \mathrm{eig}(\hat{I}, -\hat{\varepsilon}^{-1}\hat{s}^2)$	$\hat{E} = -\hat{\varepsilon}^{-1}\hat{s}^2\hat{E}\hat{\varepsilon}_s$
$\vec{D} = \varepsilon_s(-\hat{s}^2\hat{\varepsilon}^{-1})\vec{D}$	$[\hat{D}, \hat{\varepsilon}_s] = \mathrm{eig}(\hat{I}, -\hat{s}^2\hat{\varepsilon}^{-1})$	$\hat{D} = -\hat{s}^2\hat{\varepsilon}^{-1}\hat{D}\hat{\varepsilon}_s$
$\vec{H} = \varepsilon_s(-\hat{s}\hat{\varepsilon}^{-1}\hat{s})\vec{H}$	$[\hat{H}, \hat{\varepsilon}_s] = \mathrm{eig}(\hat{I}, -\hat{s}\hat{\varepsilon}^{-1}\hat{s})$	$\hat{H} = -\hat{s}\hat{\varepsilon}^{-1}\hat{s}\hat{H}\hat{\varepsilon}_s$
$\vec{B} = \varepsilon_s(-\hat{s}\hat{\varepsilon}^{-1}\hat{s})\vec{B}$	$[\hat{B}, \hat{\varepsilon}_s] = \mathrm{eig}(\hat{I}, -\hat{s}\hat{\varepsilon}^{-1}\hat{s})$	$\hat{B} = -\hat{s}\hat{\varepsilon}^{-1}\hat{s}\hat{B}\hat{\varepsilon}_s$

yields the electric field matrix and a diagonal eigenvalue matrix $\hat{\varepsilon}_s$ in which the nonzero elements are the ε_s's. In this method $\hat{E}$ and $\hat{\varepsilon}_s$ satisfy the equation

$$\hat{E} = -\hat{\varepsilon}^{-1}\hat{s}^2\hat{E}\hat{\varepsilon}_s. \tag{2.86}$$

A summary of similar equations for the four fields is given in Table 2.5.

Apart from the uncertainty of sign, one of the eigenvector/eigenvalue pairs for each line in Table 2.5 represents a trivial solution, and is returned as $\vec{s}$ for the eigenvector and a large value for the eigenvalue. The procedure listed in Table 2.6 recognizes and removes this pair and then reduces $\hat{E}$ and $\hat{n}_s$ to 3×2 and 2×2 matrices respectively.

To illustrate the procedure, we have included a script file **fresnel** in the *BTF Toolbox* for computing $\hat{n}_s$, $\hat{E}$, $\hat{D}$, $\hat{H}$ and $\hat{B}$. Values are calculated for the example with $n_1 = 2.4$, $n_2 = 1.55$ and $n_3 = 2.0$, $\eta = 0°$, $\psi = 40°$, $\xi = 30°$, for the propagation direction defined by $s_1 = 0.6$, $s_2 = (1 - s_1^2)^{1/2} = 0.8$, $s_3 = 0$.

Table 2.6. Eigenequation procedure for determining fields in a biaxial medium.

Given $\hat{\varepsilon}_{123}$ and $\vec{s}$ in the propagation frame

$\Downarrow$

Use rotation matrices to calculate $\hat{\varepsilon}$

$\Downarrow$

Call the eig function with

$$[\hat{E}, \hat{\varepsilon}_s] = \text{eig}(\hat{I}, -\hat{\varepsilon}^{-1}\hat{s}^2)$$

$\Downarrow$

Identify the trivial solution

$\Downarrow$

Reduce order of $\hat{E}$ to 3×2 and $\hat{\varepsilon}_s$ to 2×2

$\Downarrow$

Calculate

$\Downarrow$

$$\hat{n}_s = \hat{\varepsilon}_s^{1/2}$$

$$\hat{D} = \varepsilon_0 \hat{\varepsilon} \hat{E}$$

$$\hat{H} = \hat{s} \hat{E} \hat{n}_s / z_0$$

$$\hat{B} = \mu_0 \hat{H}$$

Chapter 3

Basis Vectors

In practice, polarizing components are used with many different types of optical source. The result obtained in a particular case depends on the characteristics of the light that is incident on the polarizing component, including the spread of wavelengths and the polarization. Thus the light from the sun is broadband and unpolarized, the light from a spectral lamp may be described as quasi-monochromatic and unpolarized, and the radiation emitted by a cw laser may be coherent and linearly polarized.

In this chapter we consider the different polarization states that are possible for light, and the mathematical representation of these states. We have divided the material to be presented into the three sections shown schematically in Fig. 3.1. From left to right the figure shows

▨ partially-polarized light represented by a four-element column vector $\vec{S}$ called a Stokes vector

▨ polarized light propagating in an isotropic medium and represented by a two-element column vector $\vec{J}$ called a Jones vector

▨ polarized light propagating in a birefringent medium and represented by a four-element column vector $\vec{F}$ called a Berreman field vector.

In each of the three cases, the polarization state of the optical fields in the medium can be described as a linear sum of a set of basis vectors and the mathematical solution of a problem, such as transmission by a multilayered birefringent coating, is simplified to the determination of the form and weighting of the basis vectors.

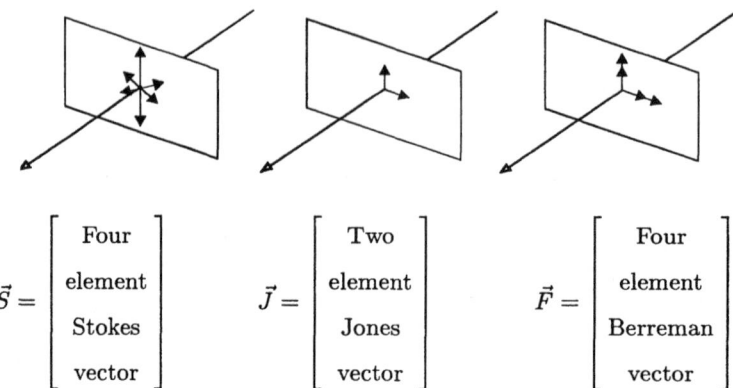

$$\vec{S} = \begin{bmatrix} \text{Four} \\ \text{element} \\ \text{Stokes} \\ \text{vector} \end{bmatrix} \qquad \vec{J} = \begin{bmatrix} \text{Two} \\ \text{element} \\ \text{Jones} \\ \text{vector} \end{bmatrix} \qquad \vec{F} = \begin{bmatrix} \text{Four} \\ \text{element} \\ \text{Berreman} \\ \text{vector} \end{bmatrix}$$

Fig. 3.1. Partially polarized light represented by a Stokes vector, polarized light in an isotropic medium represented by a Jones vector, and polarized light propagating in a birefringent layer represented by a Berreman field vector.

3.1 Partially Coherent States

3.1.1 Coherence

Fig. 3.2 shows a beam of quasimonochromatic light travelling along the x-axis. The instantaneous electric field of the light has been resolved into two fields, $E_y = |E_y| \exp(i\delta_y)$ along y and $E_z = |E_z| \exp(i\delta_z)$ along z. As time proceeds the real amplitudes $|E_y|$ and $|E_z|$ and the phase retardations δ_y and δ_z change slowly with respect to the period of the light, and in a random way.

In theory, the polarization state of the beam that is being considered here can be defined in terms of the electric fields E_y and E_z. However, in practice the frequency of an optical field is so high that E_y and E_z cannot be measured directly. From an experimental point of view it is necessary to consider the definition of polarization states in terms of irradiance, the quantity measured by practical detectors.

Due to the statistical nature of the fields, which depend on superposed light emissions from a large but finite number of atoms, repeated measurements of the irradiance will not yield exactly the same value. However, the deviations can be reduced to a negligible amount by choosing an observation time that is sufficiently long. The notation $\langle E_y E_y^* \rangle$, given here for an electric field, is used to indicate an average over such a time interval; the $*$ symbol indicates complex conjugate.

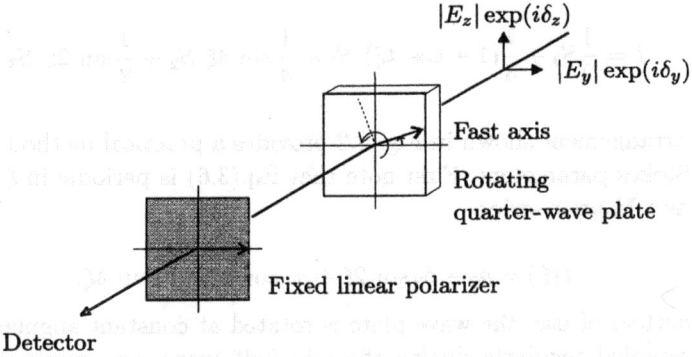

Fig. 3.2. Ellipsometer for measuring the state of polarization of a beam of light.

3.1.2 Stokes Parameters

An equation for the irradiance I at the detector in Fig. 3.2 can be derived by (i) resolving $|E_y| \exp(i\delta_y)$ and $|E_z| \exp(i\delta_z)$ along the fast and slow axes of the quarter-wave plate, (ii) multiplying the component along the slow axis by i to allow for the 90° retardation of the wave plate, resolving back along y to obtain the electric field in the transmission direction of the polarizer and, finally, determining an expression for $\langle |E|^2 \rangle$ which is the irradiance apart from a constant. After completing these algebraic steps, the result can be put in the form

$$
\begin{aligned}
I = {} & \tfrac{1}{2}(\langle |E_y|^2 \rangle + \langle |E_z|^2 \rangle) + \tfrac{1}{4}(1 + \cos 4\xi)(\langle |E_y|^2 \rangle - \langle |E_z|^2 \rangle) \\
& - \tfrac{1}{4}\sin 4\xi \, \langle 2|E_y||E_z| \cos \Delta \rangle + \tfrac{1}{2} \sin 2\xi \, \langle 2|E_y||E_z| \sin \Delta \rangle,
\end{aligned}
\tag{3.1}
$$

where $\Delta = \delta_y - \delta_z$ is the phase retardation of E_y relative to E_z.

The *Stokes parameters*,[8] defined in terms of time-averaged fields as

$$
S_0 = \langle |E_y|^2 \rangle + \langle |E_z|^2 \rangle \tag{3.2}
$$
$$
S_1 = \langle |E_y|^2 \rangle - \langle |E_z|^2 \rangle \tag{3.3}
$$
$$
S_2 = \langle 2|E_y||E_z| \cos \Delta \rangle \tag{3.4}
$$
$$
S_3 = \langle 2|E_y||E_z| \sin \Delta \rangle, \tag{3.5}
$$

quantify the state of polarization of the beam. Using the Stokes parameters, Eq.(3.1) for the irradiance I can be written as

$$I = \frac{1}{2}S_0 + \frac{1}{4}(1 + \cos 4\xi)\, S_1 - \frac{1}{4}\sin 4\xi\, S_2 + \frac{1}{2}\sin 2\xi\, S_3. \qquad (3.6)$$

The arrangement shown in Fig. 3.2 provides a practical method for measuring the Stokes parameters. First note that Eq.(3.6) is periodic in ξ and can be written as a Fourier series

$$I(\xi) = a_0 + b_2 \sin 2\xi + a_4 \cos 4\xi + b_4 \sin 4\xi. \qquad (3.7)$$

In one method of use, the wave plate is rotated at constant angular speed and $I(\xi)$ is sampled regularly during the odd half turns, i.e. while $0 \le \xi \le \pi$. During the even half turns the Fourier coefficients a_0, b_2, a_4, b_4 are calculated and the Stokes parameters are computed using

$$
\begin{aligned}
S_0 &= 2(a_0 - a_4) \\
S_1 &= 2b_2 \\
S_2 &= 2a_4 \\
S_3 &= -2b_4.
\end{aligned}
\qquad (3.8)
$$

3.1.3 Stokes Vectors

The column vector

$$\vec{S} = \begin{bmatrix} S_0 \\ S_1 \\ S_2 \\ S_3 \end{bmatrix} \qquad (3.9)$$

formed from the Stokes parameters is called a *Stokes vector*. One advantage of the Stokes vector as a descriptor of polarization state is that it can be used for partially polarized light, unpolarized light (see Fig. 3.3) and polarized light (see Figs. 3.4–3.6). Another advantage is that it provides both qualitative and quantitative measures of polarization state, through the signs and magnitudes of S_1, S_2, S_3.

The element S_0 is always positive as it represents the total irradiance, apart from a constant (see Eq.(3.2)). Equations (3.3)–(3.5) show that a positive sign for S_1 indicates a tendency for horizontal $\mathcal{P}$ (linear polarization), a positive sign for S_2 shows preference for $\mathcal{P}$ at $+45°$, and a positive sign for S_3 shows a tendency for $\mathcal{R}$ (right circular, see Sect. 3.2.3). Similarly, negative signs indicate preferences for vertical $\mathcal{P}$, $\mathcal{P}$ at $-45°$, and $\mathcal{L}$.

3.1.4 Degree of Polarization

The Stokes parameters for a beam of light satisfy the inequality,

$$0 \leq (S_1^2 + S_2^2 + S_3^2)^{1/2} \leq S_0. \tag{3.10}$$

The beam can be thought of as the superposition of a polarized part of irradiance $(S_1^2 + S_2^2 + S_3^2)^{1/2}$ and an unpolarized part of irradiance $S_0 - (S_1^2 + S_2^2 + S_3^2)^{1/2}$. Thus Eq.(3.10) is a statement of conservation of energy for the superposition.

The *degree of polarization* of the beam is defined by the equation

$$V = (S_1^2 + S_2^2 + S_3^2)^{1/2} / S_0, \tag{3.11}$$

and hence V is a number between 0 and 1.

3.1.5 Unpolarized Light

Natural or *unpolarized light* is the special case in which $S_1 = S_2 = S_3 = 0$. The normalized Stokes vector is

$$\vec{S} = \begin{bmatrix} 1 \\ 0 \\ 0 \\ 0 \end{bmatrix}, \tag{3.12}$$

and $V = 0$.

3.1.6 Partially Polarized Light

In the general case, when

$$0 < (S_1^2 + S_2^2 + S_3^2)^{1/2} < S_0, \tag{3.13}$$

the light is said to be *partially polarized*, and $0 < V < 1$.

3.1.7   Polarized Light

If

$$(S_1^2 + S_2^2 + S_3^2)^{1/2} = S_0, \tag{3.14}$$

then the light represented is completely polarized and $V = 1$.

3.1.8   Basis Vectors

The normalized Stokes vectors (see Table 3.1),

$$\begin{bmatrix} 1 \\ 0 \\ 0 \\ 0 \end{bmatrix} \quad \text{for unpolarized light,}$$

$$\begin{bmatrix} 1 \\ \pm 1 \\ 0 \\ 0 \end{bmatrix} \quad \text{for horizontal linear,}$$

$$\begin{bmatrix} 1 \\ 0 \\ \pm 1 \\ 0 \end{bmatrix} \quad \text{for linear at } \pm 45°, \text{ and}$$

$$\begin{bmatrix} 1 \\ 0 \\ 0 \\ \pm 1 \end{bmatrix} \quad \text{for right and left circular,}$$

form a complete set of basis vectors for superposition (by addition only) of *incoherent states*. Thus the Stokes vector for any partially polarized state can

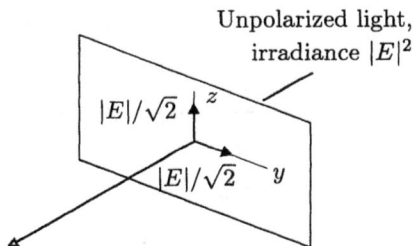

Fig. 3.3. Representation of unpolarized light.

be expressed as a linear sum of incoherent waves represented by these basis vectors, and hence the magnitude of each of S_1, S_2 and S_3 in a Stokes vector signals the relative importance of each basis state.

As an example, an unpolarized wave of unit irradiance is equivalent to the superposition of incoherent, horizontal $\mathcal{P}$ and vertical $\mathcal{P}$ states, each of one-half unit irradiance:-

$$
\begin{bmatrix} 1 \\ 0 \\ 0 \\ 0 \end{bmatrix} = \frac{1}{2} \begin{bmatrix} 1 \\ 1 \\ 0 \\ 0 \end{bmatrix} + \frac{1}{2} \begin{bmatrix} 1 \\ -1 \\ 0 \\ 0 \end{bmatrix}. \tag{3.15}
$$

This well-known representation of unpolarized light is illustrated in Fig. 3.3 for an unpolarized beam of irradiance $|E|^2$.

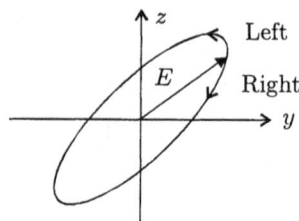

Fig. 3.4. Elliptical polarization states. The light is propagating along the x-axis, out of the plane of the diagram. At a point in space the tip of the electric vector traces an ellipse as time proceeds.

3.2 Coherent States

3.2.1 Jones Vectors

If the beam of light shown in Fig. 3.2 is coherent, then the time-average braces in Eqs.(3.1)–(3.5) can be removed and the beam can be considered as the superposition of two coherent waves E_y and E_z. A two-element column vector formed from E_y and E_z,

$$\vec{J} = \begin{bmatrix} E_y \\ E_z \end{bmatrix}, \tag{3.16}$$

can now be used as a descriptor of polarization state, and is known as a *Jones vector*.[8] With $E_y = |E_y| \exp(i\Delta)$ and $E_z = |E_z|$, where $\Delta = \delta_y - \delta_z$ is the phase lag of E_y relative to E_z, the Jones vector can be written as

$$\vec{J} = \begin{bmatrix} |E_y| e^{i\Delta} \\ |E_z| \end{bmatrix}. \tag{3.17}$$

3.2.2 Elliptical Polarization

The set of possible polarization states for coherent light can be explored by considering Jones vectors with all significant relationships between $|E_y|$, $|E_z|$ and Δ.

In the most general case, with $|E_y| \neq |E_z|$ and $\Delta \neq n\pi$, the Jones vector represents *elliptical polarization*, i.e. on a phasor diagram the tip of the electric field vector traces an ellipse as time proceeds as shown in Fig. 3.4. If the sense of rotation around the elliptical path is clockwise, then the light is said to be *right elliptical*, otherwise it is *left elliptical*.

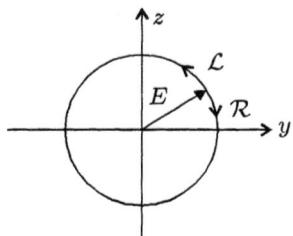

Fig. 3.5. Right $\mathcal{R}$ and left $\mathcal{L}$ circular polarization states. The light is propagating along the x-axis, out of the page. At a point in space the tip of the electric vector traces a clockwise circle for $\mathcal{R}$ and an anticlockwise circle for $\mathcal{L}$ as time proceeds.

In Fig.3.4 the principal axes of the ellipse are not aligned with the y and z axes. Coincidence will occur though, if $\Delta = (n + \frac{1}{2})\pi$. The Jones vector is then

$$\begin{bmatrix} |E_y| \\ -i|E_z| \end{bmatrix} \tag{3.18}$$

for right elliptical, and

$$\begin{bmatrix} |E_y| \\ i|E_z| \end{bmatrix} \tag{3.19}$$

for left elliptical.

3.2.3   Circular Polarization

When $|E_y| = |E_z|$ and $\Delta = (n + \frac{1}{2})\pi$, the ellipse described in the previous section degenerates into a circle (Fig. 3.5). Right and left circular polarizations are designated $\mathcal{R}$ and $\mathcal{L}$ respectively. The Jones vectors are usually given in a normalized form,

$$\vec{J}_{\mathcal{R}} = \frac{1}{\sqrt{2}} \begin{bmatrix} 1 \\ -i \end{bmatrix} \tag{3.20}$$

and

$$\vec{J}_{\mathcal{L}} = \frac{1}{\sqrt{2}} \begin{bmatrix} 1 \\ i \end{bmatrix}. \tag{3.21}$$

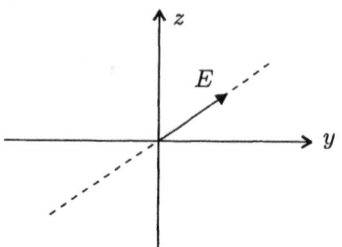

Fig. 3.6. Linear polarization. The light is propagating along the x-axis, out of the page. At a point in space the tip of the electric vector traces a straight line as time proceeds.

3.2.4 Linear Polarization

Another degeneration of the ellipse in Sect. 3.2.2 occurs when $\Delta = n\pi$, on this occasion to a straight line and hence the term *linear polarization* (Fig. 3.6). The designator $\mathcal{P}$ relates to the alternative name of *plane polarization*. As examples, the normalized Jones vector for light linearly polarized along the horizontal y-axis is

$$\vec{J}_h = \begin{bmatrix} 1 \\ 0 \end{bmatrix}, \qquad (3.22)$$

and linear polarization along the vertical z-axis is represented by

$$\vec{J}_v = \begin{bmatrix} 0 \\ 1 \end{bmatrix}. \qquad (3.23)$$

3.2.5 Basis Vectors

The Jones vectors $\vec{J}_h$ and $\vec{J}_v$ satisfy the equations

$$\vec{J}_h \cdot \vec{J}_h^* = \vec{J}_v \cdot \vec{J}_v^* = 1$$
$$\vec{J}_h \cdot \vec{J}_v^* = \vec{J}_v \cdot \vec{J}_h^* = 0, \qquad (3.24)$$

and hence form an orthonormal set. This means that any coherent polarized state can be expressed as a linear sum of $\vec{J}_h$ and $\vec{J}_v$.

3.2.6 Photons

The requirements of an orthonormal set are satisfied by $\vec{J}_\mathcal{R}$ and $\vec{J}_\mathcal{L}$, as well as by $\vec{J}_h$ and $\vec{J}_v$. Thus any coherent polarized state can be represented by an appropriate mixture of $\vec{J}_\mathcal{R}$ and $\vec{J}_\mathcal{L}$. In the quantum-mechanical description of light a photon has an intrinsic (or *spin*) angular momentum of either $-h/2\pi$, corresponding to $\mathcal{R}$, or $+h/2\pi$, corresponding to $\mathcal{L}$, where $h = 6.626176 \times 10^{-34}\,\mathrm{J\,s}$ is *Planck's constant*. Thus a plausible argument can be made for regarding $\vec{J}_\mathcal{R}$ and $\vec{J}_\mathcal{L}$ as foundation basis vectors for polarized light.

3.2.7 Ellipsometric Parameters

We have seen that the state of polarization of a coherent beam of light propagating along the x-axis can be represented by a normalized Jones vector in which the elements are based on the components of the electric field along y and z. An equivalent description, also used widely, involves a pair of values called the *ellipsometric parameters*[9] and defined by the equations

$$\Psi = \tan^{-1}(|E_y|/|E_z|) \tag{3.25}$$
$$\Delta = \mathrm{angle}(E_y/E_z). \tag{3.26}$$

Thus the first ellipsometric parameter Ψ is a measure of the ratio of the magnitudes of E_y and E_z, and the second parameter Δ is the phase retardation of E_y relative to E_z. As an example, the ellipsometric parameters for an $\mathcal{R}$ state are $\Psi = \tan^{-1}(|1|/|-i|) = \pi/4$ and $\Delta = \mathrm{angle}(1/-i) = \pi/2$.

In Sect. 3.1.2 we described apparatus for measuring the Stokes parameters of a beam of light. For a coherent beam we can express $|E_y|$, $|E_z|$, and Δ in terms of S_0, S_1, S_2, S_3,

$$|E_y| = [(S_0 + S_1)/2]^{1/2}$$
$$|E_z| = [(S_0 - S_1)/2]^{1/2}$$
$$\Delta = \tan^{-1}(S_3/S_2), \tag{3.27}$$

and hence the same apparatus can be used for measuring the components of the Jones vector or the ellipsometric parameters. A summary of Stokes vectors, Jones vectors, and ellipsometric parameters for a range of polarization states is given in Table 3.1.

Table 3.1. Polarization states.

State	Stokes vector	Jones vector	Ψ	Δ
Unpolarized	$\begin{bmatrix} 1 \\ 0 \\ 0 \\ 0 \end{bmatrix}$		$-$	$-$	$-$
Horizontal $\mathcal{P}$	$\begin{bmatrix} 1 \\ 1 \\ 0 \\ 0 \end{bmatrix}$	$\begin{bmatrix} 1 \\ 0 \end{bmatrix}$	$\frac{\pi}{2}$	0
Vertical $\mathcal{P}$	$\begin{bmatrix} 1 \\ -1 \\ 0 \\ 0 \end{bmatrix}$	$\begin{bmatrix} 0 \\ 1 \end{bmatrix}$	0	0
$\mathcal{P}$ at $45°$	$\begin{bmatrix} 1 \\ 0 \\ 1 \\ 0 \end{bmatrix}$	$\frac{1}{\sqrt{2}} \begin{bmatrix} 1 \\ 1 \end{bmatrix}$	$\frac{\pi}{4}$	0
$\mathcal{P}$ at $-45°$	$\begin{bmatrix} 1 \\ 0 \\ -1 \\ 0 \end{bmatrix}$	$\frac{1}{\sqrt{2}} \begin{bmatrix} 1 \\ -1 \end{bmatrix}$	$\frac{\pi}{4}$	π
$\mathcal{R}$	$\begin{bmatrix} 1 \\ 0 \\ 0 \\ 1 \end{bmatrix}$	$\frac{1}{\sqrt{2}} \begin{bmatrix} 1 \\ -i \end{bmatrix}$	$\frac{\pi}{4}$	$\frac{\pi}{2}$
$\mathcal{L}$	$\begin{bmatrix} 1 \\ 0 \\ 0 \\ -1 \end{bmatrix}$	$\frac{1}{\sqrt{2}} \begin{bmatrix} 1 \\ i \end{bmatrix}$	$\frac{\pi}{4}$	$-\frac{\pi}{2}$

3.3 Propagation in Layered Biaxial Media

A single plane wave propagating in the x-y plane and incident on a parallel layer of a biaxial medium will, in general, initiate four plane waves in the biaxial medium. The four waves are linearly polarized, in directions specified by the D fields, and share the same value of the Snell's law quantity

$$\beta = n \sin\theta. \qquad (3.28)$$

However the propagation directions θ of the four waves are all different in general, as are the effective refractive indices n. The problem that we wish to solve here can be stated as:- given the principal refractive indices, the angles η, ψ, ξ, and the Snell's law quantity β, how can the n's, the θ's and the field components be calculated?

Before pursuing the solution though, notice that knowledge of the four α's defined by

$$\alpha = n \cos\theta \qquad (3.29)$$

amounts to knowledge of the four n's and the four θ's, because

$$n = (\alpha^2 + \beta^2)^{1/2} \qquad (3.30)$$

and

$$\theta = \sin^{-1}(\beta/n). \qquad (3.31)$$

3.3.1 Fresnel's Quartic Equation

The n's can be obtained from Fresnel's equation. However, in this case explicit solutions are not practical because the re-cast Fresnel's equation is a quartic in α,

$$(c_{11}^2 n_1^2 + c_{12}^2 n_2^2 + c_{13}^2 n_3^2)\alpha^4$$

$$+2\beta(c_{11}c_{21}n_1^2 + c_{12}c_{22}n_2^2 + c_{13}c_{23}n_3^2)\alpha^3$$

$$+\{\beta^2[(c_{11}^2 + c_{21}^2)n_1^2 + (c_{12}^2 + c_{22}^2)n_2^2 + (c_{13}^2 + c_{23}^2)n_3^2] + (c_{11}^2 - 1)n_2^2 n_3^2$$

$$+(c_{12}^2 - 1)n_3^2 n_1^2 + (c_{13}^2 - 1)n_1^2 n_2^2\}\alpha^2$$

$$+[2\beta^3(c_{11}c_{21}n_1^2 + c_{12}c_{22}n_2^2 + c_{13}c_{23}n_3^2)$$

$$+2\beta(c_{11}c_{21}n_2^2 n_3^2 + c_{12}c_{22}n_3^2 n_1^2 + c_{13}c_{23}n_1^2 n_2^2)]\alpha$$

$$+\beta^4(c_{21}^2 n_1^2 + c_{22}^2 n_2^2 + c_{23}^2 n_3^2)$$

$$+\beta^2[(c_{21}^2 - 1)n_2^2 n_3^2 + (c_{22}^2 - 1)n_3^2 n_1^2 + (c_{23}^2 - 1)n_1^2 n_2^2] + n_1^2 n_2^2 n_3^2 = 0.$$

$$(3.32)$$

In the above equation the coefficients such as c_{11} are elements of the matrix $\hat{R}_x(-\eta)\hat{R}_z(-\psi)\hat{R}_x(-\xi)$ that transforms vectors from the material frame to the propagation frame. Three special cases of interest are considered below.

3.3.2   Propagation in the Deposition Plane

When the plane of propagation of the light corresponds to the deposition plane ($\xi = 0$) of a columnar thin film ($\eta = 0$), Fresnel's equation simplifies to

$$[(n_1^2 \cos^2 \psi + n_2^2 \sin^2 \psi)\alpha^2 + 2\beta \cos \psi \sin \psi (n_1^2 - n_2^2)\alpha$$
$$+\beta^2(n_1^2 \sin^2 \psi + n_2^2 \cos^2 \psi) - n_1^2 n_2^2] \times (\alpha^2 + \beta^2 - n_3^2) = 0. \qquad (3.33)$$

After defining

$$n_p = (\sin^2 \psi/n_1^2 + \cos^2 \psi/n_2^2)^{-1/2}, \qquad (3.34)$$

the solutions for α_1^+ and α_1^-, associated with the p polarization, can be written in the form

$$\alpha_1^\pm = -[(1 - n_p^2/n_1^2)(n_p^2/n_2^2 - 1)]^{1/2}\beta \pm n_p^2(1/n_p^2 - \beta^2/n_1^2 n_2^2)^{1/2}, \qquad (3.35)$$

and for the s polarization

$$\alpha_2^\pm = \pm(n_3^2 - \beta^2)^{1/2}. \qquad (3.36)$$

n_p is the refractive index seen by p-polarized light travelling at normal incidence, i.e. with $\beta = 0$.

3.3.3 Uniaxial Media

In a uniaxial medium, with $n_e = n_1$, $n_o = n_2 = n_3$ and $\eta = 0$, Fresnel's quartic, Eq.(3.32), becomes

$$\{(n_e^2 \cos^2 \psi + n_o^2 \sin^2 \psi)\alpha^2 + 2\beta \cos \xi \cos \psi \sin \psi (n_e^2 - n_o^2)\alpha$$
$$+\beta^2[n_e^2 \cos^2 \xi \sin^2 \psi + n_o^2(1 - \cos^2 \xi \sin^2 \psi)] - n_e^2 n_o^2\} \times (\alpha^2 + \beta^2 - n_o^2) = 0.$$
$$(3.37)$$

We consider two special cases. Crystalline wave plates are used extensively for changing the polarization of a beam of light. These plates are usually made from a uniaxial material cut in the form of a disk and with material axis-1, the optic axis, in the plane of the surface. For this application, Fresnel's equation can be simplified by substituting $\psi = 90°$ into Eq.(3.37), with the result

$$[n_o^2 \alpha^2 + \beta^2 (n_e^2 \cos^2 \xi + n_o^2 \sin^2 \xi) - n_e^2 n_o^2] \times (\alpha^2 + \beta^2 - n_o^2) = 0. \qquad (3.38)$$

A second application involves the use of a wave plate, with material axis-1 perpendicular to the surface of the plate, for offsetting phase retardation. For this case, with $\psi = 0°$ and small angles of incidence, approximate solutions to Fresnel's equation can be written in the form

$$\alpha_1^\pm = \pm n_0(1 - \frac{\beta^2}{2n_e^2})$$
$$\alpha_2^\pm = \pm n_0(1 - \frac{\beta^2}{2n_o^2}). \qquad (3.39)$$

3.3.4 Isotropic Media

In an isotropic medium, with a single refractive index n, Fresnel's equation simplifies to

$$(\alpha^2 + \beta^2 - n^2)^2 = 0, \qquad (3.40)$$

and hence the solutions are

$$\alpha_1^\pm = \alpha_2^\pm = \pm(n^2 - \beta^2)^{1/2}. \qquad (3.41)$$

3.3.5   Basis Travelling Wave Fields

We begin by writing down two of Maxwell's equations for plane, harmonic waves from Table 3.1,

$$n\hat{s}\vec{E} = z_0\vec{H}$$

$$n\hat{s}\vec{H} = -\frac{1}{z_0}\hat{\varepsilon}\vec{E}. \tag{3.42}$$

Using Eq.(2.4) for $\hat{s}$, and the equations

$$ns_x = n\cos\theta = \alpha$$

$$ns_y = n\sin\theta = \beta, \tag{3.43}$$

we can write

$$n\hat{s} = \begin{bmatrix} 0 & 0 & \beta \\ 0 & 0 & -\alpha \\ -\beta & \alpha & 0 \end{bmatrix}. \tag{3.44}$$

Substitution into Eqs. (3.42) leads to the pair of equations,

$$\begin{bmatrix} 0 & 0 & \beta \\ 0 & 0 & -\alpha \\ -\beta & \alpha & 0 \end{bmatrix} \begin{bmatrix} E_x \\ E_y \\ E_z \end{bmatrix} = z_0 \begin{bmatrix} H_x \\ H_y \\ H_z \end{bmatrix} \tag{3.45}$$

and

$$-z_0 \begin{bmatrix} 0 & 0 & \beta \\ 0 & 0 & -\alpha \\ -\beta & \alpha & 0 \end{bmatrix} \begin{bmatrix} H_x \\ H_y \\ H_z \end{bmatrix} = \begin{bmatrix} \varepsilon_{xx} & \varepsilon_{xy} & \varepsilon_{xz} \\ \varepsilon_{xy} & \varepsilon_{yy} & \varepsilon_{yz} \\ \varepsilon_{xz} & \varepsilon_{yz} & \varepsilon_{zz} \end{bmatrix} \begin{bmatrix} E_x \\ E_y \\ E_z \end{bmatrix}. \tag{3.46}$$

Six equations are implied here and two of them,

$$E_x = -(\varepsilon_{xy}E_y + \varepsilon_{xz}E_z + z_0\beta H_z)/\varepsilon_{xx} \tag{3.47}$$

and

$$H_x = (\beta/z_0)E_z, \tag{3.48}$$

may be used to eliminate the field components E_x and H_x that are normal to interfaces and not required for boundary condition matching (but may be required for power flow calculations). This leads to the eigenequation

$$
\begin{bmatrix}
-\dfrac{\beta \varepsilon_{xy}}{\varepsilon_{xx}} & z_0 - \dfrac{z_0 \beta^2}{\varepsilon_{xx}} & -\dfrac{\beta \varepsilon_{xz}}{\varepsilon_{xx}} & 0 \\[2mm]
\dfrac{\varepsilon_{yy}}{z_0} - \dfrac{\varepsilon_{xy}^2}{z_0 \varepsilon_{xx}} & -\dfrac{\beta \varepsilon_{xy}}{\varepsilon_{xx}} & \dfrac{\varepsilon_{yz}}{z_0} - \dfrac{\varepsilon_{xy}\varepsilon_{xz}}{z_0 \varepsilon_{xx}} & 0 \\[2mm]
0 & 0 & 0 & -z_0 \\[2mm]
-\dfrac{\varepsilon_{yz}}{z_0} + \dfrac{\varepsilon_{xy}\varepsilon_{xz}}{z_0 \varepsilon_{xx}} & \dfrac{\beta \varepsilon_{xz}}{\varepsilon_{xx}} & \dfrac{\beta^2}{z_0} + \dfrac{\varepsilon_{xz}^2}{z_0 \varepsilon_{xx}} - \dfrac{\varepsilon_{zz}}{z_0} & 0
\end{bmatrix}
\begin{bmatrix} E_y \\ H_z \\ E_z \\ H_y \end{bmatrix}
= \alpha
\begin{bmatrix} E_y \\ H_z \\ E_z \\ H_y \end{bmatrix}.
$$

$$(3.49)$$

Putting

$$
\hat{L} =
\begin{bmatrix}
-\dfrac{\beta \varepsilon_{xy}}{\varepsilon_{xx}} & z_0 - \dfrac{z_0 \beta^2}{\varepsilon_{xx}} & -\dfrac{\beta \varepsilon_{xz}}{\varepsilon_{xx}} & 0 \\[2mm]
\dfrac{\varepsilon_{yy}}{z_0} - \dfrac{\varepsilon_{xy}^2}{z_0 \varepsilon_{xx}} & -\dfrac{\beta \varepsilon_{xy}}{\varepsilon_{xx}} & \dfrac{\varepsilon_{yz}}{z_0} - \dfrac{\varepsilon_{xy}\varepsilon_{xz}}{z_0 \varepsilon_{xx}} & 0 \\[2mm]
0 & 0 & 0 & -z_0 \\[2mm]
-\dfrac{\varepsilon_{yz}}{z_0} + \dfrac{\varepsilon_{xy}\varepsilon_{xz}}{z_0 \varepsilon_{xx}} & \dfrac{\beta \varepsilon_{xz}}{\varepsilon_{xx}} & \dfrac{\beta^2}{z_0} + \dfrac{\varepsilon_{xz}^2}{z_0 \varepsilon_{xx}} - \dfrac{\varepsilon_{zz}}{z_0} & 0
\end{bmatrix}.
$$

$$(3.50)$$

allows us to write the eigenequation in the form

$$
\hat{L}\vec{F} = \alpha \vec{F}. \tag{3.51}
$$

We refer to $\hat{L}$ as the *auxiliary matrix*. Solutions for the four basis fields (the $\vec{F}$'s) and the four α's can be obtained by a single MATLAB call,

$$
[\hat{F}, \hat{\alpha}] = \text{eig } \hat{L}. \tag{3.52}
$$

Here the 4×4 *field matrix*

$$
\hat{F} =
\begin{bmatrix}
E_{y1}^+ & E_{y1}^- & E_{y2}^+ & E_{y2}^- \\
H_{z1}^+ & H_{z1}^- & H_{z2}^+ & H_{z2}^- \\
E_{z1}^+ & E_{z1}^- & E_{z2}^+ & E_{z2}^- \\
H_{y1}^+ & H_{y1}^- & H_{y2}^+ & H_{y2}^-
\end{bmatrix}
\tag{3.53}
$$

contains the $\vec{F}$'s as columns, and the diagonal matrix

$$\hat{\alpha} = \begin{bmatrix} \alpha_1^+ & 0 & 0 & 0 \\ 0 & \alpha_1^- & 0 & 0 \\ 0 & 0 & \alpha_2^+ & 0 \\ 0 & 0 & 0 & \alpha_2^- \end{bmatrix} \tag{3.54}$$

contains the corresponding α's in the principal diagonal positions. Note that the matrix (rather than column vector) form of Eq.(3.51) is

$$\hat{L}\hat{F} = \hat{F}\hat{\alpha}. \tag{3.55}$$

MATLAB returns the matrix $\hat{F}$ with columns normalized so that $(E_{y1}^+)^2 + (H_{z1}^+)^2 + (E_{z1}^+)^2 + (H_{y1}^+)^2 = 1$ etc, but we note here that normalization is not necessary for subsequent mathematical processes and has no physical significance.

Eq.(3.52) is the most important equation in our book. The field matrix $\hat{F}$, and the subscript/superscript notation is discussed further in the next two chapters. The relevant function in the *BTF Toolbox* for determining the field matrix is called **fmat**.

3.3.6   Power

The average Poynting flux p_x associated with each basis vector, in the direction normal to the interface, is required for the calculation of reflectance and transmittance. Four values of p_x can be calculated by applying the general equation

$$p_x = \frac{1}{2}\Re\{E_y H_z^* - E_z H_y^*\} \tag{3.56}$$

to each basis vector in turn. Thus computation of p_x requires just the elements of the field matrix $\hat{F}$. Cyclical variations of Eq.(3.56) can be written for p_y and p_z. The function **poynting** in the *BTF Toolbox* returns either p_x or a 3×4 matrix in which the rows correspond to p_x, p_y and p_z.

Chapter 4

Transfer Matrices

Many applications in optics and, indeed, other branches of physics require computation of an output such as transmittance that results from a given input to a component. A general method of approach for a linear system is to represent the input and output parameters by column vectors and determine a transfer matrix for transforming from the input column vector to the output vector, or vice-versa. A transfer matrix may apply to an entire component, to a distinct part of a component, or to any pair of reference planes within a part.

Our interest here is in transfer matrices for the Stokes vector, the Jones vector, and the Berreman vectors. As in the previous chapter, in which polarization states and basis fields are discussed in terms of column vectors, we have divided our material into three parts. Links to the previous chapter are provided by the overview given in Fig. 4.1 which illustrates

▨ the use of a 4×4 transfer matrix $\hat{M}$, called the Mueller matrix, for transferring the Stokes vector $\vec{S}$

▨ the use of a 2×2 transfer matrix $\hat{J}$, called the Jones matrix, for transferring the Jones vector $\vec{J}$

▨ the use of a 4×4 transfer matrix $\hat{M}$, called the Berreman characteristic matrix, for transferring the Berreman total field vector $\vec{m}$.

In a polarizing component such as a crystalline wave plate the optical effects of birefringence usually dominate the effects of interference and the use of the Mueller calculus or the Jones calculus is an appropriate choice for computations, depending on coherence considerations. On the other hand, interference is usually significant in and may be an essential part of the design of a multilayered birefringent coating and in such a case the Berreman calculus is used.

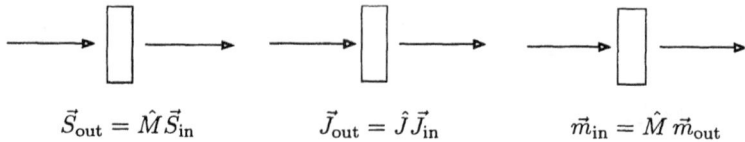

$$\vec{S}_{\text{out}} = \hat{M}\vec{S}_{\text{in}} \qquad \vec{J}_{\text{out}} = \hat{J}\vec{J}_{\text{in}} \qquad \vec{m}_{\text{in}} = \hat{M}\,\vec{m}_{\text{out}}$$

Fig. 4.1. Transfer matrices.

4.1   Mueller Calculus

The 4×4 matrix that transfers Stokes vectors from the input (in) side of an optical element to the output (out) side, as illustrated in Fig. 4.1, is called a *Mueller matrix*.[8,10] By definition the Mueller matrix $\hat{M}$ satisfies the equation

$$\vec{S}_{\text{out}} = \hat{M}\vec{S}_{\text{in}}. \tag{4.1}$$

Several Mueller matrices are listed in Table 4.1.

As an example of the use of a Mueller matrix from Table 4.1, the equation

$$\frac{1}{2}\begin{bmatrix} 1 & 1 & 0 & 0 \\ 1 & 1 & 0 & 0 \\ 0 & 0 & 0 & 0 \\ 0 & 0 & 0 & 0 \end{bmatrix}\begin{bmatrix} S_0 \\ S_1 \\ S_2 \\ S_3 \end{bmatrix} = \frac{1}{2}(S_0 + S_1)\begin{bmatrix} 1 \\ 1 \\ 0 \\ 0 \end{bmatrix} \tag{4.2}$$

shows that any light into a horizontal linear polarizer emerges with the horizontal linear polarization.

We can regard the Mueller matrices for a horizontal linear polarizer and for a retardation plate with fast axis horizontal as fundamental, and generate Mueller matrices for other orientations and series arrangements.

4.1.1   Rotated Elements

In general, if an optical element with a characteristic axis aligned with the y-axis in the y-z plane is rotated by an angle ξ about the x-axis, then the new Mueller matrix $\hat{M}(\xi)$ can be found from the original matrix $\hat{M}(0)$ by applying 4×4 rotator matrices. We have

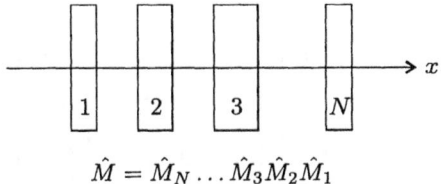

$$\hat{M} = \hat{M}_N \dots \hat{M}_3 \hat{M}_2 \hat{M}_1$$

Fig. 4.2. Mueller matrices for transmission through a series of anisotropic elements.

$$\hat{M}(\xi) = \hat{R}_m(\xi)\hat{M}(0)\hat{R}_m(-\xi) \tag{4.3}$$

where

$$\hat{R}_m(\xi) = \begin{bmatrix} 1 & 0 & 0 & 0 \\ 0 & \cos 2\xi & -\sin 2\xi & 0 \\ 0 & \sin 2\xi & \cos 2\xi & 0 \\ 0 & 0 & 0 & 1 \end{bmatrix}. \tag{4.4}$$

4.1.2   Elements in Series

For transmission through the series of N optical elements illustrated in Fig. 4.2 the transformation of the Stokes vector can be written as

$$\vec{S}_{out} = \hat{M}_N \dots \hat{M}_1 \, \vec{S}_{in}. \tag{4.5}$$

Hence the Mueller matrix for the series of elements is

$$\hat{M} = \hat{M}_N \dots \hat{M}_1; \tag{4.6}$$

the individual matrices are multiplied together in the reverse of the order that light propagates through them.

As an example of a product matrix, the combination of a horizontal linear polarizer followed by a vertical linear polarizer has the Mueller matrix

$$\frac{1}{4} \begin{bmatrix} 1 & -1 & 0 & 0 \\ -1 & 1 & 0 & 0 \\ 0 & 0 & 0 & 0 \\ 0 & 0 & 0 & 0 \end{bmatrix} \begin{bmatrix} 1 & 1 & 0 & 0 \\ 1 & 1 & 0 & 0 \\ 0 & 0 & 0 & 0 \\ 0 & 0 & 0 & 0 \end{bmatrix} = \begin{bmatrix} 0 & 0 & 0 & 0 \\ 0 & 0 & 0 & 0 \\ 0 & 0 & 0 & 0 \\ 0 & 0 & 0 & 0 \end{bmatrix} ; \qquad (4.7)$$

as expected the crossed polarizers are opaque to all polarizations. Insertion of a polarizer at +45°, between the horizontal and vertical polarizers, changes the Mueller matrix to

$$\frac{1}{4} \begin{bmatrix} 1 & 1 & 0 & 0 \\ -1 & -1 & 0 & 0 \\ 0 & 0 & 0 & 0 \\ 0 & 0 & 0 & 0 \end{bmatrix} , \qquad (4.8)$$

and hence an unpolarized input beam gives a vertical linear beam out.

4.1.3 Mueller Calculus Computations

We provide functions in the *BTF Toolbox* for Mueller calculus computations. Thus the function **rmmat** is the rotation matrix, and **mmat** is the Mueller matrix for a polarizer if one argument (ξ) is provided and the Mueller matrix for a retardation plate if two arguments (Δ and ξ) are specified. As examples, the Mueller matrix of a linear polarizer at +45° is given by **mmat**$(\pi/4)$, and the Mueller matrix for a quarter-wave plate with fast axis at +45° can be determined as **mmat**$(\pi/2, \pi/4)$.

4.2 Jones Calculus

In many applications of bulk birefringent material interference effects can be neglected and, typically, a device is characterized by the effect that it has on the polarization state of a beam of light. The surrounding medium is usually air and it is convenient to use Jones vectors to describe the state of polarization of the input and output beams.

The polarizing effect of an optical element is characterized by a 2×2 matrix called the *Jones matrix*.[8] Specifically, the Jones matrix $\hat{J}$ relates the Jones vectors of the incident and transmitted light,

$$\vec{J}_{out} = \hat{J} \vec{J}_{in}. \qquad (4.9)$$

Table 4.1. Mueller and Jones matrices.

Optical element	Mueller matrix	Jones matrix
Horizontal linear polarizer	$\frac{1}{2}\begin{bmatrix} 1 & 1 & 0 & 0 \\ 1 & 1 & 0 & 0 \\ 0 & 0 & 0 & 0 \\ 0 & 0 & 0 & 0 \end{bmatrix}$	$\begin{bmatrix} 1 & 0 \\ 0 & 0 \end{bmatrix}$
Vertical linear polarizer	$\frac{1}{2}\begin{bmatrix} 1 & -1 & 0 & 0 \\ -1 & 1 & 0 & 0 \\ 0 & 0 & 0 & 0 \\ 0 & 0 & 0 & 0 \end{bmatrix}$	$\begin{bmatrix} 0 & 0 \\ 0 & 1 \end{bmatrix}$
Linear polarizer at $+45°$	$\frac{1}{2}\begin{bmatrix} 1 & 0 & 1 & 0 \\ 0 & 0 & 0 & 0 \\ 1 & 0 & 1 & 0 \\ 0 & 0 & 0 & 0 \end{bmatrix}$	$\frac{1}{2}\begin{bmatrix} 1 & 1 \\ 1 & 1 \end{bmatrix}$
Linear polarizer at $-45°$	$\frac{1}{2}\begin{bmatrix} 1 & 0 & -1 & 0 \\ 0 & 0 & 0 & 0 \\ -1 & 0 & 1 & 0 \\ 0 & 0 & 0 & 0 \end{bmatrix}$	$\frac{1}{2}\begin{bmatrix} 1 & -1 \\ -1 & 1 \end{bmatrix}$
Retardation (Δ) plate with fast axis horizontal	$\begin{bmatrix} 1 & 0 & 0 & 0 \\ 0 & 1 & 0 & 0 \\ 0 & 0 & \cos\Delta & \sin\Delta \\ 0 & 0 & -\sin\Delta & \cos\Delta \end{bmatrix}$	$e^{-i\Delta/2}\begin{bmatrix} 1 & 0 \\ 0 & e^{i\Delta} \end{bmatrix}$
Quarter-wave plate with fast axis horizontal	$\begin{bmatrix} 1 & 0 & 0 & 0 \\ 0 & 1 & 0 & 0 \\ 0 & 0 & 0 & 1 \\ 0 & 0 & -1 & 0 \end{bmatrix}$	$e^{-i\pi/4}\begin{bmatrix} 1 & 0 \\ 0 & i \end{bmatrix}$
Quarter-wave plate with fast axis vertical	$\begin{bmatrix} 1 & 0 & 0 & 0 \\ 0 & 1 & 0 & 0 \\ 0 & 0 & 0 & -1 \\ 0 & 0 & 1 & 0 \end{bmatrix}$	$e^{i\pi/4}\begin{bmatrix} 1 & 0 \\ 0 & -i \end{bmatrix}$

A list of common Jones matrices is given in Table 4.1, and some examples are discussed below.

4.2.1 Linear Polarizer

The Jones matrix for an ideal horizontal linear polarizer is required to transfer E_y without loss and eliminate E_z. Hence the form

$$\begin{bmatrix} 1 & 0 \\ 0 & 0 \end{bmatrix}. \tag{4.10}$$

4.2.2 Retardation Plate

Now consider a retardation plate which is aligned with fast axis parallel to the y-axis. The polarizing properties of such a plate can be described by a Jones matrix of the form

$$\begin{bmatrix} 1 & 0 \\ 0 & e^{i\Delta} \end{bmatrix}. \tag{4.11}$$

The above matrix may suggest that the action of the retardation plate is to produce a phase lag of Δ in the electric field component which is in the direction of the slow axis, and leave unaltered the phase of the light component polarized in the direction of the fast axis. However, in practice the optical thickness is usually many wavelengths for propagation both along the slow axis and along the fast axis. In general a Jones matrix accounts for just the relative phase of the two waves. In the above a single phase factor can be put in either element on the principal diagonal, or it may be split between the two locations.

4.2.3 Quarter-Wave Plate

The Jones matrix for a quarter-wave plate with fast axis parallel to the y-axis can be obtained by substituting $\Delta = 90°$ into Eq.(4.11). The result is the matrix

$$\begin{bmatrix} 1 & 0 \\ 0 & i \end{bmatrix}. \tag{4.12}$$

4.2.4   Rotated Elements

If an optical element with a characteristic axis aligned with the y-axis in the y-z plane is rotated by an angle ξ about the x-axis, then the new Jones matrix $\hat{J}(\xi)$ can be found from the original matrix $\hat{J}(0)$ by applying 2×2 rotator matrices. We have

$$\hat{J}(\xi) = \hat{R}_j(\xi)\hat{J}(0)\hat{R}_j(-\xi) \tag{4.13}$$

where

$$\hat{R}_j(\xi) = \begin{bmatrix} \cos\xi & -\sin\xi \\ \sin\xi & \cos\xi \end{bmatrix}. \tag{4.14}$$

Note that

$$\hat{R}_j(\xi_2)\hat{R}_j(\xi_1) = \hat{R}_j(\xi_1 + \xi_2); \tag{4.15}$$

a rotation of ξ_1 about the x-axis followed by a rotation of ξ_2 about the same axis is equivalent to a single rotation of $\xi_1 + \xi_2$.

An advantage of the method is that a Jones matrix can be derived for the most simple geometrical orientation of an optical element, and then transformed to allow for a rotation of the element. As a trivial example, we calculate the Jones matrix for a polarizer at $90°$ starting with the matrix for the polarizer at $0°$,

$$\hat{J} = \begin{bmatrix} 0 & -1 \\ 1 & 0 \end{bmatrix} \begin{bmatrix} 1 & 0 \\ 0 & 0 \end{bmatrix} \begin{bmatrix} 0 & 1 \\ -1 & 0 \end{bmatrix}$$

$$= \begin{bmatrix} 0 & 0 \\ 0 & 1 \end{bmatrix}. \tag{4.16}$$

4.2.5   Elements in Series

Jones matrices for series arrangements of polarizing components can be formed by multiplying together the matrices representing the individual optical elements. For light travelling, in turn, through element-1, element-2, ... element-N,

$$\vec{J}_t = \hat{J}_N \ldots \hat{J}_3 \hat{J}_2 \hat{J}_1 \vec{J}_i, \tag{4.17}$$

and hence the system matrix is

$$\hat{J} = \hat{J}_N \ldots \hat{J}_3 \hat{J}_2 \hat{J}_1. \tag{4.18}$$

4.2.6 Periodic Arrangements

In some cases a transfer matrix, such as the Jones matrix for a series of wave plates, can be determined as the product of N identical, unimodular matrices. Periodic arrangements of birefringent plates fall into this category. Using $\hat{J}$ for the unimodular Jones matrix of one period gives $\hat{J}^N$ for the Jones matrix of N periods.

The following matrix identity facilitates algebraic simplification of a unimodular matrix raised to a power:-

$$\hat{J}^N = \begin{bmatrix} J_{11}U_{N-1} - U_{N-2} & J_{12}U_{N-1} \\ J_{21}U_{N-1} & J_{22}U_{N-1} - U_{N-2} \end{bmatrix} \tag{4.19}$$

where

$$U_N = \frac{\sin(N+1)\phi}{\sin\phi} \tag{4.20}$$

and

$$\phi = \cos^{-1}\left(\frac{J_{11} + J_{22}}{2}\right). \tag{4.21}$$

4.2.7 Jones Calculus Computations

We provide functions in the *BTF Toolbox* for Jones calculus computations. Thus the function **rjmat** is the rotation matrix, and **jmat** is the Jones matrix for a polarizer if one argument (ξ) is provided, and the Jones matrix for a retardation plate if two arguments (Δ and ξ) are specified. As examples, the Jones matrix of a linear polarizer at $+45°$ is given by **jmat**$(\pi/4)$, and the Jones matrix for a quarter-wave plate with fast axis at $+45°$ can be determined as **jmat**$(\pi/2, \pi/4)$.

4.3   Relationship of Mueller Calculus and Jones Calculus

Table 4.1 lists some corresponding Mueller and Jones matrices. In general an optical element that can be represented by a Jones matrix can also be represented by a Mueller matrix, but the reverse is not always true. For example an ideal depolarizer can be represented by a Mueller matrix,

$$\hat{M} = \begin{bmatrix} 1 & 0 & 0 & 0 \\ 0 & 0 & 0 & 0 \\ 0 & 0 & 0 & 0 \\ 0 & 0 & 0 & 0 \end{bmatrix}, \qquad (4.22)$$

but not by a Jones matrix.

For further discussion, and equations linking corresponding Mueller and Jones matrices, the reader is referred to an article by R.A. Chipman.[10]

4.4 Berreman Calculus

In the previous chapter we discussed a technique for determining the field matrix $\hat{F}$, the set of four basis vectors representing the waves (with the same β) that can propagate in a biaxial layer, and the four associated α's. In this chapter we discuss further properties of the field matrix, and develop a set of transfer matrices for layered biaxial media. We refer to the framework in which 4×4 matrices are used as the *Berreman calculus*.[11]

4.4.1 Field Matrix $\hat{F}$

Recall that the columns of the field matrix,

$$\hat{F} = \begin{bmatrix} E_{y1}^+ & E_{y1}^- & E_{y2}^+ & E_{y2}^- \\ H_{z1}^+ & H_{z1}^- & H_{z2}^+ & H_{z2}^- \\ E_{z1}^+ & E_{z1}^- & E_{z2}^+ & E_{z2}^- \\ H_{y1}^+ & H_{y1}^- & H_{y2}^+ & H_{y2}^- \end{bmatrix}, \qquad (4.23)$$

are the four basis vectors, and that the MATLAB call,

$$[\hat{F}, \hat{\alpha}] = \text{eig } \hat{L}, \qquad (4.24)$$

returns both $\hat{F}$ and $\hat{\alpha}$. (See the function **fmat** in the *BTF Toolbox*.)

The case of an isotropic medium requires special attention because the eigenequation (Eq.(4.24)) cannot be used. Both n and θ are known in this case, and components of the travelling wave fields are related by simple expressions. Specifically,

$$H_z^+/E_y^+ \;=\; -H_z^-/E_y^- \;=\;\;\; n/z_0\cos\theta \;=\;\; \gamma_p$$
$$H_y^+/E_z^+ \;=\; -H_y^-/E_z^- \;=\; -n\cos\theta/z_0 \;=\; \gamma_s,$$

$$(4.25)$$

where the p and s subscripts are the usual polarization designators. We use the form

$$\hat{F} = \begin{bmatrix} 1 & 1 & 0 & 0 \\ \gamma_p & -\gamma_p & 0 & 0 \\ 0 & 0 & 1 & 1 \\ 0 & 0 & \gamma_s & -\gamma_s \end{bmatrix}$$

$$(4.26)$$

for the field matrix of an isotropic medium.

Other transfer matrices which we shall use require the reciprocal of $\hat{F}$, and for an isotropic medium this has the form

$$\hat{F}^{-1} = \frac{1}{2} \begin{bmatrix} 1 & 1/\gamma_p & 0 & 0 \\ 1 & -1/\gamma_p & 0 & 0 \\ 0 & 0 & 1 & 1/\gamma_s \\ 0 & 0 & 1 & -1/\gamma_s \end{bmatrix}.$$

$$(4.27)$$

4.4.2 Field Coefficients $\vec{a}$

In a biaxial layer four waves propagate as a result of multiple reflections and transmissions. Each of these waves is just a basis wave multiplied by a complex coefficient. We use the symbols a_1^+, a_1^-, a_2^+, a_2^- for the coefficients, arrange them as a column vector

$$\vec{a} = \begin{bmatrix} a_1^+ \\ a_1^- \\ a_2^+ \\ a_2^- \end{bmatrix},$$

$$(4.28)$$

and refer to the column vector $\vec{a}$ as the *field coefficients*.

4.4.3 Total Field $\vec{m}$

Transformation across an interface requires the *total field*,

$$\vec{m} = \begin{bmatrix} E_y \\ H_z \\ E_z \\ H_y \end{bmatrix} = \begin{bmatrix} a_1^+ E_{y1}^+ + a_1^- E_{y1}^- + a_2^+ E_{y2}^+ + a_2^- E_{y2}^- \\ a_1^+ H_{z1}^+ + a_1^- H_{z1}^- + a_2^+ H_{z2}^+ + a_2^- H_{z2}^- \\ a_1^+ E_{z1}^+ + a_1^- E_{z1}^- + a_2^+ E_{z2}^+ + a_2^- E_{z2}^- \\ a_1^+ H_{y1}^+ + a_1^- H_{y1}^- + a_2^+ H_{y2}^+ + a_2^- H_{y2}^- \end{bmatrix}, \tag{4.29}$$

to be conserved. In practice this can be assured by transforming the travelling wave fields in one medium, say just to the right of an interface, to the total field at the interface, followed by a transformation of the total field to travelling wave fields in the second medium. This is the 4×4 equivalent of the Heavens 2×2 transfer matrix method[12] for isotropic thin films.

The total field at any point in the medium can be determined from $\hat{F}$ and $\vec{a}$. Writing out the product $\hat{F}\vec{a}$ gives

$$\hat{F}\vec{a} = \begin{bmatrix} E_{y1}^+ & E_{y1}^- & E_{y2}^+ & E_{y2}^- \\ H_{z1}^+ & H_{z1}^- & H_{z2}^+ & H_{z2}^- \\ E_{z1}^+ & E_{z1}^- & E_{z2}^+ & E_{z2}^- \\ H_{y1}^+ & H_{y1}^- & H_{y2}^+ & H_{y2}^- \end{bmatrix} \begin{bmatrix} a_1^+ \\ a_1^- \\ a_2^+ \\ a_2^- \end{bmatrix} = \begin{bmatrix} a_1^+ E_{y1}^+ + a_1^- E_{y1}^- + a_2^+ E_{y2}^+ + a_2^- E_{y2}^- \\ a_1^+ H_{z1}^+ + a_1^- H_{z1}^- + a_2^+ H_{z2}^+ + a_2^- H_{z2}^- \\ a_1^+ E_{z1}^+ + a_1^- E_{z1}^- + a_2^+ E_{z2}^+ + a_2^- E_{z2}^- \\ a_1^+ H_{y1}^+ + a_1^- H_{y1}^- + a_2^+ H_{y2}^+ + a_2^- H_{y2}^- \end{bmatrix}, \tag{4.30}$$

and hence

$$\hat{F}\vec{a} = \vec{m}. \tag{4.31}$$

In words, Eq.(4.31) states that, *the field matrix transforms the field coefficients to the total field at the same point in a medium.*

A rearrangement of Eq.(4.31) yields

$$\hat{F}^{-1}\vec{m} = \vec{a}, \tag{4.32}$$

and hence, *the inverse of the field matrix transforms the total field to the field coefficients at the same point.*

4.4.4 Phase Matrix $\hat{A}_d$

The four travelling wave fields in a biaxial layer change phase linearly with displacement in the x-direction, but at different rates. Along the same path, which is assumed to be always in the same layer, the absolute values of the a's

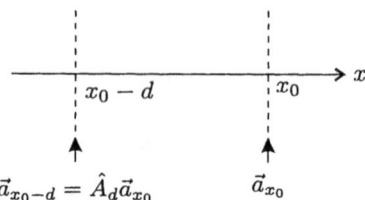

Fig. 4.3. The phase matrix $\hat{A}_d$ transforms the field coefficients $\vec{a}$ between two planes in the same layer.

remain constant – just the phases change. For a right-to-left displacement, from $x = x_0$ to $x = x_0 - d$, these phase changes are accounted for by the *phase matrix*

$$
\hat{A}_d =
\begin{bmatrix}
\exp[-i\phi_1^+] & 0 & 0 & 0 \\
0 & \exp[-i\phi_1^-] & 0 & 0 \\
0 & 0 & \exp[-i\phi_2^+] & 0 \\
0 & 0 & 0 & \exp[-i\phi_2^-]
\end{bmatrix},
\tag{4.33}
$$

where

$$
\phi_{1,2}^\pm = k\alpha_{1,2}^\pm d.
\tag{4.34}
$$

The transformation property of $\hat{A}_d$ can be written as

$$
\vec{a}_{x_0-d} = \hat{A}_d \, \vec{a}_{x_0},
\tag{4.35}
$$

and stated in words as, *the phase matrix transforms the field coefficients between two points in the same layer.* This property is illustrated in Fig. 4.3.

In the *BTF Toolbox* the function **pmat** is used to determine the phase matrix.

4.4.5 Characteristic Matrix $\hat{M}$

From the properties of $\hat{F}$ and $\hat{A}_d$ it is evident that the *characteristic matrix*,

$$
\hat{M} = \hat{F}\hat{A}_d\hat{F}^{-1},
\tag{4.36}
$$

transforms the total field between the two planes that are specified in $\hat{A}_d$ and are within the same biaxial layer,

$$
\vec{m}_{x_0-d} = \hat{M}\vec{m}_{x_0}.
\tag{4.37}
$$

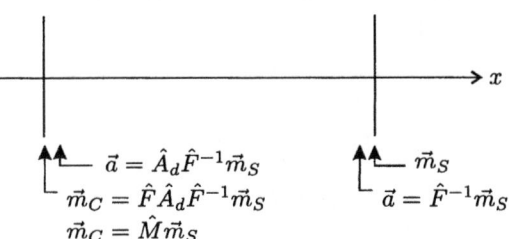

Fig. 4.4. The characteristic matrix of a layer, $\hat{M} = \hat{F}\hat{A}_d\hat{F}^{-1}$, transforms the total field $\vec{m}$ across the layer.

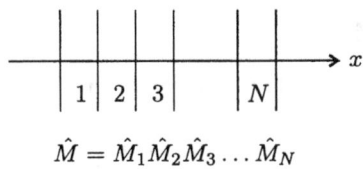

$$\hat{M} = \hat{M}_1\hat{M}_2\hat{M}_3\ldots\hat{M}_N$$

Fig. 4.5. Stack of biaxial layers.

$\hat{M}$ can be defined for any two planes, but the most common use for $\hat{M}$ is to transform the total field from the right-hand interface of a layer to the left-hand interface, and in such a case $\hat{M}$ is called the characteristic matrix of the layer. Fig. 4.4 illustrates in detail the transformation performed by the characteristic matrix of a single layer. Starting on the right-hand side, the total field $\vec{m}_S$ on the substrate interface is transformed by the matrix $\hat{F}^{-1}$ to give the field coefficients $\vec{a}$ just inside the layer, then $\vec{a}$ is transformed across the layer by the phase matrix $\hat{A}_d$, and finally, transformed by the matrix $\hat{F}$ to give the total field $\vec{m}_C$ at the cover. The overall transformation is equivalent to

$$\vec{m}_C = \hat{M}\vec{m}_S. \tag{4.38}$$

The characteristic matrix of a stack of N layers surrounded by a cover C and substrate S, as illustrated in Fig. 4.5, is

$$\hat{M} = \hat{M}_1\hat{M}_2\ldots\hat{M}_N. \tag{4.39}$$

In detail, the transformation from substrate to cover has the form

$$\vec{m}_C = \hat{F}_1\hat{A}_{d1}\hat{F}_1^{-1}\hat{F}_2\hat{A}_{d2}\hat{F}_2^{-1}\ldots\hat{F}_N\hat{A}_{dN}\hat{F}_N^{-1}\vec{m}_S. \tag{4.40}$$

Computation of the characteristic matrix for a layer or a stack of layers can be achieved by using the function **cmat** in the *BTF Toolbox*.

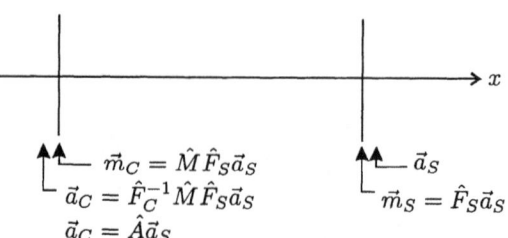

Fig. 4.6. The system matrix $\hat{A} = \hat{F}_C^{-1}\hat{M}\hat{F}_S$ transforms the field coefficients $\vec{a}$ from the substrate to the cover.

4.4.6   System Matrix $\hat{A}$

Finally, consider the 4×4 matrix $\hat{A}$ formed for a general stack of layers as

$$\hat{A} = \hat{F}_C^{-1}\hat{M}\hat{F}_S, \tag{4.41}$$

where $\hat{F}_C$ is the field matrix for the cover medium and $\hat{F}_S$ is the field matrix for the substrate. If we start with the field coefficients $\vec{a_S}$ at a point just to the right of the last layer/substrate interface, as illustrated in Fig. 4.6, then we see that the three matrices transform in turn, field coefficients just in the substrate to total fields at the substrate, to total fields at the cover, to the field coefficients just in the cover,

$$\vec{a}_C = \hat{A}\vec{a}_S. \tag{4.42}$$

The combined transformation requires properties of the cover, films and substrate, i.e. the complete system, and we shall see later that $\hat{A}$ as defined above is useful for determining system properties such as reflectance, transmittance, and the condition for a mode in a planar waveguide. For this reason we call $\hat{A}$ the *system matrix*, and the appropriate function in the *BTF Toolbox* is called **smat**. In practice the "system" can be defined by planes other than the cover and the substrate. As an example, the phase matrix A_d is a special case of the system matrix.

4.4.7   Properties of $\hat{M}$

Symmetries of $\hat{M}$ for a Single Film

The characteristic matrix of a single biaxial film has 16 elements, but only eight parameters, n_1, n_2, n_3, η, ψ, ξ, β and d are required for computation of the matrix. Of the sixteen elements in the matrix only ten are different irrespective of sign, as displayed in the equation

$$\hat{M} = \begin{bmatrix} A & B & C & D \\ E & A & F & G \\ -G & -D & H & I \\ -F & -C & J & H \end{bmatrix}. \qquad (4.43)$$

Determinant of $\hat{M}$

In the general case of a single biaxial layer the determinant of the characteristic matrix is

$$\begin{aligned} \det \hat{M} &= \det(\hat{F}\hat{A}_d\hat{F}^{-1}) \\ &= \det \hat{A}_d \\ &= \exp[-i(\phi_1^+ + \phi_1^- + \phi_2^+ + \phi_2^-)], \qquad (4.44) \end{aligned}$$

and hence

$$|\det \hat{M}| = 1. \qquad (4.45)$$

For a stack of biaxial layers $\det \hat{M}$ is the product of the determinants of the individual characteristic matrices, and hence Eq.(4.45) holds for a general stack of biaxial layers.

The phase angles in Eq.(4.44) add to zero for an isotropic layer, $\phi_1^+ + \phi_1^- = 0$ and $\phi_2^+ + \phi_2^- = 0$. Thus, for a single isotropic layer, or for a stack of isotropic films,

$$\det \hat{M} = 1. \qquad (4.46)$$

4.4.8 Computation of Film Parameters from $\hat{M}$

The procedure for calculating $\hat{M}$ from the film parameters n_1, n_2, n_3, $\eta = 0$, ψ, ξ, d of a biaxial film has a unique reversal. To show this we begin with Eq.(4.36) which defines the characteristic matrix, $\hat{M} = \hat{F}\hat{A}_d\hat{F}^{-1}$, and rearrange it in the form of an eigenequation,

$$\hat{M}\hat{F} = \hat{F}\hat{A}_d. \qquad (4.47)$$

Thus the matrices $\hat{F}$ and $\hat{A}_d$ can be determined from the matrix $\hat{M}$ by using the MATLAB eig function,

$$[\hat{F}, \hat{A}_d] = \text{eig } \hat{M}. \qquad (4.48)$$

The four α's can be found from $\hat{F}$, using one of the three equations implied by Eq.(3.45), $\alpha = -z_0 H_y/E_z$. The four E_z's required here are given by the third row of $\hat{F}$ and the four H_y's are located in the fourth row. Hence we can construct $\hat{\alpha}$ using

$$
\hat{\alpha} = \begin{bmatrix} -z_0 F_{41}/F_{31} & 0 & 0 & 0 \\ 0 & -z_0 F_{42}/F_{32} & 0 & 0 \\ 0 & 0 & -z_0 F_{43}/F_{33} & 0 \\ 0 & 0 & 0 & -z_0 F_{44}/F_{34} \end{bmatrix}. \tag{4.49}
$$

Next consider the matrix version of Eq.(3.51), $\hat{L}\hat{F} = \hat{F}\hat{\alpha}$. Rearranging this equation into the form

$$
\hat{L} = \hat{F}\hat{\alpha}\hat{F}^{-1} \tag{4.50}
$$

provides a method for determining $\hat{L}$, which is defined by and related to β and the propagation frame ε's by Eqs.(3.49) and (3.51). The solution to these equations can be expressed as

$$
\beta = [z_0 L_{21} + z_0 L_{43} - z_0 L_{23}(L_{11}/L_{13} - L_{13}/L_{11})]^{1/2} \tag{4.51}
$$

and

$$
\begin{aligned}
\varepsilon_{xx} &= \beta^2/(1 - L_{12}/z_0) \\
\varepsilon_{xy} &= -\varepsilon_{xx} L_{11}/\beta \\
\varepsilon_{xz} &= -\varepsilon_{xx} L_{13}/\beta \\
\varepsilon_{yz} &= z_0 L_{23} + \varepsilon_{xy}\varepsilon_{xz}/\varepsilon_{xx} \\
\varepsilon_{yy} &= z_0 L_{21} + \varepsilon_{xy}^2/\varepsilon_{xx} \\
\varepsilon_{zz} &= \beta^2 + \varepsilon_{xz}^2/\varepsilon_{xx} - z_0 L_{43}.
\end{aligned} \tag{4.52}
$$

Next Eq.(2.78) is solved for the film parameters:-

$$
\begin{aligned}
\xi &= \tan^{-1}(\varepsilon_{xz}/\varepsilon_{xy}) \\
\psi &= \frac{1}{2}\tan^{-1}[2(\varepsilon_{xy}\cos\xi + \varepsilon_{xz}\sin\xi)/(\varepsilon_{xx} - \varepsilon_{yy}\cos^2\xi - \varepsilon_{yz}\sin 2\xi - \varepsilon_{zz}\sin^2\xi)] \\
n_1 &= [\varepsilon_{xx} + (\varepsilon_{xy}\cos\xi + \varepsilon_{xz}\sin\xi)\tan\psi]^{1/2} \\
n_3 &= (\varepsilon_{yy} - \varepsilon_{yz}\varepsilon_{xy}/\varepsilon_{xz})^{1/2} \\
n_2 &= (\varepsilon_{xx} + \varepsilon_{yy} + \varepsilon_{zz} - n_1^2 - n_3^2)^{1/2}.
\end{aligned} \tag{4.53}
$$

The final step requires determination of d from the non-zero elements of $\hat{A}_d$, which we write in the form $A_{djj} = \exp(2\pi\alpha_{jj}d/\lambda)$. We can obtain a value of the thickness, d_j, that satisfies this equation for a particular j by taking the natural logarithm and rearranging. However, $d_j + m_j\lambda/\alpha_{jj}$ (with m_j an integer) will be a solution as well, and we have four equations

$$d_j = \frac{-\lambda\log(A_{djj})}{2i\pi\alpha_{jj}} + \frac{m_j\lambda}{\alpha_{jj}}. \qquad (4.54)$$

The true value of d can be found by a method reminiscent of the method of exact fractions used with the Fabry-Perot interferometer. A possible value of d_1 is chosen, with m_1 an integer, and Eq.(4.54) is checked to see if it is satisfied with integral values of m_j for $j = 2, 3, 4$. In practice we determine d in a MATLAB loop, using a range of integers m_1 and finding the best fit to the desired condition. For further details see the script file **reverse** in the *BTF Toolbox*.

4.5 Abelès and Heavens Calculus

4.5.1 Isotropic Layer

It is instructive to see how the 4×4 characteristic matrix $\hat{M}$ simplifies for an isotropic layer, i.e. when $n_1 = n_2 = n_3 = n$. From Eq.(3.41) we see that the four solutions to Fresnel's equation can be expressed as $\alpha, -\alpha, \alpha, -\alpha$ where α is the positive square root of $n^2 - \beta^2$. Hence the four phases required for $\hat{A}_d$ are $\phi, -\phi, \phi, -\phi$, where $\phi = k\alpha d$, and the field ratios are $\gamma_p = n^2/z_0(n^2 - \beta^2)^{1/2}$, $\gamma_s = -(n^2 - \beta^2)^{1/2}/z_0$. Writing out $\hat{M} = \hat{F}\hat{A}_d\hat{F}^{-1}$ with these simplifications gives

$$\hat{M} = \tfrac{1}{2} \begin{bmatrix} 1 & 1 & 0 & 0 \\ \gamma_p & -\gamma_p & 0 & 0 \\ 0 & 0 & 1 & 1 \\ 0 & 0 & \gamma_s & -\gamma_s \end{bmatrix} \times$$

$$\begin{bmatrix} \exp[-i\phi] & 0 & 0 & 0 \\ 0 & \exp[+i\phi] & 0 & 0 \\ 0 & 0 & \exp[-i\phi] & 0 \\ 0 & 0 & 0 & \exp[+i\phi] \end{bmatrix} \begin{bmatrix} 1 & 1/\gamma_p & 0 & 0 \\ 1 & -1/\gamma_p & 0 & 0 \\ 0 & 0 & 1 & 1/\gamma_s \\ 0 & 0 & 1 & -1/\gamma_s \end{bmatrix} \tag{4.55}$$

$$= \begin{bmatrix} \hat{F}_p & \cdot \\ \cdot & \hat{F}_s \end{bmatrix} \begin{bmatrix} \hat{A}_{dp} & \cdot \\ \cdot & \hat{A}_{ds} \end{bmatrix} \begin{bmatrix} \hat{F}_p^{-1} & \cdot \\ \cdot & \hat{F}_s^{-1} \end{bmatrix}.$$

Here F_p, F_s, A_{dp} and A_{ds} are the Heavens[12] 2×2 matrices for isotropic thin films. These can be used for transferring travelling wave fields through interfaces and across layers for p or s polarized light in the same way that the 4×4 matrices are used.

Equation (4.55) for $\hat{M}$ can be put in the form of a partitioned matrix,

$$\hat{M} = \begin{bmatrix} \hat{M}_p & \cdot \\ \cdot & \hat{M}_s \end{bmatrix} \tag{4.56}$$

where

$$\hat{M}_p = \begin{bmatrix} \cos\phi & -i\gamma_p^{-1}\sin\phi \\ -i\gamma_p\sin\phi & \cos\phi \end{bmatrix} \tag{4.57}$$

and

$$\hat{M}_s = \begin{bmatrix} \cos\phi & -i\gamma_s^{-1}\sin\phi \\ -i\gamma_s\sin\phi & \cos\phi \end{bmatrix}. \tag{4.58}$$

The 2×2 matrices $\hat{M}_p$ and $\hat{M}_s$ are the Abelès[13] characteristic matrices for p and s polarized light. Note that

$$\det\hat{M}_p = \det\hat{M}_s = 1, \tag{4.59}$$

and see Eq.(4.46).

Table 4.2. Heavens matrices ($\hat{F}_p$, $\hat{F}_s$, $\hat{A}_{dp}$, $\hat{A}_{ds}$) and Abelès matrices ($\hat{M}_p$, $\hat{M}_s$) for propagation in isotropic layers.

p-polarization	*s*-polarization

$$\hat{F}_p = \begin{bmatrix} 1 & 1 \\ \gamma_p & -\gamma_p \end{bmatrix} \qquad\qquad \hat{F}_s = \begin{bmatrix} 1 & 1 \\ \gamma_s & -\gamma_s \end{bmatrix}$$

$$\hat{A}_{dp} = \begin{bmatrix} e^{-i\phi} & 0 \\ 0 & e^{i\phi} \end{bmatrix} \qquad\qquad \hat{A}_{ds} = \begin{bmatrix} e^{-i\phi} & 0 \\ 0 & e^{i\phi} \end{bmatrix}$$

$$\hat{M}_p = \begin{bmatrix} \cos\phi & -i\gamma_p^{-1}\sin\phi \\ -i\gamma_p\sin\phi & \cos\phi \end{bmatrix} \qquad \hat{M}_s = \begin{bmatrix} \cos\phi & -i\gamma_s^{-1}\sin\phi \\ -i\gamma_s\sin\phi & \cos\phi \end{bmatrix}$$

$$\alpha = (n^2 - \beta^2)^{1/2} \qquad\qquad \alpha = (n^2 - \beta^2)^{1/2}$$

$$\phi = kd\alpha \qquad\qquad\qquad \phi = kd\alpha$$

$$\gamma_p = n^2/z_0\alpha \qquad\qquad\qquad \gamma_s = -\alpha/z_0$$

The Abelès matrix method is used extensively for thin film calculations in which the film materials are isotropic. Table 4.2 lists the Heavens and Abelès matrices together with equations needed for computing them from the film parameters.

4.5.2 Deposition Plane

The general 4×4 matrix $\hat{M}$ simplifies for propagation in a principal plane, such as the deposition plane of a tilted columnar thin film ($\xi = 0$). In this case the *p* and *s* polarizations are decoupled with n_1 and n_2 associated with *p*, and n_3 associated with *s*. Matrices similar to $\hat{F}_p$, $\hat{F}_s$, $\hat{A}_{dp}$, $\hat{A}_{ds}$, $\hat{M}_p$ and $\hat{M}_s$ for an isotropic medium can be used.

We can write the product $\hat{F}\hat{A}_d\hat{F}^{-1}$ in the form

$$\hat{M} = \tfrac{1}{2} \begin{bmatrix} 1 & 1 & 0 & 0 \\ \gamma_p & -\gamma_p & 0 & 0 \\ 0 & 0 & 1 & 1 \\ 0 & 0 & \gamma_s & -\gamma_s \end{bmatrix} \times$$

$$\begin{bmatrix} \exp[-i\phi_p^+] & 0 & 0 & 0 \\ 0 & \exp[-i\phi_p^-] & 0 & 0 \\ 0 & 0 & \exp[-i\phi_s^+] & 0 \\ 0 & 0 & 0 & \exp[-i\phi_s^-] \end{bmatrix} \begin{bmatrix} 1 & 1/\gamma_p & 0 & 0 \\ 1 & -1/\gamma_p & 0 & 0 \\ 0 & 0 & 1 & 1/\gamma_s \\ 0 & 0 & 1 & -1/\gamma_s \end{bmatrix}.$$

$$(4.60)$$

Expanding and putting

$$\begin{aligned} \phi_p^+ &= \phi_p + \phi_p' \\ \phi_p^- &= -\phi_p + \phi_p' \end{aligned} \tag{4.61}$$

$$\begin{aligned} \phi_s^+ &= \phi_s \\ \phi_s^- &= -\phi_s, \end{aligned} \tag{4.62}$$

leads to the partitioned matrix

$$\hat{M} = \begin{bmatrix} \hat{M}_p & \cdot \\ \cdot & \hat{M}_s \end{bmatrix}, \tag{4.63}$$

where

$$\hat{M}_p = e^{i\phi_p'} \begin{bmatrix} \cos\phi_p & -i\gamma_p^{-1}\sin\phi_p \\ -i\gamma_p\sin\phi_p & \cos\phi_p \end{bmatrix} \tag{4.64}$$

and

$$\hat{M}_s = \begin{bmatrix} \cos\phi_s & -i\gamma_s^{-1}\sin\phi_s \\ -i\gamma_s\sin\phi_s & \cos\phi_s \end{bmatrix}. \tag{4.65}$$

The term $e^{i\phi'_p}$ cancels out (or disappears when multiplied by its complex conjugate) in many equations that involve elements of the matrix $\hat{M}_p$, such as expressions for the reflectance, transmittance, and the modal condition for a planar waveguide. In such circumstances

$$\hat{M}'_p = \begin{bmatrix} \cos\phi_p & -i\gamma_p^{-1}\sin\phi_p \\ -i\gamma_p\sin\phi_p & \cos\phi_p \end{bmatrix} \tag{4.66}$$

can be used as the Abelès matrix for p-polarized light. Thus 2×2 matrix algebra can be used to derive solutions to both single layer and multilayered birefringent stacks, provided the propagation of light is always in a deposition plane. For this purpose it is convenient to complete the specification of the elements of the 2×2 matrices in terms of the parameters of the birefringent layer.

First of all, for the matrix $\hat{M}_s$, the field ratio is $\gamma_s = -(n_3^2 - \beta^2)^{1/2}/z_0$ and $\phi_s = -kdz_0\gamma_s$. As well, from Eq.(4.61) it follows that

$$\phi_p = (\phi_p^+ - \phi_p^-)/2 \tag{4.67}$$

and

$$\phi'_p = (\phi_p^+ + \phi_p^-)/2, \tag{4.68}$$

and hence the phase thicknesses ϕ_p^+ and ϕ_p^- can be expressed in terms of the solutions of Fresnel's equation that are listed in Eq.(3.35). The results are

$$\phi_p = kn_p^2 d/z_0\gamma_p, \tag{4.69}$$

$$\phi'_p = kd\beta[(1 - n_p^2/n_1^2)(n_p^2/n_2^2 - 1)]^{1/2}, \tag{4.70}$$

and

$$\gamma_p = 1/z_0(1/n_p^2 - \beta^2/n_1^2 n_2^2)^{1/2}. \tag{4.71}$$

A summary for the special case of propagation in the deposition plane of a biaxial layer is given in Table 4.3.

4.5.3 Berreman Calculus Computations

General Birefringent Coatings

To facilitate calculation of the characteristic matrix $\hat{M}$ and the system matrix $\hat{A}$ in Berreman calculus, we add to the list of input arguments used by the *BTF Toolbox*, so that the complete set becomes –

Table 4.3. Heavens matrices ($\hat{F}_p$, $\hat{F}_s$, $\hat{A}_{dp}$, $\hat{A}_{ds}$) and Abelès matrices ($\hat{M}_p$, $\hat{M}_s$) for propagation in the deposition plane.

$$p\text{-polarization} \qquad\qquad\qquad\qquad s\text{-polarization}$$

$$\hat{F}_p = \begin{bmatrix} 1 & 1 \\ \gamma_p & -\gamma_p \end{bmatrix} \qquad\qquad \hat{F}_s = \begin{bmatrix} 1 & 1 \\ \gamma_s & -\gamma_s \end{bmatrix}$$

$$\hat{A}_{dp} = e^{-i\phi'_p} \begin{bmatrix} e^{-i\phi_p} & 0 \\ 0 & e^{i\phi_p} \end{bmatrix} \qquad\qquad \hat{A}_{ds} = \begin{bmatrix} e^{-i\phi_s} & 0 \\ 0 & e^{i\phi_s} \end{bmatrix}$$

$$\hat{M}_p = e^{i\phi'_p} \begin{bmatrix} \cos\phi_p & -i\gamma_p^{-1}\sin\phi_p \\ -i\gamma_p\sin\phi_p & \cos\phi_p \end{bmatrix} \qquad \hat{M}_s = \begin{bmatrix} \cos\phi_s & -i\gamma_s^{-1}\sin\phi_s \\ -i\gamma_s\sin\phi_s & \cos\phi_s \end{bmatrix}$$

$$n_p = (\sin^2\psi/n_1^2 + \cos^2\psi/n_2^2)^{-1/2}$$

$$\gamma_p = 1/z_0(1/n_p^2 - \beta^2/n_1^2 n_2^2)^{1/2} \qquad\qquad \gamma_s = -(n_3^2 - \beta^2)^{1/2}/z_0$$

$$\phi_p = kn_p^2 d/z_0\gamma_p \qquad\qquad\qquad \phi_s = -kdz_0\gamma_s$$

$$\phi'_p = kd\beta[(1 - n_p^2/n_1^2)(n_p^2/n_2^2 - 1)]^{1/2}$$

$$indices = [n_1 \; n_2 \; n_3]$$

$$angles = [\eta \; \psi \; \xi]$$

$$material = [indices \; angles]$$

$$layer = [material \; d/\lambda]$$

$$cover, \; substrate = [material \; NaN]$$

$$stack = \begin{bmatrix} layer1 \\ layer2 \\ \cdots\cdots \\ layerN \end{bmatrix}$$

$$system = \begin{bmatrix} cover \\ stack \\ substrate \end{bmatrix}.$$

We have included here the definition of *layer* as a 7-element row vector, with the 7th element equal to the ratio of the layer thickness and the wavelength of the light. As well, definitions are given for *cover* and *substrate*; the "not-a-number" symbol *NaN* distinguishes these from *layer*.

The most straightforward way to proceed to the characteristic matrix $\hat{M}$ for a stack of layers is to compile the matrix *stack* and then use the line

$$\hat{M} = \mathbf{cmat}(stack, \; \beta).$$

However this will be inefficient if the system contains the same film or period of films repeated several times. In such a case the characteristic matrices of the different films should be determined, $\hat{M}_1 = \mathbf{cmat}(layer1, \; \beta)$ etc, and used to form the characteristic matrix of the stack.

Similarly, it is possible to proceed to $\hat{A}$ directly,

$$\hat{A} = \mathbf{smat}(system, \; \beta),$$

or in stages as

$$\hat{A} = \mathbf{smat}(\hat{F}_C, \; \hat{M}, \; \hat{F}_S).$$

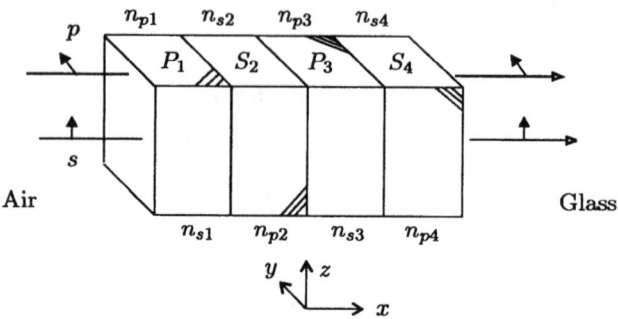

Fig. 4.7. Typical PS coating.

PS Coatings

Simplifications of the Abelès matrices and the 4×4 matrix $\hat{M}$ occur when the light is incident normally on a layer $(\beta = 0)$ which is aligned so that $\xi = 0$, $\pi/2$, π, or $3\pi/2$. A stack of four such layers is shown schematically in Fig. 4.7. The columnar structure of each layer is represented by three parallel lines drawn on one edge. This indicates both the deposition plane of the layer and the part of the coating that was closest to the source during deposition.

Two external beams of light with polarizations labelled p and s are shown in the figure. For these external beams the labels p and s are referenced to the usual y and z axes. As the upper beam propagates through the coating it continues to vibrate in the same plane. When the beam is in the first layer the vibration is parallel to the deposition plane of the layer, in the second layer the vibration is perpendicular to the deposition plane, in the third layer it is parallel, and in the fourth it is perpendicular. The appropriate refractive indices are labelled n_{p1}, n_{s2}, n_{p3}, n_{s4} – within a layer p and s are referenced to the deposition plane. We refer to the first and third layers as P layers and to the second and fourth as S layers. The coating formed with P and S layers is called a PS coating and is represented symbolically as

$$aP_1S_2P_3S_4g,$$

where a stands for air and g stands for glass. Notice that the beam emerges with unchanged polarization. A characteristic feature of PS coatings is the absence of cross-coupling of polarizations.

Next consider the stack as seen by the lower external beam, labelled s in Fig. 4.7. In this case the refractive indices are n_{s1}, n_{p2}, n_{s3}, n_{p4}, but the d/λ's are unchanged, in both cases equal to d_1/λ, d_2/λ, d_3/λ, d_4/λ. The structure could be represented as $aS_1P_2S_3P_4g$.

This second perceived structure is a "dual" of the first, obtained by interchanging P and S in the symbolic representation, and interchanging p and s in the set of refractive indices. Thus it is only necessary to use one symbolic representation to define a PS stack and we use the first one. As well, the characteristic matrices of two layers which are identical apart from ξ values of 0 and π, or $\pi/2$ and $3\pi/2$, are the same. Thus a computer routine does not have to distinguish the orientation of layers such as the first and third layers in Fig. 4.7. The practical reason for depositing the layers as shown in the figure is to avoid the accumulation of wedging through the coating.

From the above discussion it follows that a P layer or an S layer is completely specified by two refractive indices and d/λ. To avoid possible confusion, due to the multiple use of p and s polarization descriptors, we refer to the pair of refractive indices that define the oriented material as n_y and n_z. The *BTF Toolbox* uses the input arguments

$$psmaterial = [n_y \ n_z]$$

$$pslayer = [psmaterial \ d/\lambda]$$

$$pscover = [n_C \ n_C \ NaN]$$

$$pssubstrate = [n_S \ n_S \ NaN]$$

$$psstack = \begin{bmatrix} pslayer1 \\ pslayer2 \\ \cdots\cdots \\ pslayerN \end{bmatrix}$$

$$pssystem = \begin{bmatrix} pscover \\ psstack \\ pssubstrate \end{bmatrix},$$

for computations with PS coatings. In particular, the function **cmat** accepts *pslayer*, *psstack*, and *pssystem* as single input arguments for computation of the 4×4 characteristic matrix $\hat{M}$. One advantage here is that explicit expressions are used for the elements of $\hat{M}$, and hence computation time is much shorter.

PS layers are discussed further in Sect. 5.5.2 and in Sect. 16.4. In the latter we show that the example PS coating illustrated in Fig. 4.7 can be designed to

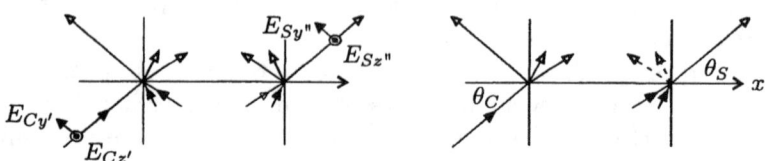

Fig. 4.8. Transmission through a wave plate, taking interference into account (left) and neglecting interference (right).

be a birefringent antireflection coating, passing the p-polarized beam without loss but reflecting some of the s-polarized beam.

4.6   Relationship of Jones and Berreman Calculus

Finally in this chapter we explore the relationship between the Jones matrix $\hat{J}$ and the system matrix $\hat{A}$ by working through a particular example, a birefringent plate separating a cover medium and a substrate medium. Two cases are considered, the Jones matrix including interference, illustrated in the left-hand side of Fig. 4.8, and the Jones matrix including primary interfacial reflections but not interference (Fig. 4.8, right).

4.6.1   Jones Matrix with Interference

For the case in which interference is included, the matrix

$$\hat{A} = \hat{F}_C^{-1} \hat{F} \hat{A}_d \hat{F}^{-1} \hat{F}_S \tag{4.72}$$

transforms tangential travelling wave fields from the right-hand side of the plate to the left-hand side,

$$
\begin{bmatrix} E_{Cy}^+ \\ E_{Cy}^- \\ E_{Cz}^+ \\ E_{Cz}^- \end{bmatrix} = \hat{A} \begin{bmatrix} E_{Sy}^+ \\ 0 \\ E_{Sz}^+ \\ 0 \end{bmatrix}. \tag{4.73}
$$

Now, for the formation of the Jones matrix, we require only the electric field components with the + superscript, and write

$$
\begin{bmatrix} E_{Cy}^+ \\ E_{Cz}^+ \end{bmatrix} = \begin{bmatrix} A_{11} & A_{13} \\ A_{31} & A_{33} \end{bmatrix} \begin{bmatrix} E_{Sy}^+ \\ E_{Sz}^+ \end{bmatrix}.
$$
(4.74)

Rearranging, and then transforming the column vectors to axes aligned with the incoming and outgoing rays gives

$$
\begin{bmatrix} E_{Sy''}^+ \\ E_{Sz''}^+ \end{bmatrix} = \begin{bmatrix} \cos\theta_S & 0 \\ 0 & 1 \end{bmatrix}^{-1} \begin{bmatrix} A_{11} & A_{13} \\ A_{31} & A_{33} \end{bmatrix}^{-1} \begin{bmatrix} \cos\theta_C & 0 \\ 0 & 1 \end{bmatrix} \begin{bmatrix} E_{Cy'}^+ \\ E_{Cz'}^+ \end{bmatrix}.
$$
(4.75)

Hence the Jones matrix with interference is

$$
\hat{J} = \begin{bmatrix} \cos\theta_S & 0 \\ 0 & 1 \end{bmatrix}^{-1} \begin{bmatrix} A_{11} & A_{13} \\ A_{31} & A_{33} \end{bmatrix}^{-1} \begin{bmatrix} \cos\theta_C & 0 \\ 0 & 1 \end{bmatrix}.
$$
(4.76)

4.6.2 Jones Matrix with Reflections but without Interference

When a narrow laser beam passes through a relatively thick birefringent plate at an oblique angle, as shown in the right-hand side of Fig. 4.8, the beams that are multiply reflected do not overlap, and hence interference does not occur. Even so, the matrix method can be used to trace the travelling wave fields from the right-hand side of the plate to the left-hand side. Application of $\hat{F}^{-1}\hat{F}_S$ leads to the set of four waves in the plate at the interface with the substrate (Fig. 4.8), but it is assumed that the pair shown with broken lines can be disregarded. The matrix

$$
\hat{A}_d^+ = \begin{bmatrix} \exp[-i\phi_1^+] & 0 & 0 & 0 \\ 0 & 0 & 0 & 0 \\ 0 & 0 & \exp[-i\phi_2^+] & 0 \\ 0 & 0 & 0 & 0 \end{bmatrix},
$$
(4.77)

transfers the phase of the other pair to the cover interface, and finally the application of $\hat{F}_C^{-1}\hat{F}$ leaves the incident and reflected waves in the cover. The matrix $\hat{A}^+$ for the complete operation has a similar form to $\hat{A}$,

$$
\hat{A}^+ = \hat{F}_C^{-1}\hat{F}\hat{A}_d^+\hat{F}^{-1}\hat{F}_S,
$$
(4.78)

and an argument similar to the one used in the previous section leads to the equation

$$\hat{J} = \begin{bmatrix} \cos\theta_S & 0 \\ 0 & 1 \end{bmatrix}^{-1} \begin{bmatrix} A_{11}^+ & A_{13}^+ \\ A_{31}^+ & A_{33}^+ \end{bmatrix}^{-1} \begin{bmatrix} \cos\theta_C & 0 \\ 0 & 1 \end{bmatrix} \tag{4.79}$$

for the Jones matrix with reflections but no interference.

Chapter 5

Reflection and Transmission

In this chapter we consider methods for computing the optical reflection and transmission coefficients from bulk and layered biaxial media. Our objectives are to enhance understanding of this problem and to provide simple, yet general algorithms for computation.

▨ We begin by solving the general case in which the cover, the layers, and the substrate may all be birefringent and light can be incident on the layers from the cover side and from the substrate side. The solution makes use of a matrix $\hat{r}$ that contains ratios of the field coefficients. The case of light incident on a crystal/crystal interface is considered, first as a special case of the general equations and then with "stand alone" equations.

▨ The solutions for the reflection and transmission coefficients require identification of the individual basis vectors in an anisotropic cover or substrate, and this is achieved by sorting the columns of the field matrix $\hat{F}$.

▨ Finally we show that, when the cover medium, the layers and the substrate are all isotropic, the general equations for the reflectance and the transmittance can be rearranged into the standard equations derived from 2×2 matrices.

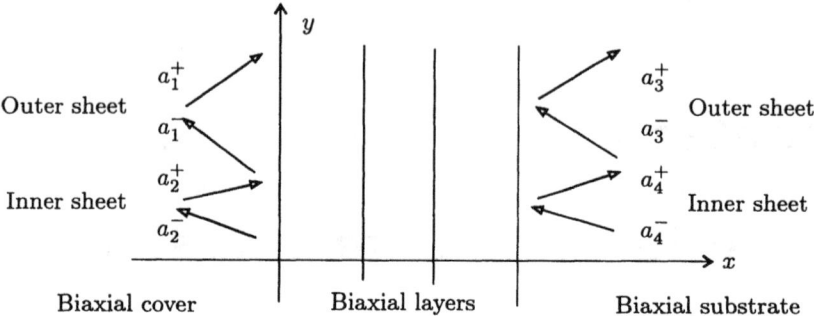

Fig. 5.1. Labelling scheme used for the amplitudes of the four basis vectors that propagate in the cover and the amplitudes of the four basis vectors in the substrate.

5.1 General Case – All Media Biaxial

The arrangement of cover, layers, and substrate is illustrated in Fig. 5.1. The a's are the field coefficients for the incident and outgoing waves. As in previous sections, the + superscript indicates right-going waves and the negative superscript indicates left-going waves. In this chapter we use the subscripts 1 and 2 to apply to the cover and the subscripts 3 and 4 to apply to the substrate.

Computation of the reflectance and transmittance requires the field strengths in the cover and in the substrate, i.e. both the field matrix $\hat{F}$ and the field coefficients $\vec{a}$ are required for the cover and substrate. The set of field coefficients

$$
\begin{bmatrix}
a_1^+ \\
a_2^+ \\
a_3^- \\
a_4^-
\end{bmatrix}
\tag{5.1}
$$

for waves travelling towards the layers may be regarded as inputs to the problem, and the "missing link" is the set of field coefficients,

$$
\begin{bmatrix}
a_1^- \\
a_2^- \\
a_3^+ \\
a_4^+
\end{bmatrix},
\tag{5.2}
$$

for the output waves.

The "balance" between input and output fields is set by the boundary conditions. At the interfaces of a stack of anisotropic films sandwiched between an anisotropic cover and an anisotropic substrate, the boundary conditions are satisfied, provided the total field $\hat{F}_C \vec{a_C}$ at the cover is equal to the result of transferring the total field $\hat{F}_S \vec{a_S}$ in the substrate to the cover, i.e. $\hat{F}_C \vec{a_C} = \hat{M} \hat{F}_S \vec{a_S}$. Rearranging and using the system matrix $\hat{A} = \hat{F}_C^{-1} \hat{M} \hat{F}_S$ yields the condition $\vec{a_C} = \hat{A} \vec{a_S}$, i.e.

$$
\begin{bmatrix} a_1^+ \\ a_1^- \\ a_2^+ \\ a_2^- \end{bmatrix} = \begin{bmatrix} A_{11} & A_{12} & A_{13} & A_{14} \\ A_{21} & A_{22} & A_{23} & A_{24} \\ A_{31} & A_{32} & A_{33} & A_{34} \\ A_{41} & A_{42} & A_{43} & A_{44} \end{bmatrix} \begin{bmatrix} a_3^+ \\ a_3^- \\ a_4^+ \\ a_4^- \end{bmatrix} . \tag{5.3}
$$

Next we rearrange Eq.(5.3) to put the field coefficients associated with the output waves on the left-hand side and the field coefficients associated with the input waves on the right-hand side,

$$
\begin{bmatrix} 0 & 0 & -A_{11} & -A_{13} \\ 1 & 0 & -A_{21} & -A_{23} \\ 0 & 0 & -A_{31} & -A_{33} \\ 0 & 1 & -A_{41} & -A_{43} \end{bmatrix} \begin{bmatrix} a_1^- \\ a_2^- \\ a_3^+ \\ a_4^+ \end{bmatrix} = \begin{bmatrix} -1 & 0 & A_{12} & A_{14} \\ 0 & 0 & A_{22} & A_{24} \\ 0 & -1 & A_{32} & A_{34} \\ 0 & 0 & A_{42} & A_{44} \end{bmatrix} \begin{bmatrix} a_1^+ \\ a_2^+ \\ a_3^- \\ a_4^- \end{bmatrix} . \tag{5.4}
$$

Then we can write

$$
\begin{bmatrix} a_1^- \\ a_2^- \\ a_3^+ \\ a_4^+ \end{bmatrix} \equiv \hat{r} \begin{bmatrix} a_1^+ \\ a_2^+ \\ a_3^- \\ a_4^- \end{bmatrix} , \tag{5.5}
$$

where

$$
\hat{r} \equiv
\begin{bmatrix}
r_{11} & r_{12} & t_{13} & t_{14} \\
r_{21} & r_{22} & t_{23} & t_{24} \\
t_{31} & t_{32} & r_{33} & r_{34} \\
t_{41} & t_{42} & r_{43} & r_{44}
\end{bmatrix}
=
\begin{bmatrix}
0 & 0 & -A_{11} & -A_{13} \\
1 & 0 & -A_{21} & -A_{23} \\
0 & 0 & -A_{31} & -A_{33} \\
0 & 1 & -A_{41} & -A_{43}
\end{bmatrix}^{-1}
\begin{bmatrix}
-1 & 0 & A_{12} & A_{14} \\
0 & 0 & A_{22} & A_{24} \\
0 & -1 & A_{32} & A_{34} \\
0 & 0 & A_{42} & A_{44}
\end{bmatrix}.
$$

$$(5.6)$$

Here $\hat{r}$ is to be regarded as an intermediary matrix, because its elements are ratios of the field coefficients rather than ratios of actual fields. Specifically,

$$
a_2^+ = a_3^- = a_4^- = 0 \quad a_1^+ = a_3^- = a_4^- = 0 \quad a_1^+ = a_2^+ = a_4^- = 0 \quad a_1^+ = a_2^+ = a_3^- = 0
$$

$$
\begin{array}{llll}
r_{11} = a_1^-/a_1^+ & r_{12} = a_1^-/a_2^+ & t_{13} = a_1^-/a_3^- & t_{14} = a_1^-/a_4^- \\
r_{21} = a_2^-/a_1^+ & r_{22} = a_2^-/a_2^+ & t_{23} = a_2^-/a_3^- & t_{24} = a_2^-/a_4^- \\
t_{31} = a_3^+/a_1^+ & t_{32} = a_3^+/a_2^+ & r_{33} = a_3^+/a_3^- & r_{34} = a_3^+/a_4^- \\
t_{41} = a_4^+/a_1^+ & t_{42} = a_4^+/a_2^+ & r_{43} = a_4^+/a_3^- & r_{44} = a_4^+/a_4^-.
\end{array}
$$

$$(5.7)$$

The (irradiance) reflectance and transmittance coefficients, which are to be held in the matrix

$$
\hat{R} \equiv
\begin{bmatrix}
R_{11} & R_{12} & T_{13} & T_{14} \\
R_{21} & R_{22} & T_{23} & T_{24} \\
T_{31} & T_{32} & R_{33} & R_{34} \\
T_{41} & T_{42} & R_{43} & R_{44}
\end{bmatrix},
$$

$$(5.8)$$

are defined in terms of ratios of power flow in the x-direction. Thus $R_{12} = P_1^-/P_2^+ = |a_1^-|^2|p_1^-|/|a_2^+|^2 p_2^+ = |r_{12}|^2|p_1^-/p_2^+|$ etc., where the p's are the powers carried by the basis vectors along the x-axis and may be calculated using Eq.(3.56). The reflectance and transmittance coefficients are given by

$$
\hat{R} =
\begin{bmatrix}
|r_{11}|^2|p_1^-/p_1^+| & |r_{12}|^2|p_1^-/p_2^+| & |t_{13}|^2|p_1^-/p_3^-| & |t_{14}|^2|p_1^-/p_4^-| \\
|r_{21}|^2|p_2^-/p_1^+| & |r_{22}|^2|p_2^-/p_2^+| & |t_{23}|^2|p_2^-/p_3^-| & |t_{24}|^2|p_2^-/p_4^-| \\
|t_{31}|^2|p_3^+/p_1^+| & |t_{32}|^2|p_3^+/p_2^+| & |r_{33}|^2|p_3^+/p_3^-| & |r_{34}|^2|p_3^+/p_4^-| \\
|t_{41}|^2|p_4^+/p_1^+| & |t_{42}|^2|p_4^+/p_2^+| & |r_{43}|^2|p_4^+/p_3^-| & |r_{44}|^2|p_4^+/p_4^-|
\end{bmatrix}.
$$

$$(5.9)$$

5.1.1 Crystal-Crystal Interface

In the absence of films $\hat{M} = \hat{I}$ and hence $\hat{A} = \hat{F}_C^{-1}\hat{F}_S$. Thus the general equations developed above are applicable to the crystal-crystal interface. Alternatively, the boundary conditions for the crystal–crystal interface can be expressed by the equation $\hat{F}_C\vec{a}_C = \hat{F}_S\vec{a}_S$, i.e.

$$
\begin{bmatrix}
E_{y1}^+ & E_{y1}^- & E_{y2}^+ & E_{y2}^- \\
H_{z1}^+ & H_{z1}^- & H_{z2}^+ & H_{z2}^- \\
E_{z1}^+ & E_{z1}^- & E_{z2}^+ & E_{z2}^- \\
H_{y1}^+ & H_{y1}^- & H_{y2}^+ & H_{y2}^-
\end{bmatrix}
\begin{bmatrix}
a_1^+ \\
a_1^- \\
a_2^+ \\
a_2^-
\end{bmatrix}
=
\begin{bmatrix}
E_{y3}^+ & E_{y3}^- & E_{y4}^+ & E_{y4}^- \\
H_{z3}^+ & H_{z3}^- & H_{z4}^+ & H_{z4}^- \\
E_{z3}^+ & E_{z3}^- & E_{z4}^+ & E_{z4}^- \\
H_{y3}^+ & H_{y3}^- & H_{y4}^+ & H_{y4}^-
\end{bmatrix}
\begin{bmatrix}
a_3^+ \\
a_3^- \\
a_4^+ \\
a_4^-
\end{bmatrix},
\tag{5.10}
$$

and then a procedure similar to that used above leads to

$$
\hat{r} =
\begin{bmatrix}
E_{y1}^- & E_{y2}^- & -E_{y3}^+ & -E_{y4}^+ \\
H_{z1}^- & H_{z2}^- & -H_{z3}^+ & -H_{z4}^+ \\
E_{z1}^- & E_{z2}^- & -E_{z3}^+ & -E_{z4}^+ \\
H_{y1}^- & H_{y2}^- & -H_{y3}^+ & -H_{y4}^+
\end{bmatrix}^{-1}
\begin{bmatrix}
-E_{y1}^+ & -E_{y2}^+ & E_{y3}^- & E_{y4}^- \\
-H_{z1}^+ & -H_{z2}^+ & H_{z3}^- & H_{z4}^- \\
-E_{z1}^+ & -E_{z2}^+ & E_{z3}^- & E_{z4}^- \\
-H_{y1}^+ & -H_{y2}^+ & H_{y3}^- & H_{y4}^-
\end{bmatrix}.
\tag{5.11}
$$

Thus the only difference is that, in the direct method for an interface, $\hat{r}$ is defined in terms of the columns of $\hat{F}$ rather than the columns of $\hat{A}$.

5.2 Sorting Columns of $\hat{F}$

In general it is not necessary to sort the basis vectors associated with anisotropic layers, because the characteristic matrix $\hat{M}$ for a film does not depend on the order of the columns of $\hat{F}$. However, a minimum sort of the cover and substrate basis vectors is necessary because the equations leading to the reflectance and transmittance coefficients require identification of the positive-going (+) and negative-going (-) basis waves. In the remaining part of this section we explain the various situations that arise, and need to be addressed, by considering the most complicated numerical example.

Consider first the plot of α versus β shown in the left-hand part of Fig. 5.2 for an anisotropic substrate specified by $n_1 = 2.4$, $n_2 = 1.55$, $n_3 = 2.0$, $\eta = 0$, $\psi = -45°$, $\xi = 0°$. In this case the eigenvectors are decoupled and propagate with p and s polarizations. Thus in this special situation ($\xi = 0°$) it would

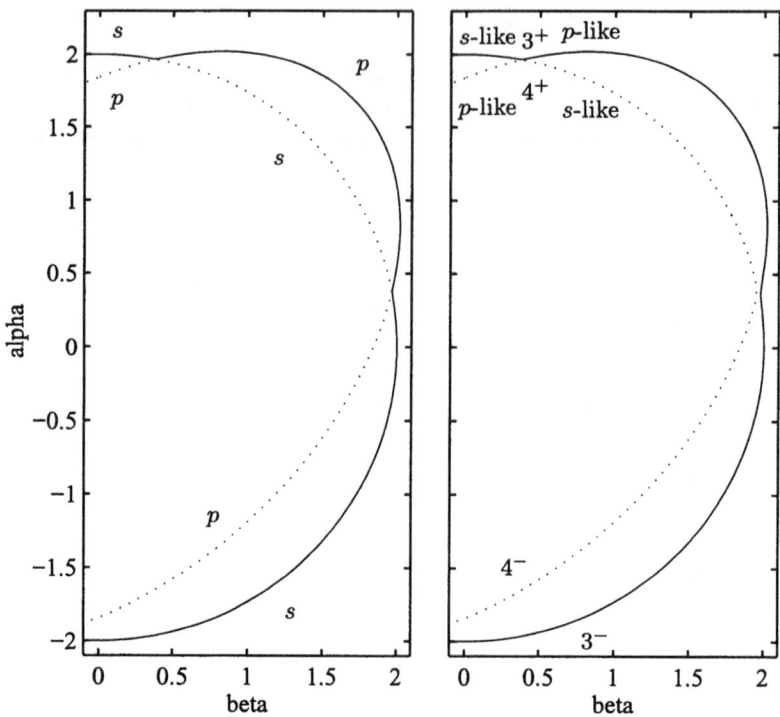

Fig. 5.2. Plots of $\alpha = n\cos\theta$ versus $\beta = n\sin\theta$ for a biaxial medium with $n_1 = 2.4$, $n_2 = 1.55$, $n_3 = 2.0$, $\eta = 0$, $\psi = -45°$ and $\xi = 0°$ (left), $\xi = 2°$ (right). The cusp in the outer sheet of the refractive index surface provides the most complicated example for sorting and matching α's determined as eigenvalues with optical features. (Adapted from I.J. Hodgkinson, S. Kassam and Q.H. Wu, *Journal of Computational Physics* 133, 75, 1997. Copyright © 1997 Academic Press. Reprinted with permission.)

be natural to sort the α's according to polarization. However, for refractive index sections in which ξ is not exactly zero the inner and outer sheets of the refractive index surface do not touch, and a sort based on polarization leads to discontinuities in plotted curves of reflectance and transmittance as functions of angle of incidence or β.

Fig. 5.2 (right) shows the outer sheet (solid line) and inner sheet (broken line) for the anisotropic substrate with $\xi = 2°$. For a given β the four associated values of α can be determined by drawing a vertical line in the figure, and the directions of the Poynting vector (indicating power flow) obtained by drawing normals to the curves. The positive α direction in Fig. 5.2 corresponds to the x-axis shown normal to the substrate in Fig. 5.1. It is clear that the sign of α (and hence the sign of the x-component of the wave vector) is not a reliable indicator of the sense of power flow along the x-axis. For this reason we take the terms positive-going (+) and negative-going (-) to refer to positive and negative senses of power flow along the x-axis for non-evanescent waves. In the case of evanescent waves, which carry no average power along x, the terms positive-going and negative-going are conveniently associated with the sign of the imaginary part of α, as this implies exponentially decreasing field strengths for waves moving away from the interface(s).

Apart from the necessary sort of cover and substrate basis vectors considered above, matching of the subscript pairs 1,2 and 3,4 with optical characteristics of the cover and substrate media is desirable to prevent fragmentation in plotted curves such as R_{11} versus θ. To illustrate suitable procedures we consider the above substrate (with $\xi = 2°$) together with an air ($n_1 = n_2 = n_3 = 1$) cover. For small values of β the "optical characteristic" used is simply association with the refractive index outer sheet (label 1 for an anisotropic cover and label 3 for the substrate) or the inner sheet (label 2 for an anisotropic cover and label 4 for the substrate).

In this particular example the cover is isotropic and normal practice dictates that the basis vectors should represent p and s polarizations. In such a case we use the matrix $\hat{F}$ defined by Eqs.(4.25) and (4.26). The subscripts 1 and 2 in previous equations translate to p and s in the cover, and 3 and 4 would translate to p and s in an isotropic substrate.

The first column of the matrix $\hat{R}$ is plotted in Fig. 5.3 as a function of β, for the range $0 \leq \beta \leq 1$ corresponding to $\theta_C \leq 0 \leq 90°$. For each of these curves the incident light is the 1+ (p) wave in the cover. The upper part of the figure shows a Brewster angle reflection for $R_{11} \equiv R_{pp}$. For small values of β the incident light excites p-like (4+) waves in the substrate, and hence T_{41} is large. The sudden fall in T_{41} and the corresponding rapid rise in T_{31} is caused by the switches in polarization character from p-like to s-like and s-like to p-like shown by the labels on Fig. 5.2 (right).

Unfortunately, pairs of α's cannot always be identified with the outer and inner sheets of the refractive index surface, and our example has been chosen to illustrate this point. Suppose that a line of constant β is moved from the left-hand side to the right-hand side of Fig. 5.4, in which both real and imaginary parts of α are plotted as functions of β. The intersections made can be classed as (i) outer sheet (two real), inner sheet (two real), (ii) outer sheet (two real), inner sheet (pair of complex conjugates), (iii) outer sheet (four real), (iv) outer sheet (two pairs of complex conjugates). In each case the four positions on the refractive index surface can be identified by considering the numerical order of the real parts of α, together with the sign of the x-component of the Poynting vector or the sign of the imaginary part of α. The labels on Fig. 5.4, which

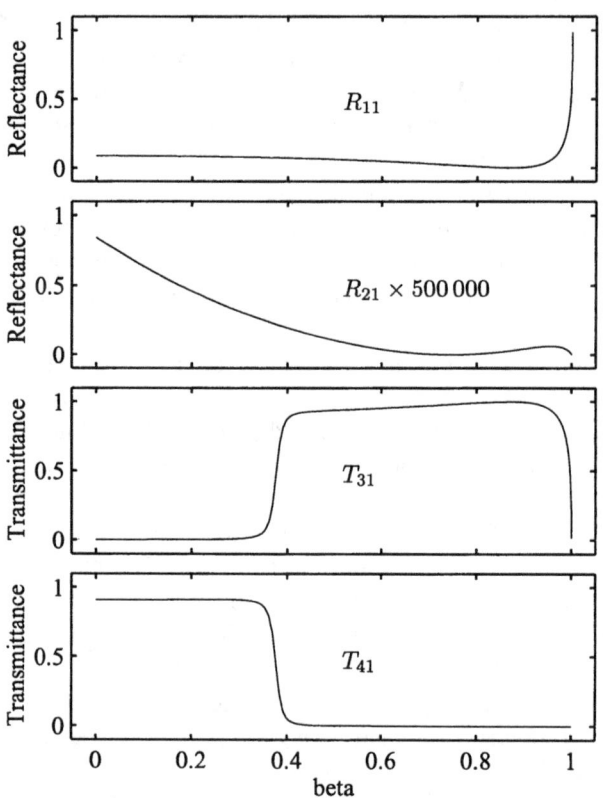

Fig. 5.3. Four of the sixteen R–T coefficients plotted as functions of $\beta = n \sin \theta$ for an air cover medium and the biaxial substrate specified by $n_1 = 2.4$, $n_2 = 1.55$, $n_3 = 2.0$, $\psi = -45°$ and $\xi = 2°$. (Adapted from I.J. Hodgkinson, S. Kassam and Q.H. Wu, *Journal of Computational Physics* 133, 75, 1997. Copyright © 1997 Academic Press. Reprinted with permission.)

result from such a sorting procedure, ensure both identification and continuity of reflectance and transmittance curves for this complicated example. In specific (and more usual) cases in which a cusp is not present in the outer sheet, sorting is correspondingly simpler.

5.3 Isotropic Cover and Substrate

In most applications of optical coatings the cover and the substrate that surround the coating are both isotropic. For this reason it is appropriate to consider equations that relate to this special case.

For each isotropic bounding medium the field matrix $\hat{F}$ has the simple form given in Eqs.(4.25) and (4.26). The 1's in the field matrix are positioned so that the elements of the $\vec{a}$'s are just the electric field components. Thus for the cover we can write

$$
\vec{a}_C = \begin{bmatrix} a_1^+ \\ a_1^- \\ a_2^+ \\ a_2^- \end{bmatrix} = \begin{bmatrix} E_{Cy}^+ \\ E_{Cy}^- \\ E_{Cz}^+ \\ E_{Cz}^- \end{bmatrix}, \tag{5.12}
$$

and for the substrate

$$
\vec{a}_S = \begin{bmatrix} a_3^+ \\ a_3^- \\ a_4^+ \\ a_4^- \end{bmatrix} = \begin{bmatrix} E_{Sy}^+ \\ E_{Sy}^- \\ E_{Sz}^+ \\ E_{Sz}^- \end{bmatrix}. \tag{5.13}
$$

Now we make an additional assumption, that light is incident from the cover side only, as this leads to simpler equations without loss of generality. This assumption means that $E_{Sy}^- = 0$ and $E_{Sz}^- = 0$. When the new expressions for $\vec{a}_C$ and $\vec{a}_S$ are substituted into the boundary condition $\vec{a}_C = \hat{A}\vec{a}_S$, the four equations that are implied can be rearranged to highlight the reflected and transmitted electric fields,

$$
E_{Cy}^- = \frac{(A_{21}A_{33} - A_{23}A_{31})E_{Cy}^+ + (A_{11}A_{23} - A_{13}A_{21})E_{Cz}^+}{A_{11}A_{33} - A_{13}A_{31}}
$$

$$
E_{Cz}^- = \frac{(A_{33}A_{41} - A_{31}A_{43})E_{Cy}^+ + (A_{11}A_{43} - A_{13}A_{41})E_{Cz}^+}{A_{11}A_{33} - A_{13}A_{31}}
$$

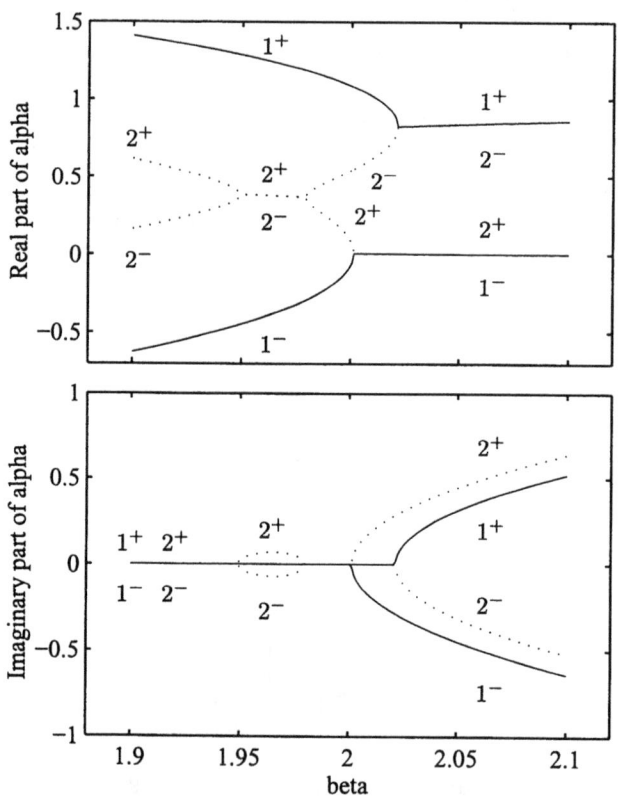

Fig. 5.4. Real and imaginary parts of α near the cusp described in Fig. 5.2. The labelling scheme both satisfies the minimum sorting requirement and prevents fragmentation in plotted reflectance and transmittance curves. (Adapted from I.J. Hodgkinson, S. Kassam and Q.H. Wu, *Journal of Computational Physics* 133, 75, 1997. Copyright © 1997 Academic Press. Reprinted with permission.)

$$E_{Sy}^+ = \frac{A_{33}E_{Cy}^+ - A_{13}E_{Cz}^+}{A_{11}A_{33} - A_{13}A_{31}}$$

$$E_{Sz}^+ = \frac{-A_{31}E_{Cy}^+ + A_{11}E_{Cz}^+}{A_{11}A_{33} - A_{13}A_{31}}. \tag{5.14}$$

As well, equations relating the elements of $\hat{A}$ and $\hat{M}$ can be obtained from the equation $\hat{A} = \hat{F}_C^{-1}\hat{M}\hat{F}_S$, i.e.

$$\hat{A} = \frac{1}{2}\begin{bmatrix} 1 & 1/\gamma_{Cp} & 0 & 0 \\ 1 & -1/\gamma_{Cp} & 0 & 0 \\ 0 & 0 & 1 & 1/\gamma_{Cs} \\ 0 & 0 & 1 & -1/\gamma_{Cs} \end{bmatrix}\begin{bmatrix} M_{11} & M_{12} & M_{13} & M_{14} \\ M_{21} & M_{22} & M_{23} & M_{24} \\ M_{31} & M_{32} & M_{33} & M_{34} \\ M_{41} & M_{42} & M_{43} & M_{44} \end{bmatrix}\begin{bmatrix} 1 & 1 & 0 & 0 \\ \gamma_{Sp} & -\gamma_{Sp} & 0 & 0 \\ 0 & 0 & 1 & 1 \\ 0 & 0 & \gamma_{Ss} & -\gamma_{Ss} \end{bmatrix}. \tag{5.15}$$

Equating coefficients for the first and third column elements of $\hat{A}$ (the elements used in Eqs. (5.14)) gives

$$\begin{aligned} A_{11} &= (M_{11} + \gamma_{Sp}M_{12} + M_{21}/\gamma_{Cp} + \gamma_{Sp}M_{22}/\gamma_{Cp})/2 \\ A_{21} &= (M_{11} + \gamma_{Sp}M_{12} - M_{21}/\gamma_{Cp} - \gamma_{Sp}M_{22}/\gamma_{Cp})/2 \\ A_{31} &= (M_{31} + \gamma_{Sp}M_{32} + M_{41}/\gamma_{Cs} + \gamma_{Sp}M_{42}/\gamma_{Cs})/2 \\ A_{41} &= (M_{31} + \gamma_{Sp}M_{32} - M_{41}/\gamma_{Cs} - \gamma_{Sp}M_{42}/\gamma_{Cs})/2, \end{aligned} \tag{5.16}$$

and

$$\begin{aligned} A_{13} &= (M_{13} + \gamma_{Ss}M_{14} + M_{23}/\gamma_{Cp} + \gamma_{Ss}M_{24}/\gamma_{Cp})/2 \\ A_{23} &= (M_{13} + \gamma_{Ss}M_{14} - M_{23}/\gamma_{Cp} - \gamma_{Ss}M_{24}/\gamma_{Cp})/2 \\ A_{33} &= (M_{33} + \gamma_{Ss}M_{34} + M_{43}/\gamma_{Cs} + \gamma_{Ss}M_{44}/\gamma_{Cs})/2 \\ A_{43} &= (M_{33} + \gamma_{Ss}M_{34} - M_{43}/\gamma_{Cs} - \gamma_{Ss}M_{44}/\gamma_{Cs})/2. \end{aligned} \tag{5.17}$$

5.3.1 Amplitude Reflection and Transmission Coefficients

The amplitude reflection and transmission coefficients can now be expressed in terms of the elements of $\hat{A}$. Thus, for reflection in the cover, we can write

$$\begin{bmatrix} r_{pp} & r_{ps} \\ r_{sp} & r_{ss} \end{bmatrix} \equiv \begin{bmatrix} r_{11} & r_{12} \\ r_{21} & r_{22} \end{bmatrix}$$

$$= \begin{bmatrix} E_{Cy}^-/E_{Cy}^+ & E_{Cy}^-/E_{Cz}^+ \\ E_{Cz}^-/E_{Cy}^+ & E_{Cz}^-/E_{Cz}^+ \end{bmatrix}$$

$$= \begin{bmatrix} A_{21}A_{33} - A_{23}A_{31} & A_{11}A_{23} - A_{13}A_{21} \\ A_{33}A_{41} - A_{31}A_{43} & A_{11}A_{43} - A_{13}A_{41} \end{bmatrix} \Big/ \begin{vmatrix} A_{11} & A_{13} \\ A_{31} & A_{33} \end{vmatrix}. \quad (5.18)$$

Similarly, the cover-to-substrate amplitude transmission coefficients are given by

$$\begin{bmatrix} t_{pp} & t_{ps} \\ t_{sp} & t_{ss} \end{bmatrix} \equiv \begin{bmatrix} t_{31} & t_{32} \\ t_{41} & t_{42} \end{bmatrix}$$

$$= \begin{bmatrix} E_{Sy}^+/E_{Cy}^+ & E_{Sy}^+/E_{Cz}^+ \\ E_{Sz}^+/E_{Cy}^+ & E_{Sz}^+/E_{Cz}^+ \end{bmatrix}$$

$$= \begin{bmatrix} A_{33} & -A_{13} \\ -A_{31} & A_{11} \end{bmatrix} \Big/ \begin{vmatrix} A_{11} & A_{13} \\ A_{31} & A_{33} \end{vmatrix}$$

$$= \begin{bmatrix} A_{11} & A_{13} \\ A_{31} & A_{33} \end{bmatrix}^{-1}. \quad (5.19)$$

5.3.2 Irradiance Reflectance Coefficients

The irradiance reflectance coefficients can be calculated from the corresponding amplitude coefficients. Thus the reflectance coefficients for the cover are

$$\begin{bmatrix} R_{pp} & R_{ps} \\ R_{sp} & R_{ss} \end{bmatrix} \equiv \begin{bmatrix} R_{11} & R_{12} \\ R_{21} & R_{22} \end{bmatrix}$$

$$= \begin{bmatrix} |r_{pp}|^2 & -|r_{ps}|^2\gamma_{Cp}/\gamma_{Cs} \\ -|r_{sp}|^2\gamma_{Cs}/\gamma_{Cp} & |r_{ss}|^2 \end{bmatrix}, \quad (5.20)$$

and the cover-to-substrate transmittance coefficients are

$$\begin{bmatrix} T_{pp} & T_{ps} \\ T_{sp} & T_{ss} \end{bmatrix} \equiv \begin{bmatrix} T_{31} & T_{32} \\ T_{41} & T_{42} \end{bmatrix}$$

$$= \begin{bmatrix} |t_{pp}|^2\gamma_{Sp}/\gamma_{Cp} & -|t_{ps}|^2\gamma_{Sp}/\gamma_{Cs} \\ -|t_{sp}|^2\gamma_{Ss}/\gamma_{Cp} & |t_{ss}|^2\gamma_{Ss}/\gamma_{Cs} \end{bmatrix}. \tag{5.21}$$

5.4 All Media Isotropic

When the films as well as the bounding media are isotropic, the matrix equations for the amplitude reflectance and transmittance coefficients simplify, and the non-zero elements can be expressed in terms of the elements of $\hat{M}$. Thus

$$\begin{bmatrix} r_p & 0 \\ 0 & r_s \end{bmatrix} \equiv \begin{bmatrix} r_{11} & r_{12} \\ r_{21} & r_{22} \end{bmatrix}$$

$$= \begin{bmatrix} A_{21}/A_{11} & 0 \\ 0 & A_{43}/A_{33} \end{bmatrix}, \tag{5.22}$$

so that

$$r_p = \frac{\gamma_{Cp}M_{11} + \gamma_{Cp}\gamma_{Sp}M_{12} - M_{21} - \gamma_{Sp}M_{22}}{\gamma_{Cp}M_{11} + \gamma_{Cp}\gamma_{Sp}M_{12} + M_{21} + \gamma_{Sp}M_{22}} \tag{5.23}$$

$$r_s = \frac{\gamma_{Cs}M_{33} + \gamma_{Cs}\gamma_{Ss}M_{34} - M_{43} - \gamma_{Ss}M_{44}}{\gamma_{Cs}M_{33} + \gamma_{Cs}\gamma_{Ss}M_{34} + M_{43} + \gamma_{Ss}M_{44}}. \tag{5.24}$$

Similarly

$$\begin{bmatrix} t_p & 0 \\ 0 & t_s \end{bmatrix} \equiv \begin{bmatrix} t_{31} & t_{32} \\ t_{41} & t_{42} \end{bmatrix}$$

$$= \begin{bmatrix} 1/A_{11} & 0 \\ 0 & 1/A_{33} \end{bmatrix} \tag{5.25}$$

and hence

$$t_p = \frac{2\gamma_{Cp}}{\gamma_{Cp}M_{11} + \gamma_{Cp}\gamma_{Sp}M_{12} + M_{21} + \gamma_{Sp}M_{22}} \tag{5.26}$$

$$t_s = \frac{2\gamma_{Cs}}{\gamma_{Cs}M_{33} + \gamma_{Cs}\gamma_{Ss}M_{34} + M_{43} + \gamma_{Ss}M_{44}}. \tag{5.27}$$

5.4.1   Phase Changes on Reflection and Transmission

The phase of a reflected or transmitted beam is generally not the same as the phase of the input beam. The difference, δ = (phase of output - phase of input), is referred to either as a *phase change on reflection* or a *phase change on transmission*. In our notation a phase factor $\exp(i\delta)$ in an output corresponds to a phase lag. The phase factors are incorporated in the elements of the matrix $\hat{r}$, and the complete set of phase factors can be determined as the angle of $\hat{r}$,

$$\hat{\delta} = \text{angle}\,\hat{r}. \tag{5.28}$$

The phase changes given by Eq. (5.28) are referred to the basis field vectors as they are defined in this book. Some adjustments are needed to give the values associated with "conventional" positive directions for the electric field. Thus, for isotropic media the conventional phase changes on reflection (or on total internal reflection) are $\pi - \text{angle}\,\hat{r}_{11}$ for p and $-\text{angle}\,\hat{r}_{22}$ for s.

5.5   Computations Using the BTF Toolbox

5.5.1   General Birefringent Coating

The function **reflect** in the *BTF Toolbox* contains a compact algorithm for implementation of the equations required for calculating the reflectance and transmittance coefficients for a general stack of birefringent layers.

The most straightforward way to proceed is to compile the matrix *system* and then use either

$$\hat{R} = \textbf{reflect}(system,\ \beta),$$

or

$$[\hat{R},\ \hat{r}] = \textbf{reflect}(system,\ \beta).$$

However, this will be inefficient if the system contains the same film or period of films repeated several times. In such a case the characteristic matrices of the different films should be determined and used to calculate the characteristic matrix $\hat{M}$ of the stack. The field matrices for the cover, $\hat{F}_C = \textbf{fmat}(cover)$, and for the substrate, $\hat{F}_S = \textbf{fmat}(substrate)$, are required as well. Finally

$$\hat{R} = \textbf{reflect}(\hat{F}_C,\ \hat{M},\ \hat{F}_S),$$

or

$$[\hat{R},\ \hat{r}] = \textbf{reflect}(\hat{F}_C,\ \hat{M},\ \hat{F}_S).$$

5.5.2 PS Coatings

The *BTF Toolbox* function **reflect** accepts *psstack*, and *pssystem* as arguments for the computation of $\hat{R}$. The formats include

$$\hat{R} = \mathbf{reflect}(pssystem),$$

and

$$\hat{R} = \mathbf{reflect}(n_C, \hat{M}, n_S).$$

Chapter 6

Guided Waves

A biaxial thin film deposited onto a glass substrate, and of sufficient thickness, can act as an optical planar waveguide. Light that enters the film may be trapped by total internal reflection at both the film/cover interface and the film/substrate interface. One useful property of waveguides is that very large power densities can be achieved. For example, a laser beam can be coupled into a film of thickness about $1\,\mu$m.

Here we are interested in understanding the intrinsic properties of anisotropic planar waveguides so that they can be used for determining the principal refractive indices of biaxial thin film media.

We consider

▨ conditions for a mode, including modal cutoffs

▨ visualization of modal contours

▨ modal field structure and polarization

▨ prism couplers.

Numerical examples are used to illustrate and compare the characteristic properties of isotropic and anisotropic guides. Two single layer waveguides defined by the parameters in Table 6.1 are used for this purpose, and are referred to as the isotropic waveguide and the anisotropic waveguide.

Table 6.1. Planar waveguides.

Parameter	Symbol	Isotropic Waveguide	Anisotropic Waveguide
Cover index	n_C	1	1
Substrate index	n_S	1.516	1.516
Layer indices	n_1, n_2, n_3	1.7	1.8, 1.55, 1.7
Column angle	ψ	-	39°
Layer thickness	d	1.5 μm	1.5 μm
Wavelength	λ	632.8 nm	632.8 nm

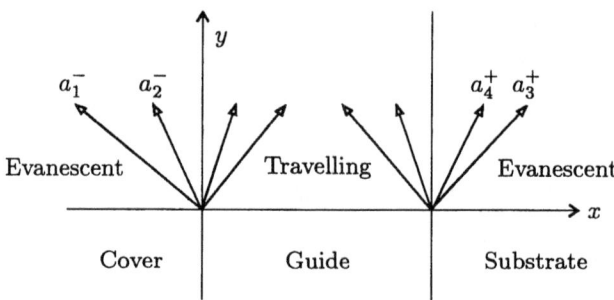

Fig. 6.1. Outward-going evanescent fields in the bounding media of a free biaxial waveguide.

6.1   Modal Condition

6.1.1   General Case

We begin by considering the most general case of a free waveguide, in which the semi-infinite cover, the layers of the waveguide, and the semi-infinite substrate may all be anisotropic. Figure 6.1 illustrates some of the conditions that must be satisfied by the fields of such a waveguide. In the cover and in the substrate the fields are outward going and evanescent. The first point means that no additional sources are irradiating the guide, and hence the field coefficients a_1^+, a_2^+ in the cover and a_3^-, a_4^- in the substrate are all equal to zero. The requirement of evanescent cover and substrate fields means that energy flowing is trapped in the guide, and we assume that it is travelling in the y direction as shown in the figure.

In general, waveguiding involves round-trip constructive interference of the travelling wave fields in the guide, and for this reason occurs only for discrete values of β, called the modal values. From a theoretical point of view the central problem is computation of the β values, since the matrix method developed earlier can be used to complete the solution, trace all field profiles etc.

Now, even in the more elementary case of isotropic layers and bounding media, it is not possible to write an analytical expression for the modal β values. As well, matrix methods proceed most easily when β is one of the specified values, and for this reason the usual method for determining the modes involves a modal condition that is tested over a range of β values. Thus we require an expression, a function of β, that is easy to compute and signifies the presence of a mode.

Returning to Fig. 6.1, we note that the cover and substrate field coefficients shown there are required to satisfy the general equation

$$\vec{a}_C = \hat{A}\vec{a}_S, \tag{6.1}$$

i.e.

$$\begin{bmatrix} 0 \\ a_1^- \\ 0 \\ a_2^- \end{bmatrix} = \hat{A} \begin{bmatrix} a_3^+ \\ 0 \\ a_4^+ \\ 0 \end{bmatrix} \tag{6.2}$$

where $\hat{A}$ is the system matrix. Four equations in which a_1^-, a_2^-, a_3^+ and a_4^+ are the unknowns are implied here, and inspection shows that the condition for a non-trivial solution is $A_{11}A_{33} - A_{13}A_{31} = 0$. This could be used as the modal condition, but in general the elements of the system matrix are complex, and it is preferable to use

$$|A_{11}A_{33} - A_{13}A_{31}| = 0 \tag{6.3}$$

as the modal condition in its most general form.[14]

6.1.2 Isotropic Cover and Substrate

When the bounding media are isotropic, the modal condition can be expressed in terms of the elements of the characteristic matrix $\hat{M}$. We have for the total field at the cover interface

$$\vec{m}_C \equiv \begin{bmatrix} E_{Cy} \\ H_{Cz} \\ E_{Cz} \\ H_{Cy} \end{bmatrix} = \begin{bmatrix} E_{Cy}^- \\ -\gamma_{Cp}E_{Cy}^- \\ E_{Cz}^- \\ -\gamma_{Cs}E_{Cz}^- \end{bmatrix}, \tag{6.4}$$

and at the substrate interface

$$\vec{m}_S \equiv \begin{bmatrix} E_{Sy} \\ H_{Sz} \\ E_{Sz} \\ H_{Sy} \end{bmatrix} = \begin{bmatrix} E_{Sy}^+ \\ \gamma_{Sp}E_{Sy}^+ \\ E_{Sz}^+ \\ \gamma_{Ss}E_{Sz}^+ \end{bmatrix}. \tag{6.5}$$

Thus the general equation

$$\vec{m}_C = \hat{M}\vec{m}_S \tag{6.6}$$

becomes

$$\begin{bmatrix} E_{Cy}^- \\ -\gamma_{Cp}E_{Cy}^- \\ E_{Cz}^- \\ -\gamma_{Cs}E_{Cz}^- \end{bmatrix} = \hat{M} \begin{bmatrix} E_{Sy}^+ \\ \gamma_{Sp}E_{Sy}^+ \\ E_{Sz}^+ \\ \gamma_{Ss}E_{Sz}^+ \end{bmatrix}. \tag{6.7}$$

Elimination of the E's leads to the modal condition in the form

$$\begin{aligned} |(\gamma_{Cp}M_{11} + \gamma_{Cp}\gamma_{Sp}M_{12} + M_{21} + \gamma_{Sp}M_{22}) \times \\ (\gamma_{Cs}M_{33} + \gamma_{Cs}\gamma_{Ss}M_{34} + M_{43} + \gamma_{Ss}M_{44}) - \\ (\gamma_{Cp}M_{13} + \gamma_{Cp}\gamma_{Ss}M_{14} + M_{23} + \gamma_{Ss}M_{24}) \times \\ (\gamma_{Cs}M_{31} + \gamma_{Cs}\gamma_{Sp}M_{32} + M_{41} + \gamma_{Sp}M_{42})| = 0. \end{aligned}$$

Alternatively, substitution of Eqs.(5.16) and (5.17) into Eq.(6.3) leads to the same expression.

Table 6.2. Modal conditions on the elements of the system matrix.

General Hybrid Modes	p-modes	s-modes
$|A_{11}A_{33} - A_{13}A_{31}| = 0$	$|A_{11}| = 0$	$|A_{33}| = 0$

6.1.3 Uncoupled Modes

In the special cases of an isotropic waveguide or propagation in a common deposition plane of an anisotropic waveguide, the matrices $\hat{A}$ and $\hat{M}$ have the forms

$$\hat{A} = \begin{bmatrix} A_{11} & A_{12} & 0 & 0 \\ A_{21} & A_{22} & 0 & 0 \\ 0 & 0 & A_{33} & A_{34} \\ 0 & 0 & A_{43} & A_{44} \end{bmatrix} \tag{6.8}$$

$$\hat{M} = \begin{bmatrix} M_{11} & M_{12} & 0 & 0 \\ M_{21} & M_{22} & 0 & 0 \\ 0 & 0 & M_{33} & M_{34} \\ 0 & 0 & M_{43} & M_{44} \end{bmatrix}, \tag{6.9}$$

and the modal condition can be separated into two expressions, one for p-polarized modes and one for s-polarized modes. In terms of the elements of $\hat{A}$ the modal conditions are

$$|A_{11}| = 0 \quad (p\text{-modes}) \tag{6.10}$$

and

$$|A_{33}| = 0 \quad (s\text{-modes}). \tag{6.11}$$

A summary of modal conditions based on the elements of the system matrix $\hat{A}$ is given in Table 6.2.

Equivalent modal expressions, based on the elements of $\hat{M}$, for the special cases discussed in this section are

$$|\gamma_{Cp}M_{11} + \gamma_{Cp}\gamma_{Sp}M_{12} + M_{21} + \gamma_{Sp}M_{22}| = 0 \quad (p\text{-modes}) \qquad (6.12)$$

$$|\gamma_{Cs}M_{33} + \gamma_{Cs}\gamma_{Ss}M_{34} + M_{43} + \gamma_{Ss}M_{44})| = 0 \quad (s\text{-modes}). \qquad (6.13)$$

However, 2×2 matrix algebra can be used for these special cases and the modal conditions can be expressed in terms of the elements of the smaller matrices. Such modal conditions have the form of Eq.(6.12) with appropriate p and s γ's. Thus, for an isotropic guide the matrices given in Eqs.(4.57) and (4.58) can be used, and for propagation in the deposition plane of a birefringent waveguide, the matrices $\hat{M}_p$ and $\hat{M}_s$ given by Eq.(4.64) (or (4.66)) and Eq.(4.65) are appropriate,

$$|\gamma_{Cp}(M_p)_{11} + \gamma_{Cp}\gamma_{Sp}(M_p)_{12} + (M_p)_{21} + \gamma_{Sp}(M_p)_{22}| = 0 \quad (p\text{-modes}) \qquad (6.14)$$

$$|\gamma_{Cs}(M_s)_{11} + \gamma_{Cs}\gamma_{Ss}(M_s)_{12} + (M_s)_{21} + \gamma_{Ss}(M_s)_{22}| = 0 \quad (s\text{-modes}). \qquad (6.15)$$

6.1.4 Poles of R

Notice that the modal expressions occur in the denominator of the corresponding expressions for the reflection coefficients r_{CG} from the cover to the layers and r_{SG} from the substrate to the layers. Thus the use of these equations would yield infinities for r_{CG} and r_{SG} at the β values corresponding to the modes, and for this reason we may state *the modes of a planar waveguide occur at the poles of the reflection coefficient*. A simple physical explanation can be given to explain poles in the reflection coefficients – the guide has outgoing fields but no incoming fields.

6.1.5 Examples

As examples we have determined the modes of the isotropic waveguide described in Table 6.1, and modes of the anisotropic guide (described in the same table) for the cases $\xi = 0°$ and $\xi = 45°$. In Fig. 6.2 the log of the test expression is plotted as a function of β for the anisotropic guide with $\xi = 45$. Each sharp minimum in Fig. 6.2 indicates the presence of a mode. After refinement to four decimal places, the values shown in Table 6.3 were obtained. See Sect. 6.7 for a discussion of modal order (m) and modal designation.

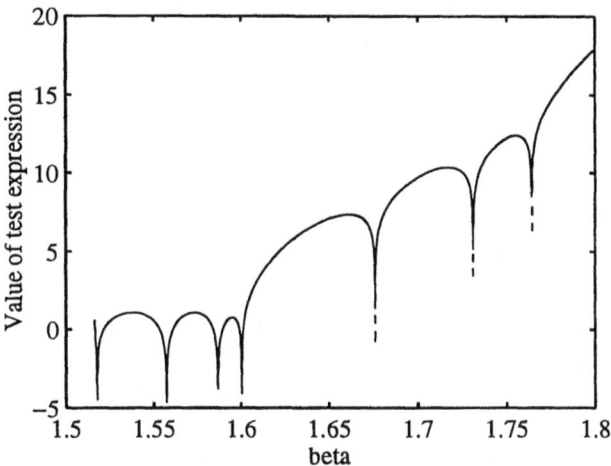

Fig. 6.2. Log plot of modal test expression $|A_{11}A_{33} - A_{13}A_{31}|$.

Table 6.3. Bound modes of planar waveguides.

Isotropic		Biaxial $\xi = 0°$		Biaxial $\xi = 45°$		
Mode	β	Mode	β	Mode	β	$m_1^+ m_1^- m_2^+ m_2^-$
TE_0	1.6899	$TM_{0,0}$	1.6935	$TM_{0,45}$	1.7639	0 1 0 0
TM_0	1.6889	$TE_{0,0}$	1.6899	$TE_{0,45}$	1.7310	1 2 0 0
TE_1	1.6593	$TE_{1,0}$	1.6593	$TE_{1,45}$	1.6760	2 3 0 0
TM_1	1.6556	$TM_{1,0}$	1.6578	$TM_{1,45}$	1.6003	3 3 0 0
TE_2	1.6081	$TE_{2,0}$	1.6081	$TE_{2,45}$	1.5868	3 4 0 1
TM_2	1.6004	$TM_{2,0}$	1.5986	$TM_{2,45}$	1.5573	3 4 1 2
TE_3	1.5381	$TE_{3,0}$	1.5381	$TE_{3,45}$	1.5178	4 4 1 2
TM_3	1.5282	$TM_{3,0}$	1.5225	–	–	–

6.2   Modal Cutoffs

The set of modal values β for a planar waveguide is subject to a lower limiting value, to ensure that total internal reflection occurs at both the cover interface and the substrate interface, and an upper limit to ensure that the internal fields in the layers are not all evanescent. For a multilayered guide with isotropic layers the limits can be stated as

$$n_C, \, n_S < \beta < \text{maximum layer index} \quad \text{(isotropic guide)}. \tag{6.16}$$

Thus for the single layer isotropic guide, used here as an example, the modal β values are constrained to lie within the range 1.516 to 1.7.

Similar principles apply to anisotropic guides, but the upper limiting value of β then depends on the geometry, i.e. on the value of ξ as well as the principal refractive indices. For propagation in the deposition plane, i.e. when $\xi = 0$, α and β are related by Eq.(3.33). The expression in the square brackets of this equation can be written as a quadratic in α, and then the condition on β for the solution to change from real to complex can be determined. This is the upper modal cutoff for the p-polarization. The upper cutoff for the s-polarization is just n_3, and we have

$$n_C, \, n_S < \beta < [n_1^2 \cos^2 \psi + n_2^2 \sin^2 \psi]^{1/2} \quad (p\text{-modes}) \tag{6.17}$$

$$n_C, \, n_S < \beta < n_3 \quad (s\text{-modes}). \tag{6.18}$$

A plot of α versus β, computed using Eq.(3.33) provides a useful method for determining the upper limit of β in a particular case. Figure 6.3 provides such a plot for the anisotropic guide that we are using as an example, and shows that the inner sheet and outer sheet basis waves are evanescent for $\beta > 1.598$ and $\beta > 1.775$ respectively. Thus the maximum allowed value of β for hybrid modes in the anisotropic waveguide is 1.775.

6.3   Modal Contours

For a given film thickness d, the set of possible modes for a planar waveguide can be investigated by computing β values for various values of the azimuthal angle ξ between the deposition plane and the propagation plane. We wish to display the set of possible β values and note that the difference between the β's of a pair of neighbouring modes may be considerably smaller than the average of the β's. That is, the fractional differences may be small. For this reason we increase sensitivity and enhance visualization by plotting $(\beta - n_S) \sin \xi$ versus

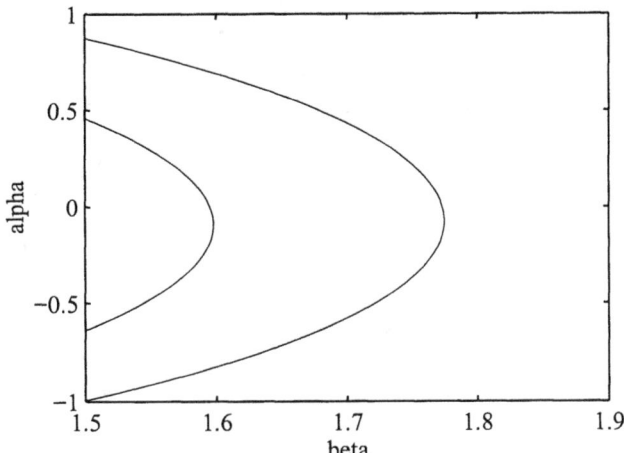

Fig. 6.3. The largest (right-most) value of β in the plot determines the upper modal cutoff.

$(\beta - n_S) \cos \xi$. Such polar plots effectively show modal contours superposed on the plane of the waveguide.

Consider first the isotropic waveguide defined in Table 6.1. Clearly the modes of an isotropic guide do not depend on the azimuthal angle ξ and hence the modal contours are concentric circles, as shown in the upper part of Fig. 6.4.

In the lower part of Fig. 6.4 the dots plotted at intervals of $\Delta \xi = 1°$ indicate the modes that were determined for the anisotropic waveguide. Modal contours formed by joining the dots have interesting characteristic features. First of all we note the resemblance to a "head-with-mask" shape, which we call βatman. Further investigations show the relationship of βatman's features to the refractive indices. For example Fig. 6.5, in which the inner and outer curves indicate the onset of evanescence for basis waves associated with the inner and outer sheets of the refractive index surface of the biaxial material, shows similar shapes. Despite the complicated structure of hybrid modal patterns, we can conclude that βatman's "head" in the pattern is associated with the outer sheet and his "mask" is related to the inner sheet.

Another property of the modal contour lines is that they may approach one another closely in some areas of the pattern but do not cross.

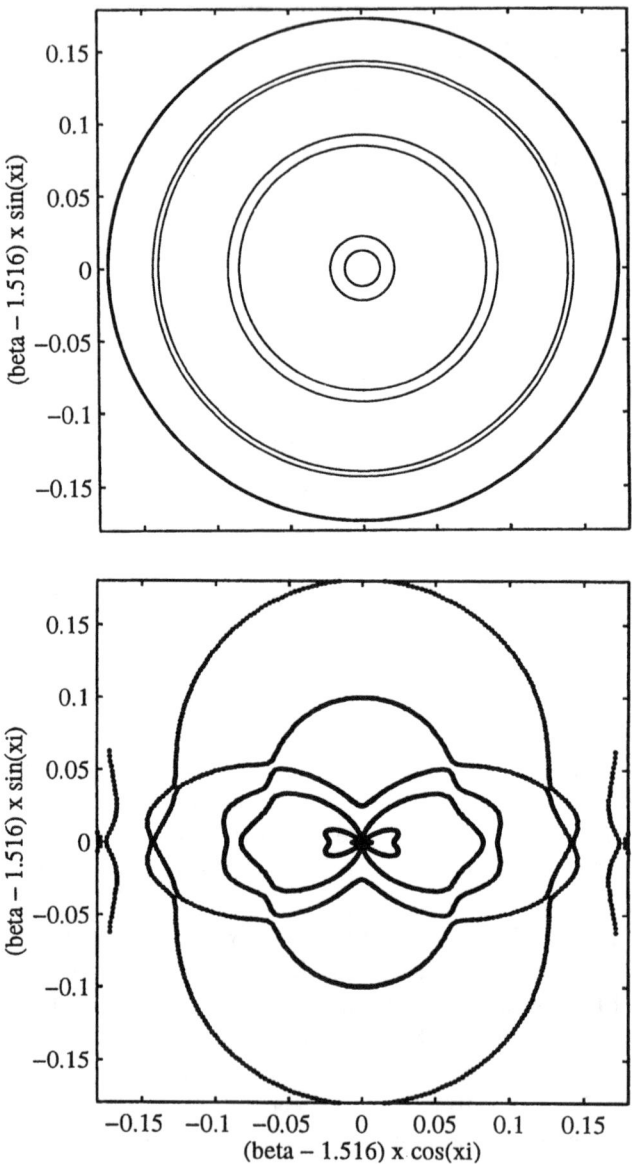

Fig. 6.4. Modal contours (polar plots of $(\beta - n_S)\sin\xi$ versus $(\beta - n_S)\cos\xi$) form concentric circles for the isotropic waveguide (upper) and a characteristic βatman shape for the anisotropic waveguide (lower). (Adapted from I.J. Hodgkinson, S. Kassam, J. Hazel, S.J. Cloughley and Q.H. Wu, *Applied Optics* 35, 5569, 1996. Copyright © Optical Society of America, 1995. Reprinted with permission.)

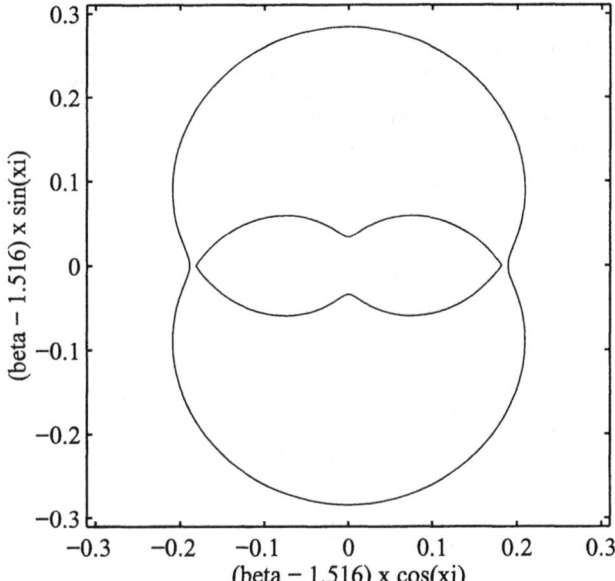

Fig. 6.5. Curves marking the onset of evanescence for basis waves associated with the inner sheet of the refractive index surface (inside curve) and the outer sheet (outside curve). (Adapted from I.J. Hodgkinson, S. Kassam, J. Hazel, S.J. Cloughley and Q.H. Wu, *Applied Optics* 35, 5569, 1996. Copyright © Optical Society of America, 1995. Reprinted with permission.)

6.4   Modal Field Structure

The modal field structure[15] of a waveguide can be described in terms of characteristic profiles of total fields. Typically these patterns have the form of standing waves across the guide, and travel sinusoidally down the guide. Alternatively the field structure can be described in terms of the linear sum of basis vectors that develops in the anisotropic material, and of course the two descriptions are related by the 4×4 matrix method. Thus at a particular point in the guide the total field is given by the equation $\vec{m} = \hat{F}\vec{a}$ in which $\hat{F}$ contains the basis fields and $\vec{a}$ provides the coefficients for the linear sum.

Where do we start in order to plot standing wave field patterns? Clearly once we have a foothold, knowledge of $\vec{m}$ or $\vec{a}$ at some point, we can use matrices to trace fields throughout the layers and the bounding media of the guide. In fact Eq.(6.17) can be solved easily enough for the relative values of the a's at the cover and at the substrate. We use the solution forms

$$
a_C \equiv \begin{bmatrix} a_1^+ \\ a_1^- \\ a_2^+ \\ a_2^- \end{bmatrix} = \begin{bmatrix} 0 \\ A_{11}A_{23} - A_{13}A_{21} \\ 0 \\ A_{11}A_{43} - A_{13}A_{41} \end{bmatrix} \tag{6.19}
$$

$$
a_S \equiv \begin{bmatrix} a_3^+ \\ a_3^- \\ a_4^+ \\ a_4^- \end{bmatrix} = \begin{bmatrix} -A_{13} \\ 0 \\ A_{11} \\ 0 \end{bmatrix} \tag{6.20}
$$

for the all anisotropic guide. When the cover and substrate media are both isotropic, the a's can be replaced by electric field amplitudes, as in

$$
m_C \equiv \begin{bmatrix} E_{Cy} \\ H_{Cz} \\ E_{Cz} \\ H_{Cy} \end{bmatrix} = \begin{bmatrix} A_{11}A_{23} - A_{13}A_{21} \\ -\gamma_{Cp}(A_{11}A_{23} - A_{13}A_{21}) \\ A_{11}A_{43} - A_{13}A_{41} \\ \gamma_{Cs}(A_{11}A_{43} - A_{13}A_{41}) \end{bmatrix} \tag{6.21}
$$

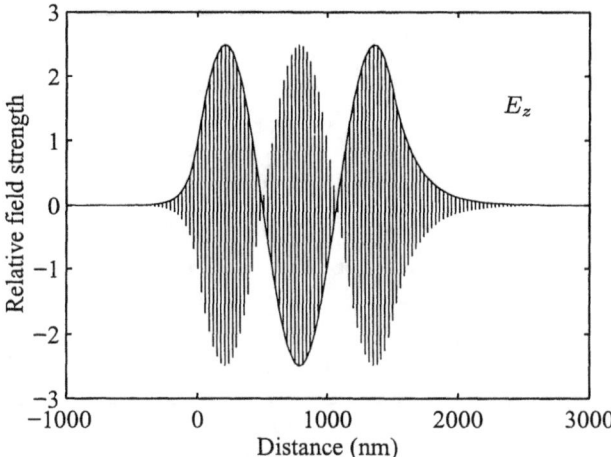

Fig. 6.6. Snapshot and excursions (vertical lines) of the electric field component E_z for the TE$_2$ mode in the isotropic waveguide.

$$m_S \equiv \begin{bmatrix} E_{Sy} \\ H_{Sz} \\ E_{Sz} \\ H_{Sy} \end{bmatrix} = \begin{bmatrix} -A_{13} \\ -\gamma_{Sp}A_{13} \\ A_{11} \\ \gamma_{Ss}A_{11} \end{bmatrix}. \tag{6.22}$$

The most striking modal field profiles, capturing the resonant nature of modes, are formed by the TE or s-polarization because the electric field component E_z is continuous in both value and gradient across an interface. The TE$_{2,0}$ mode appropriate to propagation in the deposition plane of the isotropic guide is illustrated in Fig. 6.6. Here the vertical lines show the excursion of E_z as time proceeds and the solid line is a snapshot for a particular instant of time. The middle segment of the standing wave is 180° out of phase with the outer segments including the evanescent standing waves in the cover and substrate. From the form of the profile we can reach an interesting conclusion – additional half-wave sections of material could be added to or removed from the guide without changing the β value of the particular mode.

The excursions of H_z in the TM$_{2,0}$ mode (Fig. 6.7) look similar to the previous case, but now the snapshot shows "linear phase slip" of the standing wave pattern along the x-axis, perpendicular to the plane of the guide. This is caused by the different magnitudes of α (and hence different magnitudes of the wave

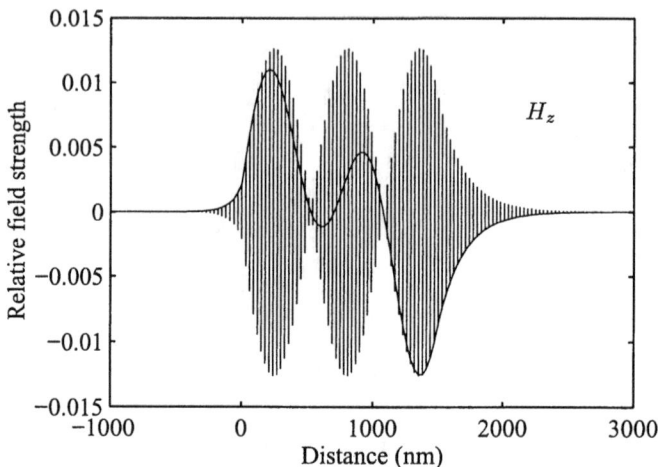

Fig. 6.7. Snapshot and excursions of H_z in the TM$_{2,0}$ mode of the anisotropic guide, showing "linear phase slip".

vector component k_x along the x-axis) of the two p-polarized basis vectors that propagate in the guide and add to give the solid line shown in the figure for an instant of time. All field vectors in the layer are influenced in the same way, so the p-polarization is retained along the x-axis. See also the discussion of phase terms in Sect. 4.5.2.

A third example, E_y and E_z plotted for TM$_{2,45}$ in Fig. 6.8, shows that the hybrid modes are even more complicated. The four basis vectors now all have different wave vector components along the x-axis, and hence the envelopes of the maximum excursions of the total fields are no longer simple harmonic in a spatial sense. As consequences, the β value cannot be maintained by adding or removing simple slices and the polarization state, as defined by the field components E_y and E_z, varies along the x-axis. In the following section we use the polarization state at the cover interface as a reference so that characteristics of different hybrid modes can be compared.

6.5   Modal Polarization

Fig. 6.9 shows the polarization of the electric field at the cover, for modes of the anisotropic guide computed at intervals of $\Delta\xi = 5°$. Each ellipse or line is centered on a modal point and is drawn appropriate to the y-axis running left to right across the page and the z-axis running toward the top of the page. Open ellipses represent the right-hand polarization sense for propagation along the

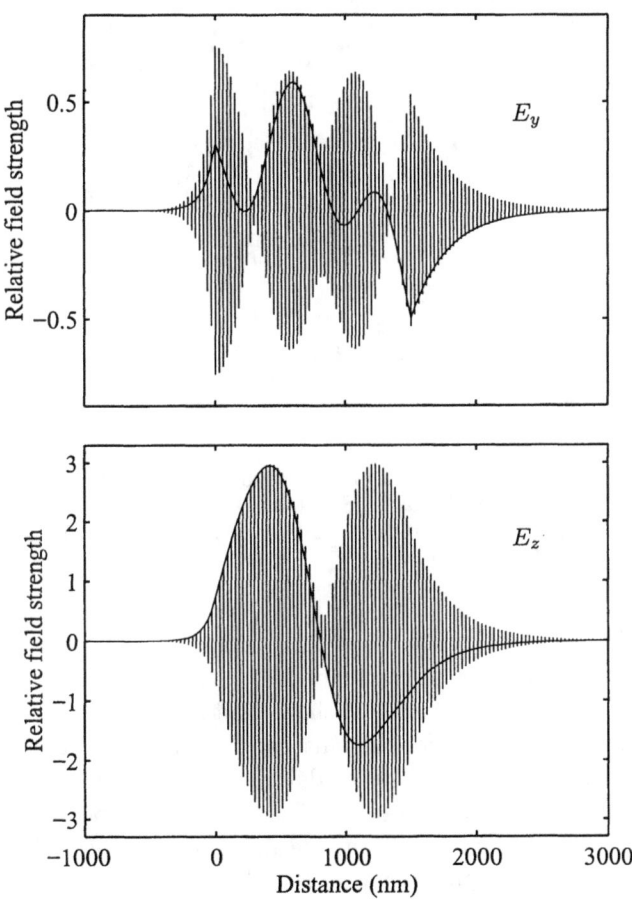

Fig. 6.8. Snapshot and excursions of E_y (upper) and E_z (lower) for the $TM_{2,45}$ hybrid mode.

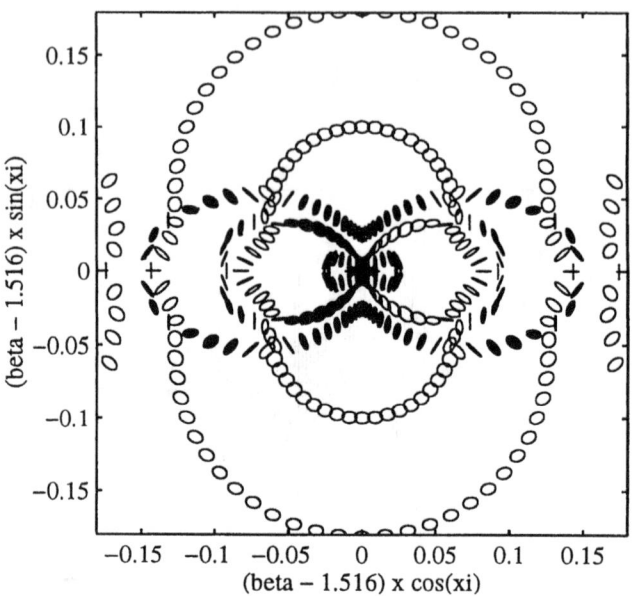

Fig. 6.9. Polarization of the modes of the anisotropic waveguide. (Adapted from I.J. Hodgkinson, S. Kassam, J. Hazel, S.J. Cloughley and Q.H. Wu, *Applied Optics* 35, 5569, 1996. Copyright © Optical Society of America, 1995. Reprinted with permission.)

negative x-axis (E_{Cy} leading E_{Cz}) and filled ellipses represent left-handedness. The reverse notation applies in the lower half of the figure.

Comparison of Figs 6.4, 6.5 and 6.9 shows that βatman's "mask" and "head" shapes, referred to previously for the anisotropic waveguide, are also associated with polarization handedness. The modal lines are formed from fragments of the "mask" and "head" shapes, and significant changes of polarization occur in regions where a modal line moves from one shape to the other. The latter point is illustrated at higher magnification in Fig. 6.10, which is plotted for a guide with thickness $d = 2.5\,\mu$m. Here the modes in the central vertical section of the figure retain the right-handed sense of the "head", whilst the modes at the sides of the figure have the left-handedness sense of the "mask".

6.6   Modal Overlap

The modal contours of the isotropic waveguide in Fig. 6.4 form concentric circles and modal overlap does not occur. Central horizontal lines in Fig. 6.4 and

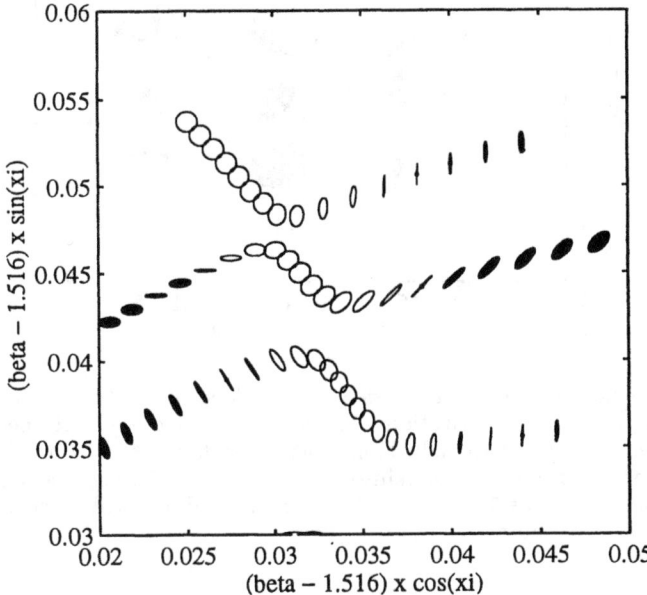

Fig. 6.10. Polarization of the modes of a waveguide with thickness $d = 2.5\,\mu$m in a region of strong coupling. Open ellipses correspond to right-hand states and filled ellipses indicate left-hand states. (Adapted from I.J. Hodgkinson, S. Kassam, J. Hazel, S.J. Cloughley and Q.H. Wu, *Applied Optics* 35, 5569, 1996. Copyright © Optical Society of America, 1995. Reprinted with permission.)

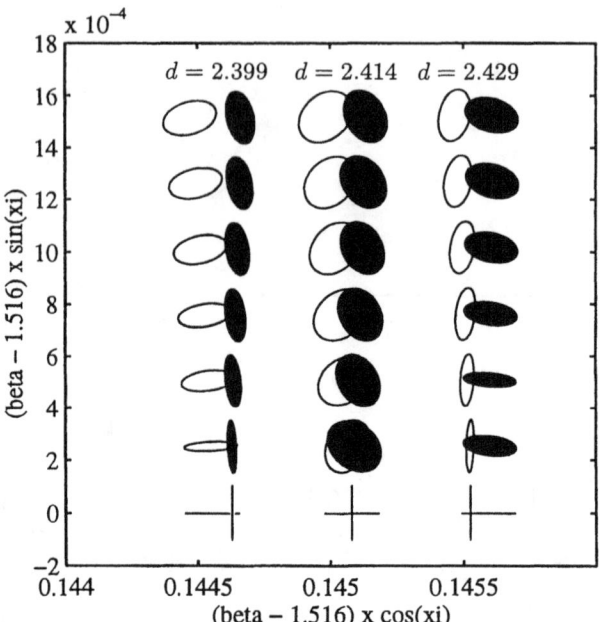

Fig. 6.11. Mechanism for modal overlap in the deposition plane. The modal lines bump together, and the polarization ellipses rotate, but the modal lines do not cut. (Open ellipses correspond to right-hand states and filled ellipses indicate left-hand states.) (Adapted from I.J. Hodgkinson, S. Kassam, J. Hazel, S.J. Cloughley and Q.H. Wu, *Applied Optics* 35, 5569, 1996. Copyright © Optical Society of America, 1995. Reprinted with permission.)

Fig. 6.9 for the anisotropic waveguide correspond to propagation in the deposition plane. In this case the four basis fields in the guide decouple into two pairs, and the polarizations are p or s. Fig. 6.9 shows overlap of the horizontal bars that represent the p-polarized modes and the vertical bars of the s-polarized modes.

In practice, modal overlap in thick biaxial guides tends to make modal identification difficult. An understanding of how modal overlap occurs, without crossing of modal contours, can be achieved by considering Fig. 6.11. The figure follows a pair of modes, as the local thickness of the guide is supposed to increase by a small amount. Overlap of the p and s modes is seen to be accompanied by steady rotation of the ellipses, as the modal lines bump together and then move apart again.[16]

6.7 Modal Order

For the isotropic guide each circular modal contour can be labelled with an integral order number m together with the p or s polarization type. Such order numbers satisfy an x-axis round-trip condition for constructive interference of the form
sum of phase thicknesses = sum of phase advances on reflection + $2m\pi$, i.e.

$$2k\alpha d = 2\phi_C + 2\phi_S + 2m\pi \qquad (6.23)$$

where $2\phi_C$ is the phase advance on reflection at the cover and similarly $2\phi_S$ applies to the substrate.

The same round-trip condition can be applied to the elliptically polarized hybrid modes of a biaxial guide. However, the situation is now complicated by two factors. Four round-trips can be defined – one consists of a basis wave associated with the inner sheet of the refractive index surface and travelling from the cover side to the substrate side followed by an outer sheet wave travelling in the opposite sense – and the m values may be different. As well, the phase relationships at an interface are not constant for a given β but depend on the relative weighting of the basis waves.

However, propagation in the deposition plane ($\xi = 0$) is similar to the isotropic case and, as modal contours do not cross, we conclude that each hybrid mode ($\xi \neq 0$) can be labelled unambiguously according to (i) the deposition plane TE or TM polarization, (ii) the deposition plane m-value, and (iii) the angle ξ. Thus, in Table 6.3, the hybrid mode designated $TM_{0,45}$ is on the same modal contour as the deposition plane mode $TM_{0,0}$.

The right-most column in Table 6.3 lists the "orders" m_1^+, m_1^-, m_2^+, m_2^-, for basis waves associated with the outer and inner sheet of the refractive index surface. The integers listed are the whole numbers of half-wavelengths in the thickness of the guide, for each basis vector.

6.8 Power Flow

At any point in a planar waveguide the time-averaged value of the Poynting vector $P_x = \Re(E_y H_z^* - E_z H_y^*)/2$ is equal to zero. In the cover, for example, the fields E_z and H_y of a TE wave are 90° out of phase, as can be seen from Eq.(6.4) in which γ_{Cs} is imaginary. During one half-period of the light wave motion energy moves outwards into the cover, but during the next half-period an equal amount returns. The process is similar to the lossless charging and discharging of a capacitor in an ac electrical circuit.

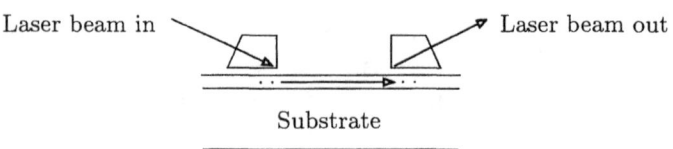

Fig. 6.12. Input and output prism couplers.

On the other hand energy travels along the y-axis, both in the layers of the guide and in the bounding media close to the cover and substrate interfaces. The distribution of power in the guide, time-averaged at points on the x-axis, is given by

$$p_y(x) = \Re(E_z H_x^* - E_x H_z^*)/2. \tag{6.24}$$

In some cases the total power P flowing in a layer or in a bounding medium is required. For the most general anisotropic guide the total power in a particular layer is best found by numerical integration of Eq.(6.24), but analytical expressions can be derived for the total power in any layer of an isotropic guide.

6.9   Prism Couplers

Knowing that light wave paths are reversible, and that the light in a waveguide mode is trapped by total internal reflection, it is follows that an external laser beam cannot be coupled into a waveguide by simply directing it at either the cover or the substrate interface. In practice a small prism positioned close to the cover interface may be used as a coupling device. The arrangement is illustrated in Fig. 6.12, where it can be seen that a prism can act as an input coupler or an output coupler.

In the region between the waveguide and the prism the behaviour of the guide is perturbed by the presence of the prism – the evanescent tail of the fields cannot reach to infinity. As a consequence energy flows into and accumulates in the guide. Some energy flows out again, but once past the sharp edge of the prism energy is trapped in a free guide. The reverse applies in the output coupler.

A variation on the prism coupler just described is shown in Fig. 6.13. Here a prism of symmetric shape acts both as an input coupler and an output coupler, with interesting results for thin films which usually scatter a small amount of the light propagating in them. In operation, usually in the weak coupling regime

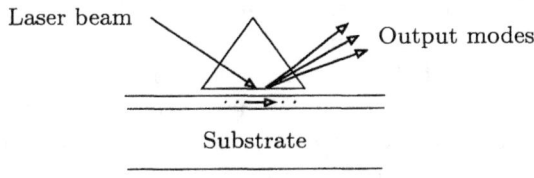

Fig. 6.13. Prism acting as simultaneous input/output coupler.

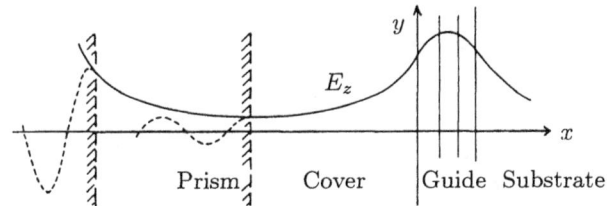

Fig. 6.14. Geometry for maximum coupling, producing the largest value of the ratio strength of E_z in the layers of the guide to strength of E_z in the prism. (Adapted from J.T. Chilwell and I.J. Hodgkinson, *Journal of the Optical Society of America A* 1, 742, 1984. Copyright © 1984 Optical Society of America. Reprinted with permission.)

where the guide is "nearly free", the angle of incidence and the polarization state of the incident laser beam are adjusted to excite a mode in the guide. Within the guide light is scattered and accumulates in other modes which may have different polarizations. An equilibrium is established whereby energy leaks out of the guide and coupler at the discrete set of angles corresponding to the modes excited by the primary and secondary processes. For a biaxial waveguide the modal display on a screen is similar and can be related to the modal contours discussed in Sect. 6.3. One application is the estimation of the three principal refractive indices of a biaxial film, by minimizing the difference between experimental and theoretical modal β values in an optimization procedure.

Anyone who has used the prism coupler as described above will have experienced frustrating moments when a laser beam refuses to cooperate and enter a waveguide (usually this happens during a demonstration for a visitor). The gap between the prism and the guide is a critical parameter, and we shall now refer to Fig. 6.14 for a physical explanation of an optimal coupling distance d_C.

Suppose that β, which is determined by the incident laser beam, is just slightly different from a modal value β_m of the free isotropic guide. A TE mode is easiest to visualize because as mentioned previously E_z is continuous both in value and slope across an interface. We can arbitrarily assign $\{0\ 0\ 1\ \gamma_{Ss}\}$ to the total field at the cover, and begin tracing E_z through the layers of the guide, through the gap, and into the prism where the incident laser beam and the totally (100%) reflected beam form a standing wave profile. The amplitude of the standing wave is twice the amplitude of the incident travelling wave laser beam. Two evanescent waves propagate in opposite senses in the gap – one is from the prism and the other is from the guide. The standing evanescent wave in the prism has an accessible minimum, as illustrated in Fig. 6.14, and that is the location of the prism face for maximum value of the ratio E_z in the guide to E_z in the prism.

Part II

Characterization of Anisotropic Films

Chapter 7

Deposition of Microstructures

Biaxial thin films are deposited in much the same way as isotropic films. The main difference is that the substrate is positioned to receive the impinging vapour at an oblique angle during the deposition of a biaxial film, and at normal incidence for an isotropic (or uniaxial) film.

The evaporant material is heated in vacuum, usually by an electron beam gun, but in some cases by a resistively heated coil or boat filament. Evaporant atoms travel from the source to the substrate where they condense and then have limited mobility. The columnar structure grows at an angle between the vapour direction and the substrate normal. Basically, self shadowing of incoming atoms and limited mobility together cause the columnar growth – condensing atoms are unable to move far enough to fill vacant positions in the shadow of existing material.[17]

The thickness required of a biaxial film for an application such as a quarter-wave plate is relatively large, perhaps twenty times the thickness of an individual film layer in an isotropic coating. For this reason the deposition parameters of a biaxial film need to be chosen with care to minimize absorbtion and scatter.

In this chapter, we

▨ describe apparatus for depositing anisotropic films, and give typical values of deposition parameters for the growth of biaxial columnar structures with low absorbtion and scatter

▨ list a range of columnar structures that have been deposited

▨ discuss computer simulation of the deposition of biaxial columnar structures, by the serial deposition of hard spheres.

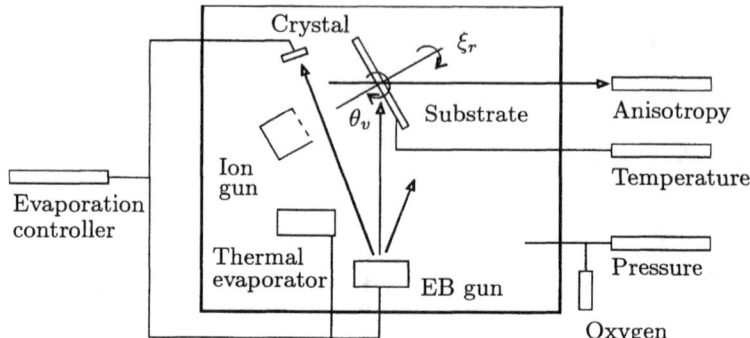

Fig. 7.1. Vacuum chamber for depositing thin film microstructures.

7.1 Vacuum Deposition

7.1.1 Apparatus

Figure 7.1 shows a typical layout of a vacuum coating chamber furnished for the deposition of thin film microstructures. The *electron beam evaporator*, the *thermal evaporator*, the *quartz crystal sensor*, the *evaporation rate and thickness controller*, the *substrate temperature controller*, the *residual gas pressure controller*, and the *ion gun* are standard optical coating accessories.

The label *anisotropy* on Fig. 7.1 refers to custom made monitors for *in situ* measurement or monitoring of specific anisotropic properties during the deposition of anisotropic coatings. In our work these have included:-

- A monitor dedicated to the recording of R_p, R_s, T_p, T_s for metal films illuminated by light incident at normal incidence. Here p and s mean, respectively, parallel and perpendicular to the deposition plane defined by the direction of the arriving metal atoms and the normal to the substrate. This monitor and the results obtained for aluminium, gold, and silver films are described in Chapt. 12.

- A transmission-mode perpendicular incidence ellipsometer for monitoring phase retardance during the deposition of thin film wave plates. This ellipsometer and the principles of perpendicular incidence ellipsometry in transmission are described in Sect. 8.1.

- Multiple-angle ellipsometers for *in situ* monitoring of the three principal refractive indices n_1, n_2, n_3 during the deposition of anisotropic dielectric films. Further details of this work can be found in Sect. 8.2.1.

- Apparatus for measuring anisotropic scatter during the growth of columnar films, as described in Sect. 10.5.

7.1.2 Deposition Parameters

Typical values of the deposition parameters required for birefringent media to make a wave plate, for example, are 1–4×10^{-4}mbar oxygen backfill pressure, 0.1 nm/s deposition rate, 300°C substrate temperature.

7.2 Columnar Structures and Effective Media

Several different types of columnar structure can be deposited, by controlling the deposition vapour angle θ_v and the substrate rotation angle ξ_r as the deposition proceeds (see Fig. 7.1). Here we consider dielectric materials only; the evolutionary growth of anisotropic microstructures during the growth and subsequent ion beam etching of thin metal films is discussed in Chapt. 12.

7.2.1 Uniaxial Media

Deposition with $\theta_v = 0$ and ξ_r constant leads to the growth of columns running normal to the substrate; the media is uniaxial, with $n_1 = n_e$ and $n_2 = n_3 = n_o$. Films deposited in this way have been considered to be isotropic, but waveguiding measurements have confirmed the uniaxial nature.

7.2.2 Biaxial Media

Deposition with $\theta_v \neq 0$ and ξ_r constant leads to a tilted columnar structure. The columns are nearly parallel and make a characteristic angle ψ (in practice a small range of angles) with the substrate. The column angle ψ is smaller than the deposition angle, and can be measured directly by electron microscopy, after the film and substrate have been fractured. If it is not practical to obtain the column angle by electron microscopy then it may be determined from multiple-angle ellipsometry, as explained in Sect. 8.2.1, or estimated using the tangent rule,[18]

$$\tan \Psi = \frac{1}{2} \tan \theta_v. \qquad (7.1)$$

A word of caution though. In our experience, the tangent rule overestimates ψ for films deposited to optimize birefringent properties. As an extreme example, the cerium oxide film shown in Fig. 7.2 was deposited at $\theta_v = 45°$, the tangent rule predicts a column angle of 27°, but the SEM micrograph indicates an angle

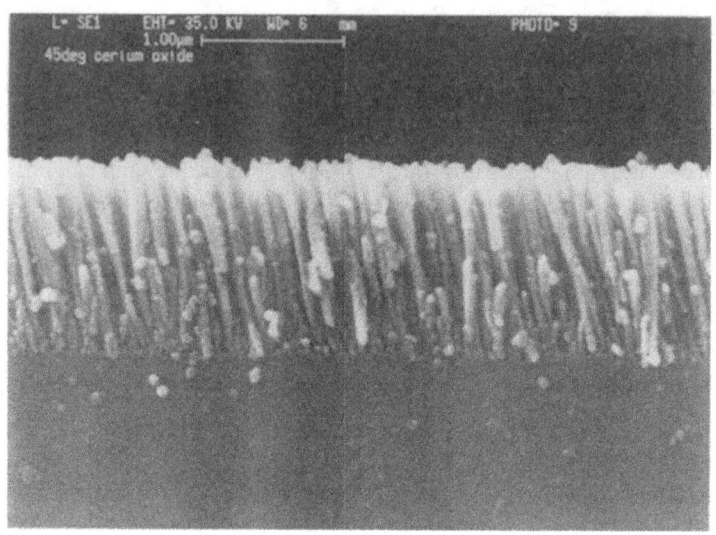

Fig. 7.2. Scanning electron micrograph of a cerium oxide film deposited at 45° and fractured in the deposition plane. The angle of the columns is ≈ 10°, significantly less than the angle of ≈ 27° predicted by the tangent rule. (Adapted from I.J. Hodgkinson, S.J. Cloughley, Q.H. Wu and S. Kassam, *Applied Optics* 35, 5563, 1996. Copyright © 1996 Optical Society of America. Reprinted with permission.)

of only 10°. Similarly, the cerium oxide film shown in Fig. 7.3 was deposited at 55° and has columns running nearly perpendicular to the substrate. The symmetry of the retardance versus optical angle of incidence graph (Fig. 7.4) is in agreement with the SEM result. As well, compare the angularly resolved retardance maps of the 55° cerium oxide film and a titanium oxide film by running the *BTF Toolbox* script file **vretard**.

The optical nature of a tilted columnar film (as determined experimentally) is biaxial with $n_1 > n_3 > n_2$. Simple modelling as form birefringence requires, for agreement between theory and experiment, the columns to be thicker or bunched preferentially in the direction perpendicular to the deposition plane.[19] Electron micrographs show both effects. Thus platelets can be seen in the fractured bi-layer of zirconium oxide shown in Fig. 7.5, and bunching of columns can be seen in the SEM micrograph of the cerium oxide film shown in Fig. 7.3. (The cerium oxide film has a large value of birefringence for light at normal incidence, $\Delta n = 0.2$, even though the column angle is nearly zero.)

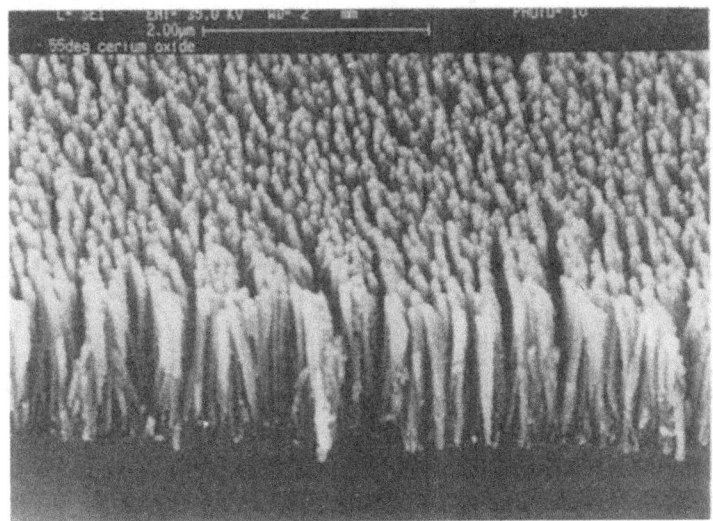

Fig. 7.3. Bunching of columns in a cerium oxide film deposited at 45° and fractured in the deposition plane. The column angle is nearly zero but the film exhibits large birefringence with $n_1 > n_3 > n_2$. (Adapted from I.J. Hodgkinson, S.J. Cloughley, Q.H. Wu and S. Kassam, *Applied Optics* 35, 5563, 1996. Copyright © 1996 Optical Society of America. Reprinted with permission.)

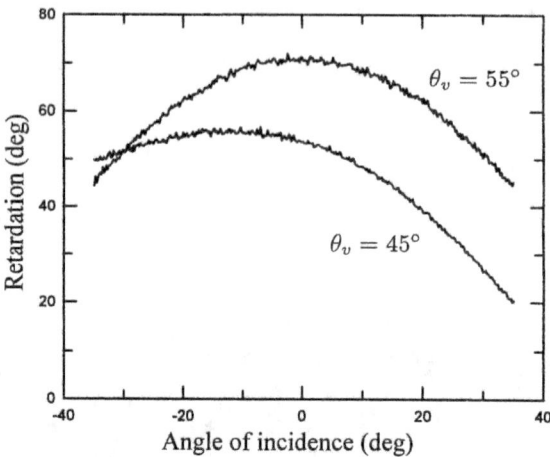

Fig. 7.4. Retardation of cerium oxide films for light incident at angle θ in the deposition plane. The turning point near $\theta = 0°$ for the film deposited at 55° indicates that a principal axis of this film is perpendicular to the substrate. (Adapted from I.J. Hodgkinson, S.J. Cloughley, Q.H. Wu and S. Kassam, *Applied Optics* 35, 5563, 1996. Copyright © 1996 Optical Society of America. Reprinted with permission.)

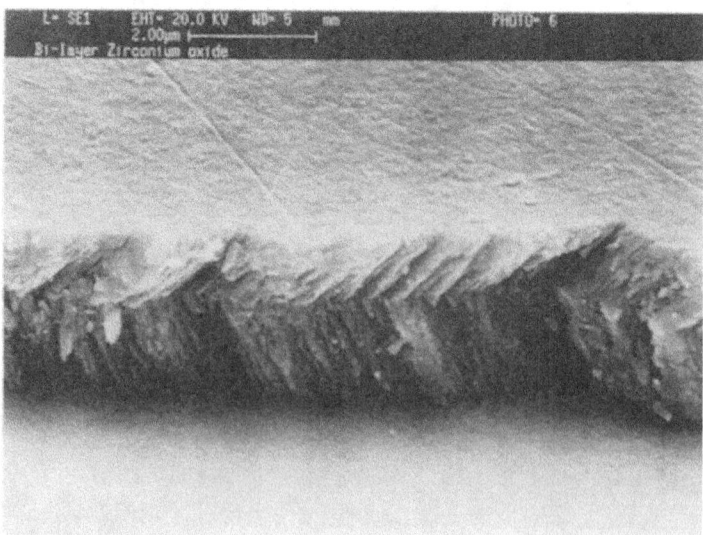

Fig. 7.5. Platelets in a zirconium oxide bi-layer. (Photograph from the authors' laboratory.)

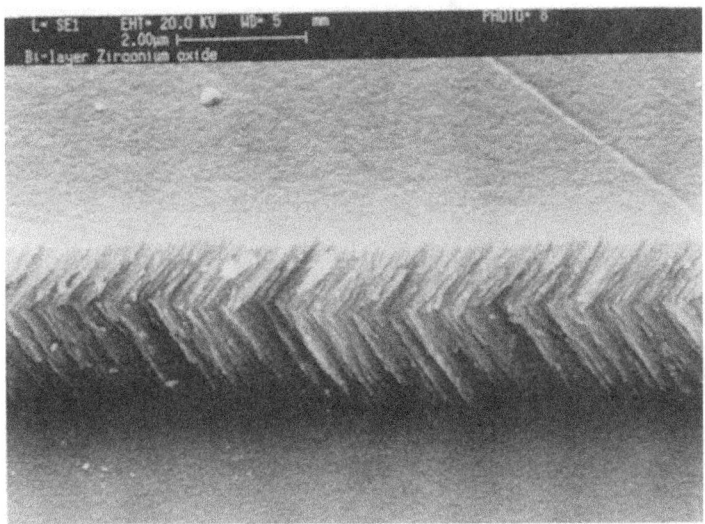

Fig. 7.6. Interface formed by two layers of zirconium oxide. (SEM photograph from authors' laboratory.)

7.2.3 Effective Anisotropic Media

Composite media, deposited as alternate thin layers of different materials or as thin layers of the same material deposited with different parameters allow control over three *effective principal indices* and three *effective material placement angles*. This is discussed in Chapt. 9.

7.2.4 Zig-Zag and Wavy Anisotropic Media

Sequential deposition from two directions can cause a zig-zag microstructure to grow. An SEM photograph showing an interface between two layers of zirconium oxide is shown in Fig. 7.6, and scatter from herring bone stacks is discussed in Sect. 10.5.2.

When the deposition angle is increased to nearly grazing incidence and the deposition temperature is kept low to reduce adatom mobility, individual zig-zag columns[20] can be grown, as shown in Fig. 7.7. In general, films with high porosity and large internal surface areas are used in devices such as gas sensors.

An oscillatory motion of the deposition angle θ_v during deposition can cause

Fig. 7.7. Individual zig-zag columns formed by sequential depositions of MgF$_2$ at $\theta_v = +85°$ and $\theta_v = -85°$. (SEM photograph supplied by K. Robbie.[21])

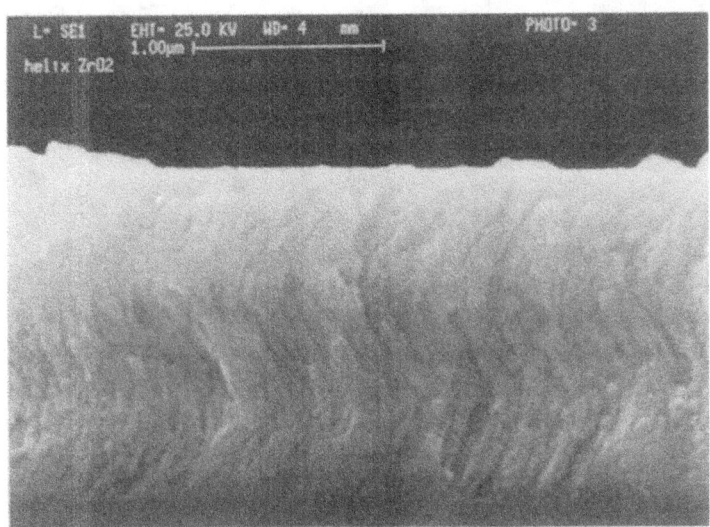

Fig. 7.8. Helical microstructure formed by depositing zirconium oxide at an oblique angle on to a substrate rotating about a normal axis. (SEM photograph from the authors' laboratory.)

a wavy microstructure. Smoothly periodic isotropic media are used currently in rugate filter designs, and future applications could include the use of wavy birefringent media.

7.2.5 Helical Microstructures

A steady rotation of the substrate about a normal axis during deposition can cause a helical microstructure to develop. N.O. Young and J. Kowal[22] reported the deposition of helical fluorite films in 1959. Left-hand and right-hand helically deposited films were found to be dextrorotatary and levorotatory, and values as large as 155°/mm were measured for the specific rotation. For comparison, the specific rotation of quartz is 21.7°/mm.

An example of a helical microstructure, for zirconium oxide deposited at an oblique angle on to a substrate rotating about a normal axis, is shown in Fig. 7.8.

Deposition at near grazing incidence on to a rotating substrate can lead to the formation of loosely-packed helical columns,[23] or individual helical columns

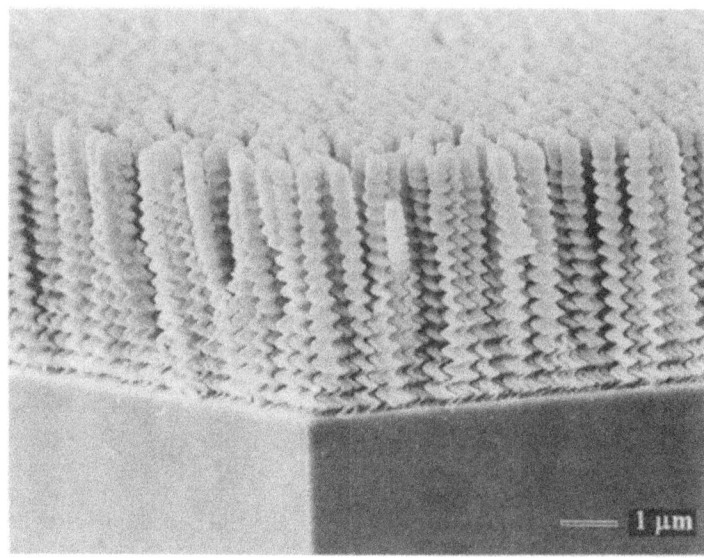

Fig. 7.9. Individual helical columns formed by depositing MgF_2 at $\theta_v = 85°$ on to a substrate rotating about a normal axis. (SEM photograph supplied by K. Robbie.[21])

as shown in Fig. 7.9.

Cholesteric liquid crystal helical media have found a wide range of optical applications and, similarly, vacuum deposited helical films are considered to have a promising future.

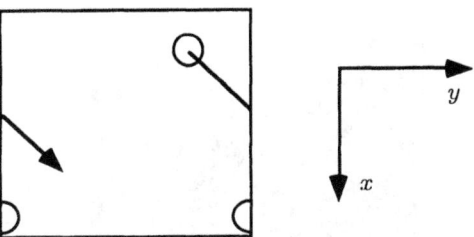

Fig. 7.10. Periodic boundary conditions. (Adapted from I.J. Hodgkinson and J.R. Gee, *Optical Interference Coatings, Proceedings of the Society of Photo-Optical Instrumentation Engineers* 2253, 1201, 1994. Copyright © 1994, SPIE. Reprinted with permission.)

7.3 Computer Modelling of Deposition

7.3.1 Serial Deposition of Hard Spheres

Computer simulations provide insight into the physical mechanisms that cause the growth of microstructural features in thin films.[24] In the HBC model (devised by D. Henderson, M.H. Brodsky and P. Chaudhari[25]) the atoms are represented by hard spheres, and the film grows in thickness as the spheres arrive serially. When a sphere impinges on the film, it relaxes into the nearest free site where it can touch three spheres, including the sphere it collided with on arrival. Thus the computer is required to store the location of all the spheres, and this limits the volume of film that can be simulated.

In our implementation of the HBC method the simulation takes place in a $50 \times 50 \times 50$ array in computer memory that corresponds to the box formed by the boundaries of the film.[26] The size of an element of the array is such that the centre of only one sphere can be in the element. A collision check is performed by considering the group of array elements around a given sphere. Periodic boundary conditions are applied to the box to simulate an infinite system. As shown in Fig. 7.10, a sphere that passes out from one side of the simulation enters the opposite side at the same angle and height.

7.3.2 Visual Analysis of Simulations

The analysis of a three-dimensional simulation is a non-trivial problem. Perhaps the best way to get an appreciation of the structure is to consider a slice of the

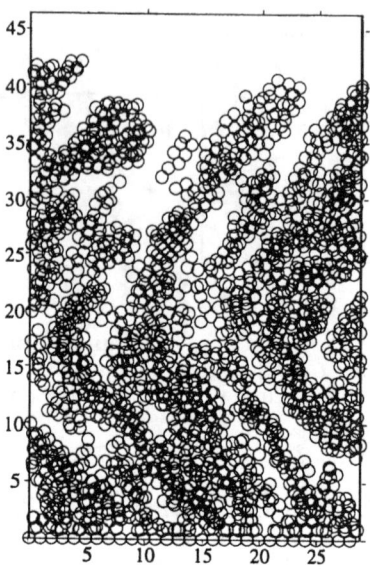

Fig. 7.11. Cross-section of sequential deposition at +50° and −50°. (Adapted from I.J. Hodgkinson and J.R. Gee, *Optical Interference Coatings, Proceedings of the Society of Photo-Optical Instrumentation Engineers* 2253, 1201, 1994. Copyright © 1994, SPIE. Reprinted with permission.)

film and plot a series of overlapping circles, one for each atom without regard for depth. The extra information conveyed about the density of particles in a region by the overlapping circles compensates for the lack of depth. This is apparent in the example, representing sequential deposition at +50° and −50°, which is reproduced in Fig. 7.11.

7.3.3 Radial Distribution Function

The radial distribution function $g(r)$ is useful for detecting crystalline structure in a simulation. We note here that in the 2–D counterpart of the HBC method, with atoms represented by disks and relaxation into the cusp formed by two disks, large areas of perfect crystallinity are inevitable. In a 3–D simulation alternation between the two close-packed structures for spheres, hexagonal close-packed and face-centred cubic,[1] leads at best to volumes of a psuedo-close-packed structure.[27,28,29]

Taking one particle in the film as an origin, the radial distribution function

Single particle separating a pair
– a sharp drop at $r = 2$

Pair of particles sharing two neighbours
– a drop at $3^{1/2} = 1.73$

Hexagonal close-packing
– singularities at $2^{1/2} = 1.41$ due to the two particles (black dots) needed to stabilize the structure, and at 1.73 and 2

Particle added to a tetrahedron
– singularity at $(2^3/3)^{1/2} = 1.63$

Two particles added to a tetrahedron
– singularity at $5/3 = 1.67$

Three particles added to a tetrahedron
– a peak at $4(2/3)^{1/2}/3 = 1.09$ and another (visualized as the addition of a particle to a ring) at $(2 \times 23^2 + 15^2 + 1^2)^{1/2}/18 = 1.99$

Tetrahedron separating a pair
– a small drop at $3^{1/2} + 2^{-1/2} = 2.44$

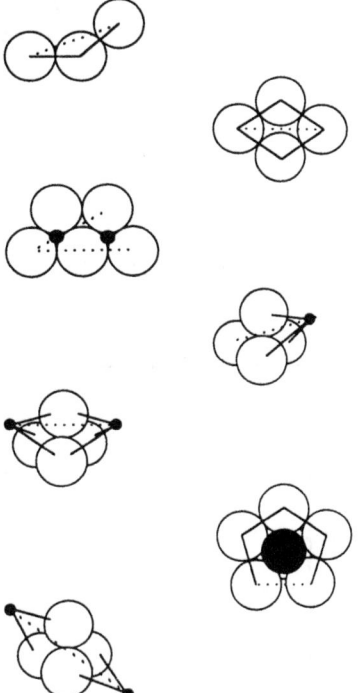

Fig. 7.12. Small clusters of spherical particles contribute to the radial distribution function. (Adapted from I.J. Hodgkinson and J.R. Gee, *Optical Interference Coatings, Proceedings of the Society of Photo-Optical Instrumentation Engineers* 2253, 1201, 1994. Copyright © 1994, SPIE. Reprinted with permission.)

gives the relative likelihood of finding another particle at a given radial distance r (expressed in units of the particle diameter). The massive peak at $r = 1$ is neglected. Several small clusters of particles and the associated features in $g(r)$ are listed in Fig. 7.12.

An example of a radial distribution function, for computer deposition at normal incidence, is given in Fig. 7.13. Four peaks, at 1.088, 1.632, 1.666 and 1.990, can be traced to the clusters listed in Fig. 7.12, and the absence of the peaks associated with crystalline structure, at 1.414, 2.236 and 2.646, indicates that the film is amorphous.

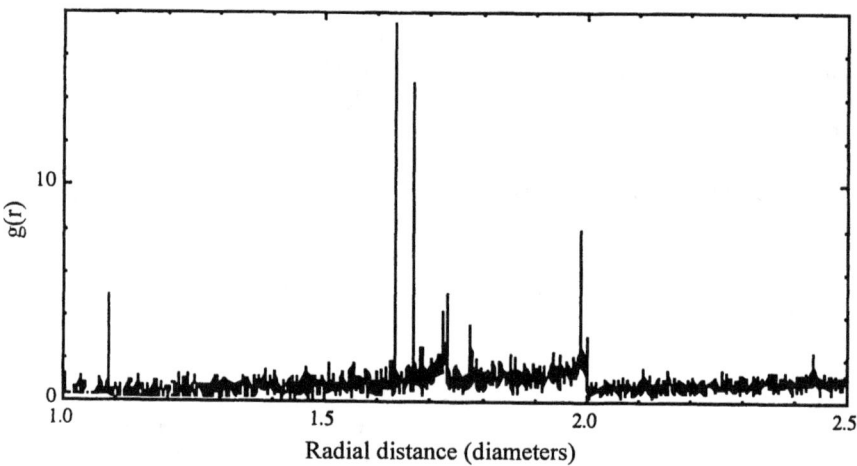

Fig. 7.13. Radial distribution function $g(r)$ for simulated deposition at $0°$. (Adapted from I.J. Hodgkinson and J.R. Gee, *Optical Interference Coatings, Proceedings of the Society of Photo-Optical Instrumentation Engineers* 2253, 1201, 1994. Copyright © 1994, SPIE. Reprinted with permission.)

7.3.4 Two-Dimensional Angular Distribution

The two-dimensional angular distribution is a projection on to a plane of the normalized vectors that connect two particles. Each vector is considered to run from the lower to the upper particle. Figure 7.14 shows two-dimensional angular distributions for a 60° simulated deposition. Here the grey and black contours are respectively at $\frac{5}{12}$ths and $\frac{7}{12}$ths of the maximum height of the distribution. The displaced centre of the upper distribution, viewed on a plane parallel to the substrate, indicates columns of material inclined at the expected angle. The lower distribution, corresponding to a view down the columns, implies elongation of the cross-section of the columns perpendicular to the deposition plane or bunching in the same direction.

7.3.5 Column Angle

The column angle in a simulation can be estimated from the variance in the relative density of the film, calculated on a grid of lines all at the same angle. For a random arrangement of particles the variance is zero for any angle, but if the particles are arranged in columns the variance is large when the grid and column directions coincide.

Figure 7.15 shows the results obtained for simulations made at a range of deposition angles. These show that the column angles obey the tangent rule for deposition angles up to 60°.

7.3.6 Birefringence

The principal refractive indices of a spherical volume of a simulated film can be determined by an iterative method in which the local electric field is calculated as the sum of an applied field and the field due to dipoles within a sufficient radius. The iterative process is terminated when the polarization stabilizes.

Sets of principal refractive indices determined in this way are found to satisfy the relationship $n_1 > n_3 > n_2$, and hence agree with the ordering found from experiment. However, the film densities and hence the values of the refractive indices, are all low relative to experimental values.

7.3.7 Conclusions from Simulations of Deposition

The simulations that we have described support the view that columnar structure is caused by self-shadowing of atoms that arrive at the substrate with limited mobility. Our most important conclusion is that *refractive anisotropy*

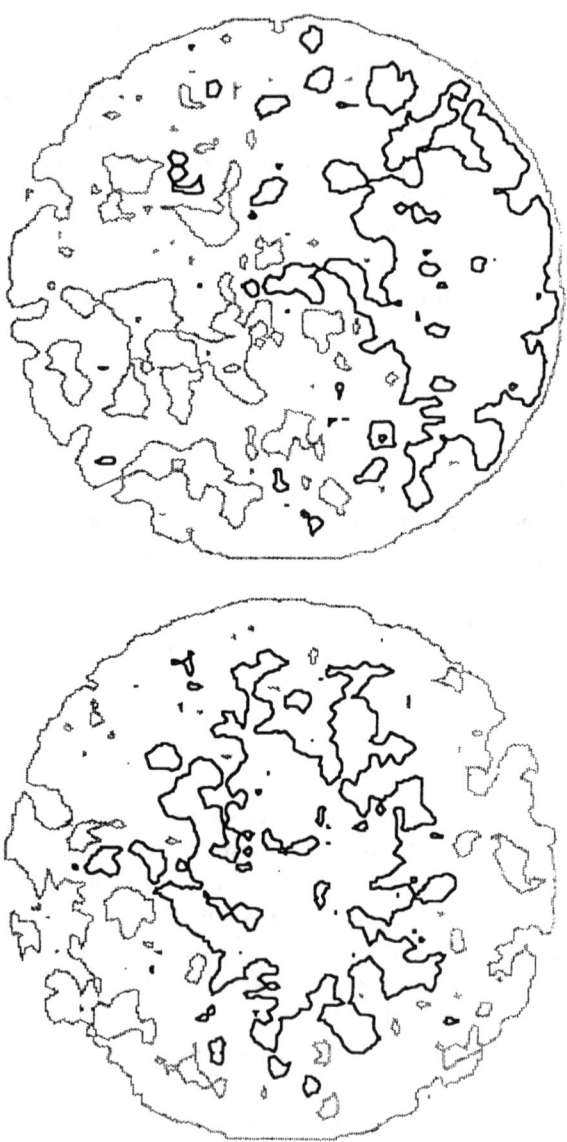

Fig. 7.14. Two-dimensional angular distributions of film deposited at 60°, viewed (upper) on a plane parallel to the substrate and (lower) on a plane perpendicular to the columns. (Adapted from I.J. Hodgkinson and J.R. Gee, *Optical Interference Coatings, Proceedings of the Society of Photo-Optical Instrumentation Engineers* 2253, 1201, 1994. Copyright © 1994, SPIE. Reprinted with permission.)

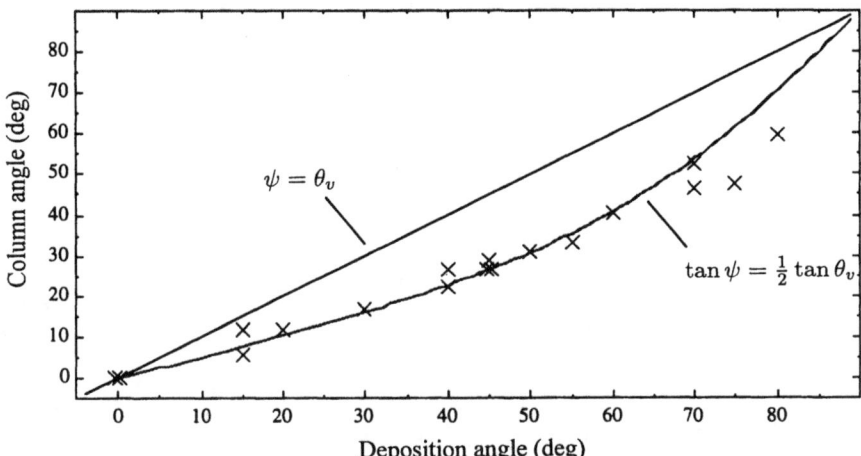

Fig. 7.15. Column angle ψ versus deposition angle θ_v. (Adapted from I.J. Hodgkinson and J.R. Gee, *Optical Interference Coatings, Proceedings of the Society of Photo-Optical Instrumentation Engineers* 2253, 1201, 1994. Copyright © 1994, SPIE. Reprinted with permission.)

with $n_1 > n_3 > n_2$ is intrinsic in a hard sphere simulation when the limit on mobility is relaxation to the nearest cusp formed by three atoms.

Chapter 8

Form Birefringence

Form birefringence is refractive anisotropy due to shape. As one example, a system of parallel layers is asymmetric in a 3–D sense and exhibits uniaxial form birefringence. As a second example, the microstructural columns in a film deposited at an oblique angle are asymmetric in shape and the film is biaxial.

For the purpose of modelling form birefringence, the columns of a biaxial film are regarded as *crystallites* (with an isotropic refractive index) embedded in a *void* or second medium (also described by an isotropic refractive index). Thus form birefringence depends on five parameters, two refractive indices and three shape factors.

In this chapter we focus on the experimental detection of form birefringence in thin films, through the measurement of phase retardance which would be zero for an isotropic film, and on the modelling of form birefringence. The former is directly applicable to the deposition of thin film wave plates, which are usually specified by the retardance of light propagating at normal incidence to the plate.

The topics covered can be summarized as

▨ the application of perpendicular incidence ellipsometry to monitoring the growth of normal incidence phase retardation during the deposition of a biaxial thin film

▨ measurement of the three principal refractive indices during the growth of a biaxial film, and after the deposition is completed

▨ modelling form birefringence with the Bragg-Pippard equations.

8.1   Perpendicular Incidence Ellipsometry

Several quantities are of interest for the deposition of thin film wave plates. Thus,

$$\Delta n = n_s - n_p, \tag{8.1}$$

where n_p and n_s are the refractive indices appropriate to light beams incident normally and polarized parallel and perpendicular to the deposition plane, is a material property.

A second material property,

$$\delta/d = 2\pi \, \Delta n/\lambda, \tag{8.2}$$

is the phase retardation per unit thickness of the birefringent film, in the absence of interference. A third quantity, Δ, is the actual phase retardation of the plate taking interference into account.

Perpendicular incidence ellipsometry, abbreviated PIE,[30] provides a method for monitoring retardation during the deposition of a wave plate. The basic principles of ellipsometry have been discussed in Chapt. 3, for the arrangement of a rotating quarter-wave plate followed by a fixed polarizer. However, several different arrangements are possible, and we shall consider now the setup illustrated in Fig. 8.1.

As indicated in Fig. 8.1 the probe light beam is circularly polarized when it reaches the sample. After passing through the sample the beam encounters a fixed analyser (linear polarizer) and a bandpass filter before being detected by a photomultiplier. An advantage of using a circularly polarized incident beam is that the photomultiplier signal is a steady direct current when the sample is isotropic. Thus the method is particularly suitable for recording small anisotropies. The method loses sensitivity as the retardation of the sample approaches 90°, but in practice the range can be extended by adding an auxiliary wave plate to produce a retardance offset.

8.1.1   Computation of Ellipsometric Parameters

The ellipsometric method yields

$$\Psi = \tan^{-1} |t_p/t_s| \tag{8.3}$$

and

$$\Delta = \text{angle}(t_p/t_s), \tag{8.4}$$

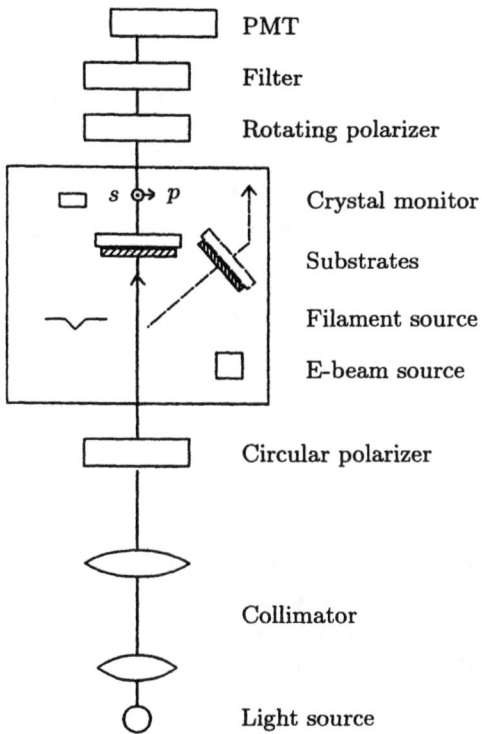

Fig. 8.1. Transmission-mode perpendicular incidence ellipsometry. (Adapted from Q.H. Wu and I.J. Hodgkinson, *Journal of Optics: Nouvelle Revue D'Optique* 25, 43, 1994. Copyright © Masson, Paris, 1994. Reprinted with permission.)

for the film (see Eqs.(3.25) and (3.26)); other birefringent effects in the light
path are subtracted in an initial zeroing operation. When irrelevant constants
are neglected, the p and s components of the light incident at the rotating
analyzer (when the film thickness is d) can be expressed as $t_s \tan \Psi \exp(i\Delta)$ and
t_s. The transmitted field is $t_s[\tan \Psi \exp(i\Delta) \cos \xi + \sin \xi]$, where ξ is the angle
between the p direction and the axis of the analyzer. The photomultiplier signal
can be obtained by multiplying the expression for the transmitted field by its
complex conjugate,

$$I(\xi) = T_s \left[\frac{1}{2}(1 + \tan^2 \Psi) + \frac{1}{2}(\tan^2 \Psi - 1) \cos 2\xi + \tan \Psi \cos \Delta \sin 2\xi\right]. \quad (8.5)$$

Thus the non-zero Fourier coefficients of $I(\xi)$ are $A_0 = T_s(1 + \tan^2 \Psi)/2$,
$A_2 = T_s(\tan^2 \Psi - 1)/2$, $B_2 = T_s \tan \Psi \cos \Delta$ and, conversely, Ψ, Δ and the
transmittances T_p, T_s can be determined from the Fourier coefficients,

$$\tan \Psi = [(A_0 + A_2)/(A_0 - A_2)]^{1/2} \qquad (8.6)$$
$$\cos \Delta = B_2/(A_0^2 - A_2^2)^{1/2} \qquad (8.7)$$
$$T_p = A_0 + A_2 \qquad (8.8)$$
$$T_s = A_0 - A_2. \qquad (8.9)$$

8.1.2   Characteristic Ellipsometric Curves

In this section we consider the theoretical shape of PIE curves, such as T v.
d, Δ v. Ψ and Δ v. d, for three representative anisotropic dielectric films.
In turn we consider:- an *ideal* non-absorbing homogeneous film, an *absorbing*
homogeneous film, and a non-absorbing *inhomogeneous* film.

Ideal Anisotropic Film

The ideal film is assumed to have refractive indices $n_p = 2.0$ and $n_s = 2.2$ at
$\lambda = 600 \, \text{nm}$. We have chosen a large value of Δn for the calculation, about
two or three times larger than practical values, so that relevant features are
displayed clearly on plotted profiles.

Figure 8.2 shows computed curves of T_p and T_s as functions of the film
thickness. In practice the refractive index of an isotropic film can be determined
from the envelope of a recorded transmittance profile and, similarly, n_p and n_s
for an anisotropic film can be estimated from the profiles T_p v. d and T_s v. d.

The Δ v. Ψ profile calculated for the ideal film is given in the left-hand
side of Fig. 8.3. This characterization of an anisotropic film has the advantage

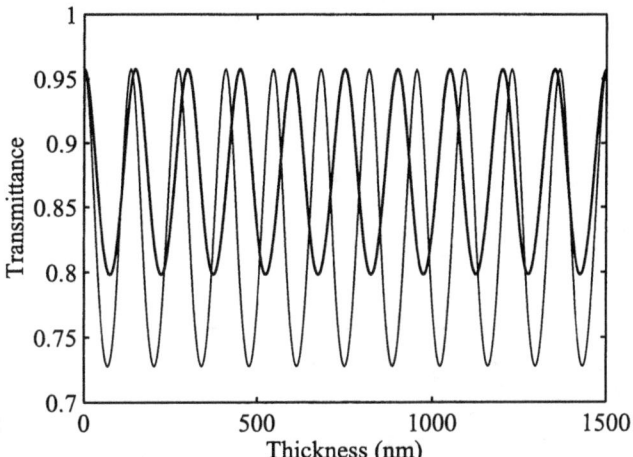

Fig. 8.2. Perpendicular incidence transmittance profiles T_p (solid line) and T_s (broken line) calculated for an ideal anisotropic dielectric film with $n_p = 2.0$ and $n_s = 2.2$ on a substrate of index 1.52. The wavelength of the incident light is 600 nm. (Adapted from Q.H. Wu and I.J. Hodgkinson, *Journal of Optics: Nouvelle Revue D'Optique* 25, 43, 1994. Copyright © Masson, Paris, 1994. Reprinted with permission.)

of being completely optical, and we consider next the information about the film that can be determined from the Δ v. Ψ profile. First we note that, in the absence of interference, the Δ v. Ψ curve would be a vertical line through $\Psi = 45°$. Further, the loops in the curve are caused by, and hence can be can be correlated with the interference modulations in the Δ v. d profile, which is shown in the right-hand side of Fig. 8.3.

The maximum positive and negative excursions of Ψ in a Δ v. Ψ profile are most closely associated with n_s. This property is emphasized in Fig. 8.3, by the superposed dots that mark the optical thickness of the film in quarter-wave steps (increments of $\lambda/4n_s$).

In practice the interference modulations in a Δ v. Ψ curve can be distinguished easily, and the number of peaks within a range of retardance,

$$N \approx \frac{\text{optical thickness}}{\text{retardation thickness}} \approx \frac{n_s}{\Delta n}, \tag{8.10}$$

provides a useful figure of merit for an anisotropic film. N can be interpreted as the number of quarter-wave film thicknesses that need to be deposited in order to make a wave plate with a phase retardation of a quarter-wave.

The broken line superposed on the Δ v. d profile in Fig. 8.3 shows $\delta = (2\pi \, \Delta n/\lambda)d$, the retardation in the absence of interference. Clearly, the material

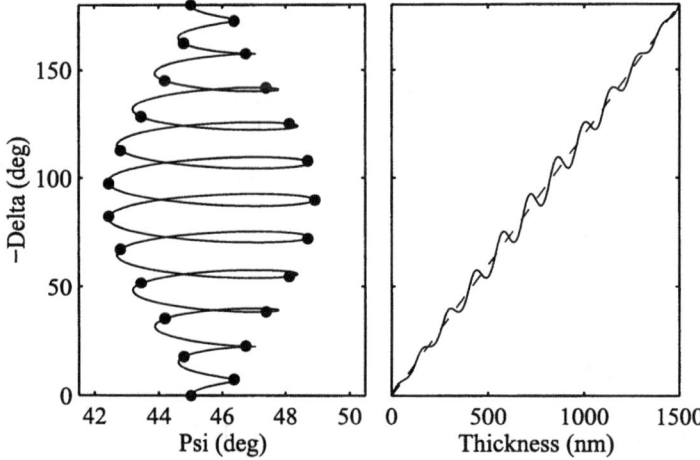

Fig. 8.3. Characteristic ellipsometric profiles Δ versus Ψ and Δ versus thickness d calculated for the ideal anisotropic dielectric film with $n_p = 2.0$ and $n_s = 2.2$. The oscillations in these profiles are most closely associated with the s refractive index. The superposed dots (left) indicate quarter-wave increments in optical thickness for the s index, and the broken line (right) which shows phase retardation in the absence of interference cuts the solid line near the quarter-wave thickness points. (Adapted from Q.H. Wu and I.J. Hodgkinson, *Journal of Optics: Nouvelle Revue D'Optique* 25, 43, 1994. Copyright © Masson, Paris, 1994. Reprinted with permission.)

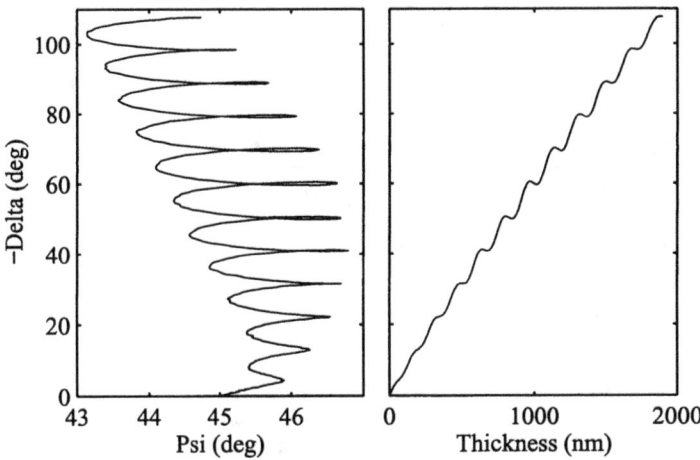

Fig. 8.4. Experimental ellipsometric profiles Δ versus Ψ and Δ versus thickness recorded with 630 nm light during the growth of a Ti_2O_3 film. About 20 quarter-wave film thicknesses of this material are required for a quarter-wave retardation plate. (Adapted from Q.H. Wu and I.J. Hodgkinson, *Journal of Optics: Nouvelle Revue D'Optique* 25, 43, 1994. Copyright © Masson, Paris, 1994. Reprinted with permission.)

property Δn is available from the gradient of a smooth line drawn through a recorded Δ v. d profile.

In practice we use both Δ v. Ψ and Δ v. d profiles to characterize our coatings. For comparison with the theoretical curves described above, experimental curves for Ti_2O_3 deposited at 330°C and 3×10^{-4} mbar oxygen are shown in Fig. 8.4. The Δ v. d curves in Figs. 8.3 and 8.4 are similar in form, but the Δ v. Ψ curves exhibit significant differences. Counting interference modulations gives $N = 20$ for this material and, from the gradient of the smoothed Δ v. d curve, $\Delta n = 0.10$ and is nearly independent of thickness.

Absorbing Anisotropic Film

For illustrative purposes we consider extreme differential absorption and large absorption coefficients. When $n_p = 2.0$ and $n_s = 2.2 + 0.02i$, for example, the absorption causes the T_s profile to fall rapidly with increasing film thickness, and this has an effect on the Δ v. Ψ curves.

Δ v. d curves for permutations of absorbing and non-absorbing indices n_p and n_s are plotted by the solid lines in Fig. 8.5. The broken line in Fig. 8.5 shows $\delta = (2\pi \, \Delta n / \lambda) d$, in this case the retardation in the absence of both absorption

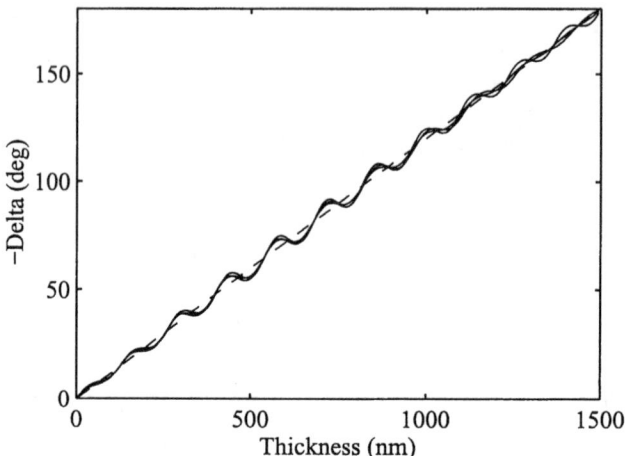

Fig. 8.5. Δ v. d profiles (solid lines) calculated for absorbing dielectric films with refractive indices $n_p = 2 + 0.02i$, $n_s = 2.2$; $n_p = 2 + 0.02i$, $n_s = 2.2 + 0.02i$; $n_p = 2.0$, $n_s = 2.2 + 0.02i$. The broken line indicates the phase retardation of the films in the absence of both absorption and interference effects. (Adapted from Q.H. Wu and I.J. Hodgkinson, *Journal of Optics: Nouvelle Revue D'Optique* 25, 43, 1994. Copyright © Masson, Paris, 1994. Reprinted with permission.)

and interference effects. Thus the material property Δn can be determined from the gradient of the smoothed Δ v. d profile, even in the presence of extreme differential absorption.

Inhomogeneous Anisotropic Film

For these calculations the refractive index inhomogeneity is assumed to be an increase of 0.3, linear with respect to film thickness. In general, a refractive index that increases with film thickness causes the peaks of a T v. d curve to increase and the troughs to decrease (and in practice the inhomogeneity coefficient can be estimated from the envelope).

Fig. 8.6 shows how inhomogeneity influences Δ. For example, in the lower, solid curve the value of n_p eventually exceeds n_s as the film grows, and the sign of the local material birefringence changes.

The broken lines in Fig. 8.6 show phase retardations in the absence of interference, calculated as the sum of δ's for thin sublayers. These lines show that, even in the case of extreme differential inhomogeneity, the material property Δn can be determined as a function of thickness d from the gradient of a smoothed Δ v. d profile.

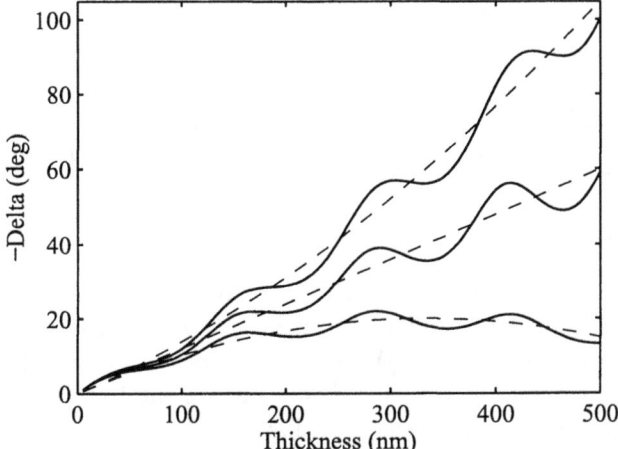

Fig. 8.6. Characteristic Δ v. d profiles calculated for inhomogeneous dielectric films. In the top curve $n_p = 2.0$ and n_s changes linearly from 2.2 at the substrate interface to 2.5 at the air interface, in the middle curve $n_p = 2.0$–2.3, $n_s = 2.2$–2.5 and in the lower curve $n_p = 2.0$–2.3, $n_s = 2.2$. The broken lines indicate the phase retardation of the films in the absence of interference effects. (Adapted from Q.H. Wu and I.J. Hodgkinson, *Journal of Optics: Nouvelle Revue D'Optique* 25, 43, 1994. Copyright © Masson, Paris, 1994. Reprinted with permission.)

Table 8.1.   Materials for birefringent coatings.   (Adapted from Q.H. Wu and I.J. Hodgkinson, *Engineering and Laboratory Notes* 2, S9, 1994. Copyright © 1994 Optical Society of America. Reprinted with permission.)

Material	Al_2O_3	Sub H1	SiO_2	Ta_2O_5	Ti_2O_3	Ti_2O_5	ZrO_2
N_{QW}	54	22	49	20	19	20	22
Bire Δn	0.027	0.079	0.026	0.082	0.104	0.093	0.087
Refrac index	1.4	1.7	<1.4	1.9	1.8	1.9	1.7
Dep temp / °C	250	330	250	330	300	330	330
Dep angle / deg	70	70	70	70	65	70	70
Dep rate / nm s^{-1}	1.3	1.0	1.3	0.8	0.7	0.6	0.6
O_2 press / 10^{-4}mbar	2	2	1	3	2	2	2
Sample ID	311	310	312	311	310	309	309

8.1.3   Experimental Values

The most important conclusions from this section are that transmission-mode PIE provides a robust method for monitoring the retardation Δ of an optical film during deposition, and the material quantity Δn can be determined from the slope of the Δ v. d graph. Some experimental values obtained using transmission-mode PIE are listed in Table 8.1.

Further values of Δn are reported in the literature. These include post-deposition values for Bi_2O_3, HfO_2, $LiNbO_3$, MoO_3, SiO_2, SnO_3, Ta_2O_5, TiO_2, $TiZrO_4$, WO_3, ZrO_2,[31] and the dispersion of Δn for HfO_2 using *in situ* PIE in reflection and in transmission.[32] Photometric methods are used as well, for example H. Wang[33] has discussed the determination of Δn for absorbing films.

8.2   Measurement of Principal Refractive Indices

Measurement of the three principal refractive indices n_1, n_2, n_3 for a biaxial film is considerably more difficult than the measurement of Δn which was discussed in Sect. 8.1. Most of the available methods make use of decoupling of the p and s polarizations for light propagating in the deposition plane of the film. This allows n_3 to be determined more or less independently of n_1 and n_2. We consider four methods.[34]

8.2.1 In Situ Measurements

It was a surprise to us to discover that measurements of the principal refractive indices of a birefringent film are easier to make *in situ* rather than *ex situ*. For the purpose of monitoring the three indices during the deposition of a film we operate two ellipsometers in transmission mode at different angles in the deposition plane. The column angle ψ is determined from an SEM micrograph of the fractured film, and the index n_3 is computed from interference modulations in T_s. The thickness d can be determined from the modulations in T_s, or from calibrated mass thickness values recorded by the quartz crystal monitor.

If δ is the phase retardation that would be recorded by an ellipsometer *in the absence of interference*, then we have $\delta = 2\pi(\alpha_p - \alpha_s)d/\lambda$. Using $\alpha_s = (n_3^2 - \beta^2)^{1/2}$ and rearranging gives $\alpha_p = (n_3^2 - \beta^2)^{1/2} + \lambda\delta/2\pi d$. Now suppose that two ellipsometers are used, and that they yield δ_1 and δ_2 at angles corresponding to β_1 and β_2. Then, for the p polarization,

$$\alpha_1 = (n_3^2 - \beta_1^2)^{1/2} + \delta_1\lambda/2\pi d \tag{8.11}$$

and

$$\alpha_2 = (n_3^2 - \beta_2^2)^{1/2} + \delta_2\lambda/2\pi d. \tag{8.12}$$

Substitution into Fresnel's equation, Eq.(3.33), gives two equations with two unknowns, n_1 and n_2. After putting

$$a = -\frac{(\alpha_1^2 - \alpha_2^2)\cos^2\psi + 2(\alpha_1\beta_1 - \alpha_2\beta_2)\cos\psi\sin\psi + (\beta_1^2 - \beta_2^2)\sin^2\psi}{(\alpha_1^2 - \alpha_2^2)\sin^2\psi - 2(\alpha_1\beta_1 - \alpha_2\beta_2)\cos\psi\sin\psi + (\beta_1^2 - \beta_2^2)\cos^2\psi}, \tag{8.13}$$

we can express the simultaneous solution in the form

$$n_2 = [\alpha_1^2(\cos^2\psi + a\sin^2\psi) + 2\alpha_1\beta_1(1 - a)\cos\psi\sin\psi + \beta_1^2(a\cos^2\psi + \sin^2\psi)]^{1/2}, \tag{8.14}$$

$$n_1 = n_2/a^{1/2}. \tag{8.15}$$

The rather subtle point that makes this realization possible is that, as explained in the previous section, δ/d can be determined as the gradient of the smoothed Δ versus d ellipsometric curve.

We note here that the column angle can be determined, together with the principal refractive indices, if ellipsometric measurements are made at three angles. In terms of the relevant β's and the α's for the p polarization,

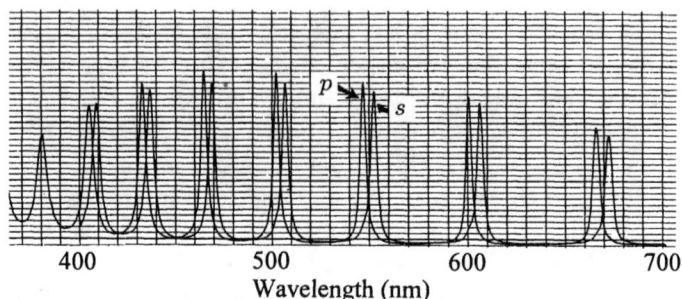

Fig. 8.7. Transmittance spectra recorded at normal incidence for a silver-silicon oxide-silver interference filter. The silver reflecting layers were deposited at normal incidence and the silicon oxide spacer layer was deposited at 60°. (Adapted from I.J. Hodgkinson, F. Horowitz, H.A. Macleod, M. Sikkens and J.J. Wharton, *Applied Optics* 24, 1568, 1985. Copyright © 1985 Optical Society of America. Reprinted with permission.)

$$\tan 2\psi =$$

$$\frac{(\alpha_1^2 - \alpha_3^2)(\beta_2^2 - \beta_3^2) - (\alpha_2^2 - \alpha_3^2)(\beta_1^2 - \beta_3^2)}{(\alpha_1^2 - \alpha_3^2 + \beta_1^2 - \beta_3^2)(\alpha_2\beta_2 - \alpha_3\beta_3) - (\alpha_2^2 - \alpha_3^2 + \beta_2^2 - \beta_3^2)(\alpha_1\beta_1 - \alpha_3\beta_3)}.$$
$$(8.16)$$

Three functions in the *BTF Toolbox*, **tao2**, **tio2** and **zro2**, provide principal indices and column angles for a given deposition angle for the materials tantalum oxide, titanium oxide and zirconium oxide. Note that **tao2**, **tio2** and **zro2** are "computer names". The actual evaporants were Ta_2O_5, Patinal titanium oxide S which is manufactured to give a stable oxygen to titanium ratio of 1.7 corresponding to a chemical composition between and Ti_3O_5 and Ti_4O_7,[35] and ZrO_2. The three functions use equations fitted to experimental measurements made for deposition angles in the range 40° to 70°, and extrapolated for deposition angles in the range 0° to 40°. Plots of principal indices and column angles, obtained by using the functions, are given later in the book, in Figs. 16.2 and 16.3.

8.2.2   Use of Narrowband Filters

In this method an interference filter is deposited, with an anisotropic spacer layer for an MDM filter (see Fig. 8.7 for typical transmittance curves), or with anisotropic H layers (including the spacer) for an all-dielectric filter. In the latter case all L and H layers are quarter-waves at the monitor wavelength λ_1 and

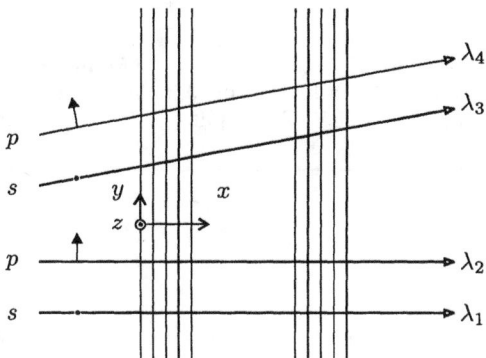

Fig. 8.8.   Measured values of the wavelengths of transmission peaks can be used for the realization of the principal indices of the anisotropic layers. (Adapted from I.J. Hodgkinson, F. Horowitz, H.A. Macleod, M. Sikkens and J.J. Wharton, *Journal of the Optical Society of America A* 2, 1693, 1985. Copyright © 1985 Optical Society of America. Reprinted with permission.)

s polarization. As shown in Fig. 8.8, three other wavelengths of peak transmittance are determined experimentally, λ_2 for p at normal incidence, λ_3 and λ_4 for s and p at an oblique angle. Values obtained for a zirconium oxide/silicon oxide filter of design $a[HL]^5 4H[HL]^5 g$ and deposited at 27° are listed in Table 8.2, together with the fitted refractive indices.

8.2.3   Photometric Method

In the photometric method the sample is illuminated with a beam of polarized light and the set (or a subset) of reflectance and transmittance coefficients, R_{pp}, R_{ps}, R_{sp}, R_{ss}, T_{pp}, T_{ps}, T_{sp}, T_{ss}, is measured at a fixed angle of incidence and for deposition plane angles in the range $0 \leq \xi \leq 180°$. In general this requires an allowance to be made for the parasitic reflections from the rear surface of the coated substrate. Typical transmittance curves are shown in Fig. 8.9; for the purpose of calculation the film column angle is assumed to be given by the tangent rule, $\psi = \tan^{-1}(\frac{1}{2} \tan \theta_v)$. The values of n_1, n_2, n_3 and d are obtained by an optimization technique, such as the simplex method.

8.2.4   Waveguide Method

In the waveguide method a laser beam is coupled into the film and values of β are measured for deposition plane angles in the range $0 \leq \xi \leq 180°$. The film must be thick enough to support several p and s modes.

Table 8.2. Refractive indices of zirconium oxide. (Adapted from I.J. Hodgkinson, F. Horowitz, H.A. Macleod, M. Sikkens and J.J. Wharton, *Journal of the Optical Society of America A* 2, 1693, 1985. Copyright © 1985 Optical Society of America. Reprinted with permission.)

Measurement	1	2	3	4
Polarization	s	p	s	p
Angle (deg)	0	0	30	30
Wavelength (nm)	627.8	624.8	601.6	599.2
Order	2	2	2	2
		Calculated refractive indices		
	n_1	n_2	n_3	n_p
	2.033	1.947	1.966	1.952

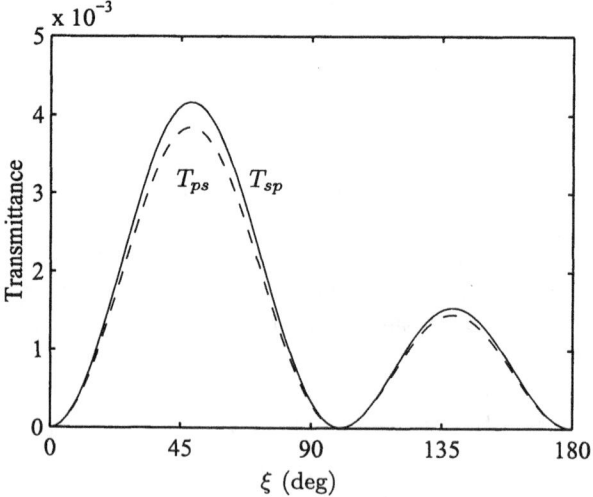

Fig. 8.9. Transmittance curves T_{ps} v. ξ and T_{sp} v. ξ computed for a TiO$_2$ layer with $n_C = 1$, $n_1 = 2.334$, $n_2 = 2.236$, $n_3 = 2.283$, $n_S = 1.516$, $\theta_v = 31.8°$, $d = 348\,\text{nm}$ and $\theta = 30°$. (Adapted from F. Flory, D. Endelema, E. Pelletier and I.J. Hodgkinson, *Applied Optics* 32, 5649, 1993. Copyright © 1993 Optical Society of America. Reprinted with permission.)

Table 8.3. Refined values of refractive indices and thicknesses of different TiO_2 layers versus the incidence angle of the material during deposition (θ_v). (Adapted from F. Flory, D. Endelema, E. Pelletier and I.J. Hodgkinson, *Applied Optics* 32, 5649, 1993. Copyright © 1993 Optical Society of America. Reprinted with permission.)

Layer	θ_v (deg)	n_1	n_2	n_3	d (nm)	Δn_{12}	Δn_{13}	Δn_{32}
1	16.1	2.356	2.289	2.306	496.0	0.067	0.050	0.017
2	20.3	2.355	2.281	2.301	444.0	0.074	0.054	0.020
3	24.5	2.349	2.269	2.297	407.6	0.080	0.052	0.028
4	31.8	2.334	2.236	2.283	291.5	0.098	0.051	0.047

A set of values refined using both photometric and waveguide data is given in Table 8.3. Here $\Delta n_{12} = n_1 - n_2$, $\Delta n_{13} = n_1 - n_3$ and $\Delta n_{32} = n_3 - n_2$. The values in the left part of the table are in agreement with the relationship $n_1 > n_3 > n_2$ that we have assumed throughout this book. Further, it can be seen that the indices n_1, n_2, n_3 decreased as the deposition angle increased. From the right part of the table we see that the differences Δn_{12} and Δn_{32} increased, but Δn_{13} stayed nearly constant.

8.3 Modelling Form Birefringence

8.3.1 Bragg-Pippard Equations

In the Bragg-Pippard (BP) method,[36] the optical medium is considered to be composed of a void or base medium of refractive index n_v in which crystallites of refractive index n_c and identical shape and orientation are embedded. When an electric field E_a is applied to the medium, the individual atoms in a crystallite become dipoles as shown schematically in Fig. 8.10, with the result that unbalanced positive and negative charges occur at the ends of the crystallite. The electric field E_d from these unbalanced charges acts against, and tends to reduce the effect of the applied field, hence the term *depolarization field*.

In the case of ellipsoidal crystallites, equations can be derived for the depolarization field and the principal refractive indices of the medium. We use the symbol p for the packing fraction, the fraction of the total volume of the medium that is occupied by crystallites. If the crystallites are distributed randomly and sparsely so that there is negligible mutual interaction of the depolarizing fields, then the principal dielectric constants are given by the equation

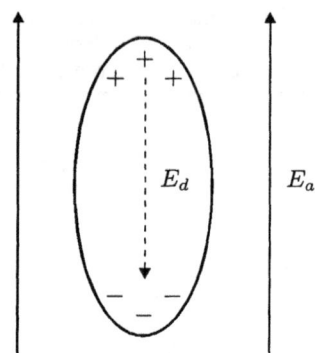

Fig. 8.10. An electric field E_a applied to a crystallite is opposed by the depolarization field E_d.

$$\varepsilon_{cdj} = \varepsilon_v + \frac{p(\varepsilon_c - \varepsilon_v)}{1 + (\varepsilon_c - \varepsilon_v)L_j/\varepsilon_v}, \quad j = 1, 2, 3. \tag{8.17}$$

In the example that we are considering the crystallites are identifiable as separate particles, and the subscript cd stands for *crystallite defined*. L_1, L_2, L_3 are called *depolarizing factors* and they have the property

$$L_1 + L_2 + L_3 = 1. \tag{8.18}$$

When the packing fraction p of the crystallites is increased so that mutual interactions of the depolarization fields need to be considered (but p is still much less than unity), replacement of the L_j's in Eq.(8.17) with $(1 - p)L_j$ leads to the crystallite-defined form of the BP equations,

$$\varepsilon_{cdj} = \varepsilon_v + \frac{p(\varepsilon_c - \varepsilon_v)}{1 + (1 - p)(\varepsilon_c - \varepsilon_v)L_j/\varepsilon_v}, \quad j = 1, 2, 3. \tag{8.19}$$

In a similar way, a second set of equations,

$$\varepsilon_{vdj} = \varepsilon_c - \frac{(1 - p)(\varepsilon_c - \varepsilon_v)}{1 - p(\varepsilon_c - \varepsilon_v)L_j/\varepsilon_c}, \quad j = 1, 2, 3, \tag{8.20}$$

can be derived for a *void-defined* (vd) medium. Here p still applies to the fraction of the material with refractive index n_c, but the depolarization factors apply to the shape of the voids. The functions **bpcd** and **bpvd** in the *BTF Toolbox* implement the BP equations.

For spherical particles we have $L_1 = L_2 = L_3 = 1/3$ and the BP equations reduce to the Maxwell-Garnett equation[37] for the effective dielectric constant of an isotropic medium,

$$\varepsilon = \varepsilon_v \frac{(1+2p)\varepsilon_c + 2(1-p)\varepsilon_v}{(1-p)\varepsilon_c + (2+p)\varepsilon_v}. \tag{8.21}$$

For cylindrical columns (with a circular cross-section, as proposed by H.K. Pulker and E. Jung[38]) we can take the depolarization factors to be $L_1 = 0$, $L_2 = 1/2$, $L_3 = 1/2$, and the BP equations have the form

$$
\begin{aligned}
\varepsilon_{cd\,1} &= (1-p)\varepsilon_v + p\varepsilon_c \\
\varepsilon_{cd\,2,3} &= \varepsilon_v \frac{(1-p)\varepsilon_v + (1+p)\varepsilon_c}{(1+p)\varepsilon_v + (1-p)\varepsilon_c} \\
\varepsilon_{vd\,1} &= \varepsilon_{cd,1} \\
\varepsilon_{vd\,2,3} &= \varepsilon_c \frac{(2-p)\varepsilon_v + p\varepsilon_c}{p\varepsilon_v + (2-p)\varepsilon_c}.
\end{aligned}
\tag{8.22}
$$

Structure Fraction

It is not uncommon for the packing density and structure of a thin film to change during deposition, and for this reason variations to the cylindrical column shape of a uniaxial film have been proposed. These include tapered cylindrical columns, which are close-packed at the substrate but decrease in diameter as the film thickness increases, and columns that expand from close-packed circular rods to close-packed hexagonal rods.[39,40] In some cases a rapid increase in refractive index occurs when, at $p = p_0 \approx 0.9$, columns begin to touch and the microstructure changes from cd to vd.[41] The effect that such a structural change has on refractive indices in a uniaxial film can be modelled by regarding the film to be a mixture of cd and vd material.[42] In general the situation can be modelled in a similar way to cases in physics in which an energy gap is involved, and a suitable equation for the effective dielectric constant may be

$$\varepsilon_2 = \frac{\varepsilon_{\text{structure1}\,2}}{1 + \exp[(p-p_0)/\Delta p]} + \frac{\varepsilon_{\text{structure2}\,2}}{1 + \exp[-(p-p_0)/\Delta p]}. \tag{8.23}$$

Here Δp is a parameter that controls the abruptness of the change in refractive index.

8.3.2 *Inversion of the Bragg-Pippard Equations*

The BP equations can be inverted, allowing n_c, p and the L's to be calculated if the principal refractive indices and the void index are known. For crystallite-defined films the inverted equations have the form

$$\varepsilon_c = \varepsilon_1 + \varepsilon_v \left[\frac{\varepsilon_1 - \varepsilon_2}{\varepsilon_2 - \varepsilon_v} + \frac{\varepsilon_1 - \varepsilon_3}{\varepsilon_3 - \varepsilon_v} \right], \tag{8.24}$$

$$p = \frac{\varepsilon_1 - \varepsilon_v}{\varepsilon_c - \varepsilon_v}, \tag{8.25}$$

$$
\begin{aligned}
L_1 &= 0, \\
L_2 &= \frac{\varepsilon_v(\varepsilon_1 - \varepsilon_2)}{(\varepsilon_c - \varepsilon_1)(\varepsilon_2 - \varepsilon_v)}, \\
L_3 &= 1 - L_1 - L_2.
\end{aligned}
\tag{8.26}
$$

Similarly, for the void-defined equations ε_c may be found from the quadratic

$$\varepsilon_c^2[\varepsilon_1 - \varepsilon_2 - \varepsilon_3 + \varepsilon_v] + \varepsilon_c[\varepsilon_2(\varepsilon_3 - \varepsilon_v) + \varepsilon_3(\varepsilon_2 - \varepsilon_v)] + \varepsilon_2\varepsilon_3(\varepsilon_v - \varepsilon_1) = 0, \tag{8.27}$$

$$p = 1 - (\varepsilon_c - \varepsilon_1)/(\varepsilon_c - \varepsilon_v), \tag{8.28}$$

and

$$
\begin{aligned}
L_1 &= 0, \\
L_2 &= [\varepsilon_c/(\varepsilon_c - \varepsilon_v) - (1 - p)\varepsilon_c/(\varepsilon_c - \varepsilon_2)]/p, \\
L_3 &= 1 - L_1 - L_2.
\end{aligned}
\tag{8.29}
$$

See the functions **bpcdi** and **bpvdi** in the *BTF Toolbox*. As examples of their use, the values given in Table 8.2 for zirconium oxide deposited at $27°$ yield, in the two models,

	cd	vd
n_c	2.09	2.18
p	0.94	0.84
L_1	0.00	0.00
L_2	0.57	0.54
L_3	0.43	0.46

$$\tag{8.30}$$

Chapter 9

Effective Media

For a given wavelength and p or s polarization the symmetric stack of isotropic films $(A/2)B(A/2)$ is known to be equivalent to a single layer. This result is evident from the form of the characteristic matrix that represents the group of layers. The refractive index of the equivalent layer is called the *Herpin index*,[43] and in practice, the method allows an index that is otherwise unavailable to be generated.

It follows that the medium ... $ABABABAB$ ... is equivalent to a homogeneous medium, and that the refractive index of the effective medium is the Herpin index. In this chapter we consider the adaptation of the Herpin index to biaxial layers. We begin by examining our expectations, by discussing the simultaneous application of the method to the p and s polarizations for isotropic media, and then move on to biaxial layers. We show that, in the most general case, the equivalent medium has three *effective principal refractive indices* n_1, n_2, n_3, and the angular position of the equivalent medium is specified by three *effective placement angles* η, ψ and ξ.

The biaxial properties of an effective medium are influenced by the anisotropy of the parallel-layered structure, and we work through selected examples to illustrate this influence. In turn, we consider cases in which

▨ A and B are isotropic materials

▨ A and B are columnar materials with a common deposition plane

▨ A and B are columnar materials with different deposition planes.

9.1   Herpin Indices for Isotropic Layers

When the components A and B that form the Herpin layer are both isotropic, the characteristic matrix $\hat{M}_{A/2}\hat{M}_B\hat{M}_{A/2}$ of the layer, which has a total thickness d, can be put in the form

$$
\hat{M} = \begin{bmatrix}
\cos\phi_p & -i\sin\phi_p/\gamma_p & 0 & 0 \\
-i\gamma_p\sin\phi_p & \cos\phi_p & 0 & 0 \\
0 & 0 & \cos\phi_s & -i\sin\phi_s/\gamma_s \\
0 & 0 & -i\gamma_s\sin\phi_s & \cos\phi_s
\end{bmatrix}. \tag{9.1}
$$

Now the various parameters of the equivalent layer can be calculated from the elements of the matrix $\hat{M}$. When this is done using the set of equations

$$
\begin{aligned}
\gamma_p &= (M_{21}/M_{12})^{1/2} \\
\gamma_s &= -(M_{43}/M_{34})^{1/2} \\
\alpha_p &= [z_0\gamma_p + (z_0^2\gamma_p^2 - 4\beta^2)^{1/2}]/2 \\
\alpha_s &= -z_0\gamma_s \\
\phi_p &= \cos^{-1}(M_{11}) \\
\phi_s &= \cos^{-1}(M_{33}) \\
d_p &= \phi_p/k\alpha_p \\
d_s &= \phi_s/k\alpha_s,
\end{aligned} \tag{9.2}
$$

it is found that the two values calculated for the thickness, d_p and d_s, are not exactly equal. This is due to interference effects in the layers and limits the use of the method to very thin periods, $d \ll \lambda$. In this regime the method is robust, in that $d_p \approx d_s \approx d$ and the equivalent p and s refractive indices, calculated using

$$
\begin{aligned}
n_p &= \alpha_p/kd_p \\
n_s &= \alpha_s/kd_s,
\end{aligned} \tag{9.3}
$$

are nearly independent of β.

The above analysis shows that the effective medium formed by alternating, thin, isotropic layers is anisotropic. Can it be described as columnar deposited media, with three refractive indices n_1, n_2, n_3 and a column angle ψ? The answer is "yes", with $n_1 = n_3 = n_p > n_2 = n_s$ and $\psi = 90°$. The effective columns are slabs parallel to the substrate. The material is uniaxial with the optic axis perpendicular to the layers, and hence we can put $n_o = n_1 = n_3$ for the ordinary index and $n_e = n_2$ for the extraordinary index. The anisotropy of the effective medium is due to form birefringence; when the electric field of the light is perpendicular to the layers (parallel to the optic axis), maximum depolarization occurs and hence $n_e < n_o$. Thus the effective medium is *negative uniaxial*.

Figure 9.1 shows the dependence of the principal refractive indices and the column angle on the material fraction f_B for an illustrative example, in which A and B are isotropic materials with $n_A = 1.35$, $n_B = 2.4$. Here f_B is the fraction of the total physical thickness of the period that is occupied by material B. A similar definition applies to f_A, and thus

$$f_A + f_B = 1. \tag{9.4}$$

Useful approximations can be derived for the ordinary and extraordinary refractive indices,

$$1/n_e^2 \approx f_A/n_A^2 + f_B/n_B^2 \tag{9.5}$$

and

$$n_o^2 \approx f_A n_A^2 + f_B n_B^2. \tag{9.6}$$

These approximations are exact in the limit of negligible film thickness (which can be stated alternatively as the long-wavelength limit). The first, Eq.(9.5), can be described as a parallel combination of dielectric constants and the second, Eq.(9.6), as a series combination.

9.2 Biaxial Layers with a Common Deposition Plane

When the layers A and B have parallel columns or a common deposition plane the parameters of the equivalent medium can be calculated using the set of equations:-

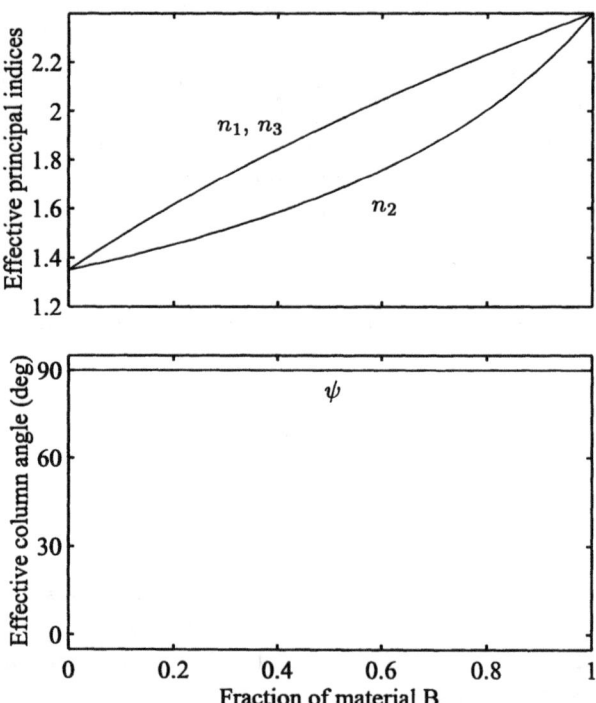

Fig. 9.1. Effective principal refractive indices and effective column angle ψ for a periodic stack of thin isotropic films with $n_A = 1.35$ and $n_B = 2.4$. The material ratio f_B is the thickness of film B divided by the total thickness of the period. The birefringence of the equivalent medium is due to depolarization caused by the parallel interfaces. (Adapted from I.J. Hodgkinson and Q.H. Wu, *Journal of the Optical Society of America A* 10, 2065, 1993. Copyright © 1993 Optical Society of America. Reprinted with permission.)

$$\gamma_p = (M_{21}/M_{12})^{1/2}$$

$$\gamma_s = -(M_{43}/M_{34})^{1/2}$$

$$\alpha_1^{\pm} = -\text{angle}[M_{11} \pm (M_{12}M_{21})^{1/2}]/kd$$

$$\alpha_2 = -z_0\gamma_s$$

$$n_{1,2}^2 = \{\alpha_1^+ - \alpha_1^- + 2\alpha_1^+\alpha_1^-/z_0/\gamma_p + 2\beta^2/z_0/\gamma_p$$
$$\pm[(\alpha_1^+ - \alpha_1^- + 2\alpha_1^+\alpha_1^-/z_0/\gamma_p + 2\beta^2/z_0/\gamma_p)^2$$
$$+4\beta^2(\alpha_1^+/z_0/\gamma_p - \alpha_1^-/z_0/\gamma_p - 2)(\alpha_1^+ - \alpha_1^-)/z_0\gamma_p]^{1/2}\}$$
$$\div 2(2 - \alpha_1^+/z_0/\gamma_p + \alpha_1^-/z_0/\gamma_p)/z_0/\gamma_p$$

$$n_3^2 = \alpha_2^2 + \beta^2$$

$$\psi = \sin^{-1}[(1/\gamma_p^2 + \beta^2/n_1^2 n_2^2 - 1/n_2^2)^{1/2}/(1/z_0^2 n_1^2 - 1/n_2^2)^{1/2}].$$

(9.7)

9.2.1 A and B Normal Columnar

As many thin films are deposited at normal incidence, and have columns running perpendicular to the substrate, the effect of such a microstructure on the anisotropic properties of the film is of interest. An example, with $n_{A1} = 1.35$, $n_{A2,3} = 1.3$, $\psi_A = 0°$, $n_{B1} = 2.4$, $n_{B2,3} = 2$ and $\psi_B = 0°$, illustrates the important points. As shown in Fig. 9.2, the medium is uniaxial for all values of f_B with $\psi = 0°$ or $90°$.

As in the previous example, the birefringence can be described by approximate expressions with the form of parallel and series combinations of the dielectric constants,

$$1/n_e^2 \approx f_A/n_{A1}^2 + f_B/n_{B1}^2, \tag{9.8}$$

and

$$n_o^2 \approx f_A n_{A2}^2 + f_B n_{B2}^2. \tag{9.9}$$

As f_B in Fig. 9.2 increases from 0 to 1, the sign of the birefringence changes twice, at values of f_B that can be determined (approximately) from the quadratic

$$(n_{A1}^2 - n_{B1}^2)(n_{A2}^2 - n_{B2}^2)f_B^2 - (n_{A1}^2 n_{A2}^2 - 2n_{A2}^2 n_{B1}^2 + n_{B1}^2 n_{B2}^2)f_B + n_{B1}^2(n_{A1}^2 - n_{A2}^2) = 0. \tag{9.10}$$

If the material of film A is isotropic, then one root moves to $f_B = 0$ and the other can be found from

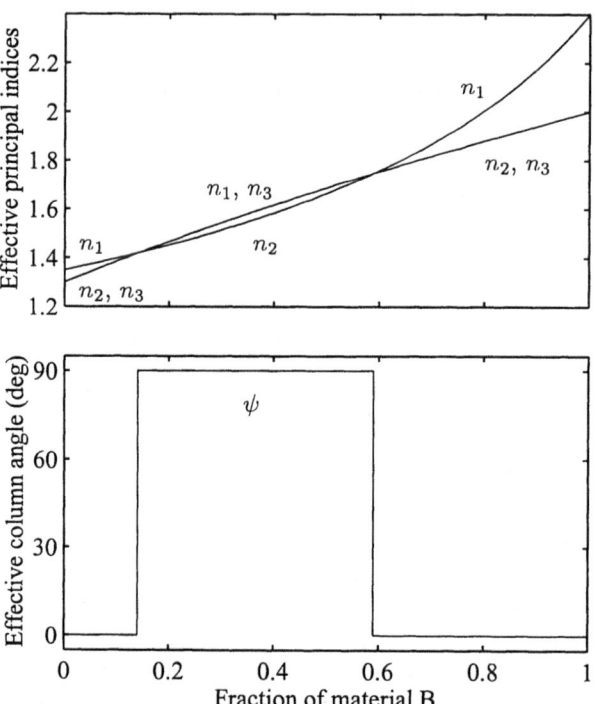

Fig. 9.2. Effective principal refractive indices and effective column angle ψ for a periodic stack formed by columnar thin film materials with $n_{A1} = 1.35$, $n_{A2,3} = 1.3$, $n_{B1} = 2.4$, $n_{B2,3} = 2$. The columns of A and B are perpendicular to the film interfaces, i.e. $\psi_A = \psi_B = 0°$. The equivalent medium behaviour is uniaxial columnar, parallel slab-like, and uniaxial columnar for small, medium and large values of the material fraction, f_B. At the two points where the refractive index curves cross the equivalent medium is isotropic. (Adapted from I.J. Hodgkinson and Q.H. Wu, *Journal of the Optical Society of America A* 10, 2065, 1993. Copyright © 1993 Optical Society of America. Reprinted with permission.)

$$f_B = n_A^2/(n_A^2 - n_{B2}^2) - n_{B1}^2/(n_A^2 - n_{B1}^2). \tag{9.11}$$

At the changeover points the three principal refractive indices are equal and the effective medium is isotropic. In the leftmost section of Fig. 9.2 the effective medium is dominated by the columnar material A, in the middle section the form birefringence of the parallel layers has the greatest effect, and in the rightmost section the properties are dominated by the columnar material B. Thus a periodic material formed by depositing alternate low and high index columnar material at normal incidence may be equivalent to an isotropic medium, a medium with columns of circular section perpendicular to the substrate, or to a medium with slab-like columns parallel to the substrate.

9.2.2 A and B Parallel, Tilted Columnar

Figure 9.3 has been plotted for the combination of two parallel tilted columnar films, with $n_{A1} = 1.35$, $n_{A2} = 1.3$, $n_{A3} = 1.31$, $\psi_A = 30°$, $n_{B1} = 2.4$, $n_{B2} = 1.95$, $n_{B3} = 2.1$, $\psi_B = 30°$. This case shows that the angular position of the effective columns is influenced by the form birefringence of the materials A and B and the form birefringence of the parallel layers. In particular, the effect of the parallel layers is to move the column axis towards the layer interfaces, i.e. parallel to the substrate.

Figure 9.4 shows the effective normal-incidence anisotropy, $\Delta n = n_3 - n_p$, as a function of f_B for the same example. By depositing alternate layers of two materials A and B, with individual normal-incidence anisotropies Δn_A and Δn_B, any value of normal-incidence anisotropy between between Δn_A and Δn_B can be engineered. In particular, it is possible to design materials with low refractive index and relatively large Δn, thus overcoming the limitation of individual low-index materials which may be required to match a substrate index, for example.

Figure 9.5 shows experimental results obtained during the deposition of silicon oxide and titanium oxide. Clearly, these confirm that equivalent anisotropic media can be formed by vacuum deposition.

The following approximations can be used for the estimation of normal-incidence anisotropy,

$$n_3^2 \approx f_A n_{A3}^2 + f_B n_{B3}^2, \tag{9.12}$$

$$n_p^2 = f_A n_{Ap}^2 + f_B n_{Bp}^2. \tag{9.13}$$

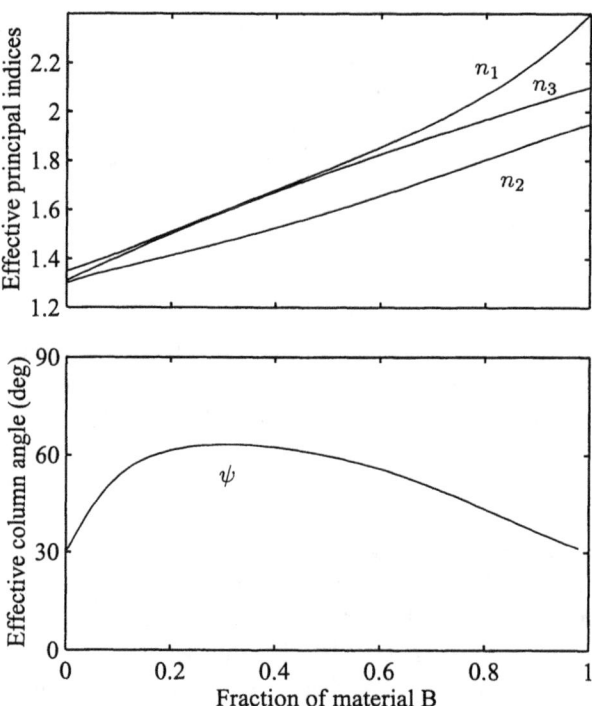

Fig. 9.3. Effective principal refractive indices and effective column angle ψ for a periodic stack formed by columnar thin film materials with $n_{A1} = 1.35$, $n_{A2} = 1.3$, $n_{A3} = 1.31$, $n_{B1} = 2.4$, $n_{B2} = 1.95$, $n_{B3} = 2.1$. The columns of A and B are parallel, with $\psi_A = \psi_B = 30°$, but the position of the effective columnar axis is rotated due to the birefringent effect of the parallel layers. (Adapted from I.J. Hodgkinson and Q.H. Wu, *Journal of the Optical Society of America A* 10, 2065, 1993. Copyright © 1993 Optical Society of America. Reprinted with permission.)

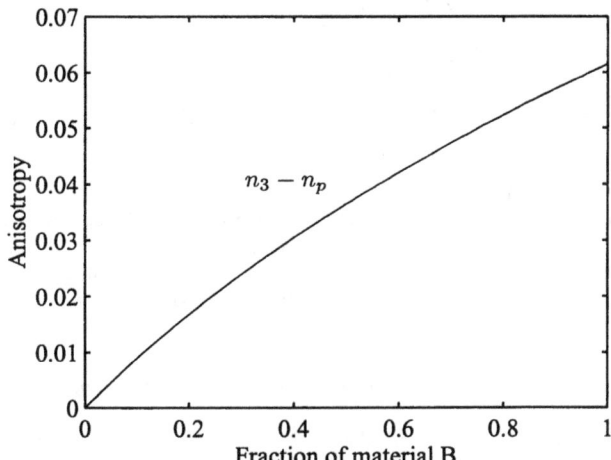

Fig. 9.4. Effective normal incidence refractive anisotropy, $n_3 - n_p$, for light traveling perpendicular to the film interfaces in the periodic medium described in Fig. 9.3. The principal axis-3 is perpendicular to the film deposition plane and the effective refractive index n_p occurs when the electric field of the light is parallel to the deposition plane. Relatively large values of normal incidence refractive anisotropy in low index films may be achieved in periodic stratified media. (Adapted from I.J. Hodgkinson and Q.H. Wu, *Journal of the Optical Society of America A* 10, 2065, 1993. Copyright © 1993 Optical Society of America. Reprinted with permission.)

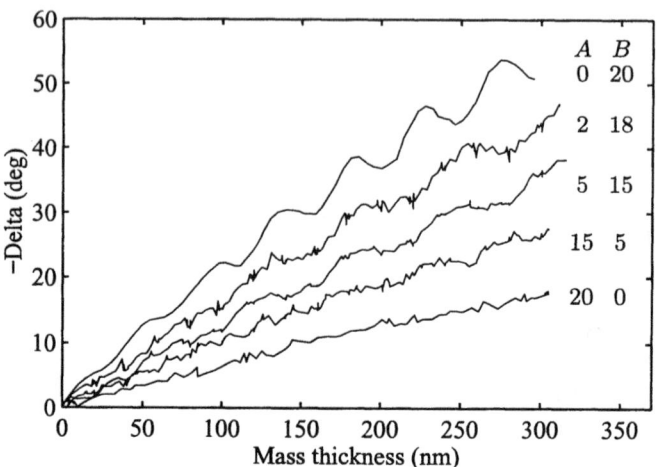

Fig. 9.5. Phase retardation recorded during the deposition of periodic stacks of silica (A) and titania (B). The labels indicate the nominal mass thicknesses in nm units of the elements forming one period.(Adapted from I.J. Hodgkinson and Q.H. Wu, *Optical Interference Coatings, Proceedings of the Society of Photo-Optical Instrumentation Engineers* 2253, 882, 1994. Copyright © 1994 SPIE. Reprinted with permission.)

9.2.3 A and B Coplanar, Tilted Columnar with $\psi_A = -\psi_B$

The geometry of deposition at an oblique angle on to a substrate from a small source leads to variations in both the thickness and the retardance of a film. Areas of the film closest to the source have the largest physical thickness and largest phase retardation. In practice non-uniformities can be reduced substantially, while maintaining the same normal-incidence birefringence, by depositing the first half of a wave plate at deposition angle θ_v and the second half at angle $-\theta_v$. The change in geometry that is required can be effected by rotating the substrate by half a turn about a normal axis.

Similar problems associated with the geometry of deposition occur for effective media formed by depositing periodic stacks of films, and can be solved in the same way, by alternating deposition angles. For this reason we consider here a relevant numerical example, defined by the parameters $n_{A1} = n_{B1} = 2.4$, $n_{A2} = n_{B2} = 1.95$, $n_{A3} = n_{B3} = 2.1$, and $\psi_A = -\psi_B = 30°$. Figure 9.6 shows that as f_B changes from 0 to 1 the effective column angle ψ changes smoothly from ψ_A to ψ_B. The parallel layer effect is small due to the similarity of the refractive indices of the two materials. Both n_3 and n_p are independent of f_B, and hence the normal-incidence birefringence $\Delta n = n_3 - n_p$ does not depend on the material fraction.

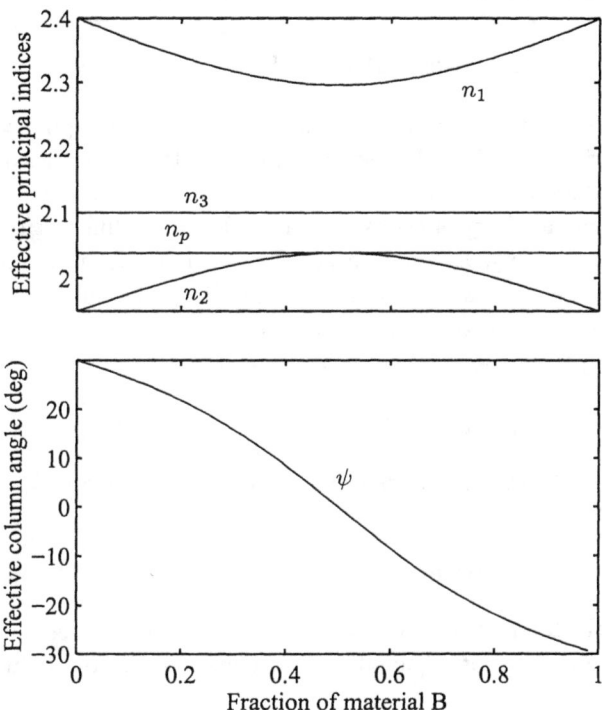

Fig. 9.6. Residual anisotropy in a periodic medium with similar constituent films deposited from opposite sides of the substrate normal ($n_{A1} = n_{B2} = 2.4$, $n_{A2} = n_{B2} = 1.95$, $n_{A3} = n_{B3} = 2.1$, $\psi_A = -\psi_B = 30°$). The indices n_3 and n_p are independent of the material fraction. Deposition of equal amounts of A and B partially compensates geometric thickness variations associated with oblique deposition without loss of normal incidence anisotropy. (Adapted from I.J. Hodgkinson and Q.H. Wu, *Journal of the Optical Society of America A* 10, 2065, 1993. Copyright © 1993 Optical Society of America. Reprinted with permission.)

At $f_B = 0.5$ in Fig. 9.6 the refractive indices satisfy $n_1 > n_3 > n_2$ but the effective column angle ψ is zero, and hence the effective medium can be described as columns of elliptical shape running perpendicular to the substrate. (Cerium oxide can be deposited so that it has this property; see Sect. 7.2.2.)

Replacement of layer B in the previous example by a double layer with $f_B = 0.5$ does not alter the curve of Δn versus f_B that is plotted in Fig. 9.4.

9.3   Biaxial Layers Deposited in Different Planes

In the general case in which A and B are biaxial layers deposited in different planes the matrix method can be used to determine the properties of the effective medium. We demonstrate, rather than offer a mathematical proof, that the effective medium has the properties of a single deposited film with three effective principal indices n_1, n_2, n_3, and three effective material placement angles η, ψ, and ξ. The key stages in a path for computing the values of the effective parameters are listed below; for further details the reader is referred to the commented script file **herpin** in the *BTF Toolbox*.

- The matrix $\hat{M}$ is computed for the period $(A/2)B(A/2)$.

- The field matrix $\hat{F}$ and the phase matrix $\hat{A}_d$ are determined using $[\hat{F}, \hat{A}_d] = \operatorname{eig} \hat{M}$.

- The matrix $\hat{\alpha}$ is determined as the log of the non-zero elements of $\hat{A}_d$.

- The matrix $\hat{L}$ for the effective medium is computed using $\hat{L} = \hat{F}\hat{\alpha}\hat{F}^{-1}$ and used for computation of the dielectric constant $\hat{\varepsilon}$ in the light propagation frame.

- The combined rotation matrix $\hat{R}_c$ that transforms the dielectric constant from the x, y, z axes to the principal axes 1, 2, 3 of the effective medium is determined by diagonalizing, $[\hat{R}_c, \hat{\varepsilon}_{123}] = \operatorname{eig} \hat{\varepsilon}$.

- The refractive indices are sorted so that they satisfy our personal preference, $n_1 \geq n_3 \geq n_2$, and corresponding changes are made to the order of the columns of $\hat{R}_c$.

- The equation $\hat{R}_x(\xi)\hat{R}_z(\psi)\hat{R}_x(\eta) = \hat{R}_c$ is solved so that the angles of the effective medium agree with our personal preference for angles. That is, starting with the material axes 1, 2, 3 aligned with the propagation axes x, y, z, the angle η is the column rotation defined by an initial rotation of the material about the x-axis, ψ is the column direction defined by a

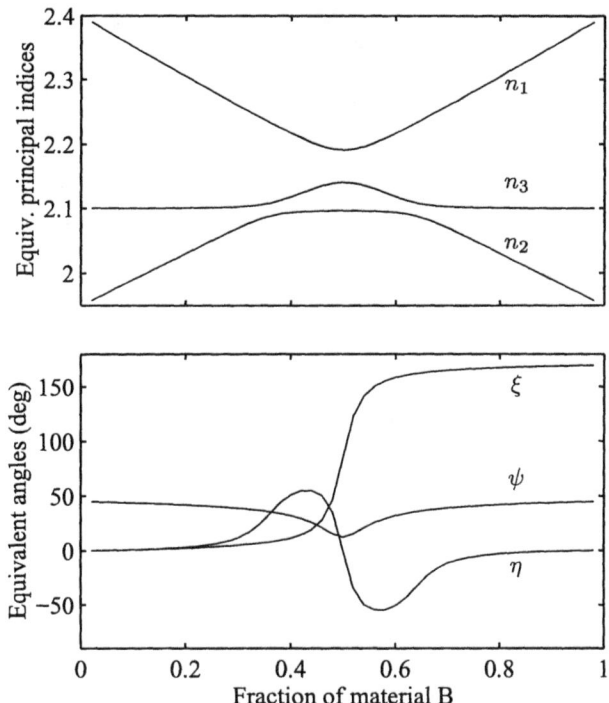

Fig. 9.7. Effective refractive indices and effective dielectric axes of a periodic medium with similar constituent films deposited in different planes ($n_{A1} = n_{B1} = 2.4$, $n_{A2} = n_{B2} = 1.95$, $n_{A3} = n_{B3} = 2.1$, $\psi_A = \psi_B = 45°$, $\xi_A = 0$, $\xi_B = 170°$). In contrast to a single deposited film, the effective medium can have all three dielectric axes tilted with respect to the substrate.

subsequent rotation of the material about the z-axis, and ξ is a subsequent rotation of the material about the x-axis.

As an example we consider a periodic medium with similar constituent films deposited in different planes, $n_{A1} = n_{B1} = 2.4$, $n_{A2} = n_{B2} = 1.95$, $n_{A3} = n_{B3} = 2.1$, $\psi_A = \psi_B = 45°$, $\xi_A = 0$, $\xi_B = 170°$. The effective refractive indices and placement angles calculated for the effective medium are plotted in Fig. 9.7. This example, which can be generated by running **herpin** from the *BTF Toolbox*, is chosen to illustrate a special feature of the effective medium – the effective principal axis-3 can be inclined at an angle to the substrate.

Chapter 10

Anisotropic Scatter

There is a rule-of-thumb in thin film physics that states "the diameter of microstructural columns is about $\frac{1}{10}$th the thickness of the film". Hence it is not surprising that very thick vacuum-deposited films can have a milky appearance due to scatter from the microstructure. Further, when the films are deposited obliquely, the scatter is found to be spatially anisotropic.

Experiments show that the deposition parameter window for the attainment of negligible or low haze in a birefringent film can be quite narrow. The goal of effective management of haze requires, at a preliminary stage, characterization of scatter distributions and later optimization of deposition parameters.

In this chapter we discuss

- observations and assessment of light scattered anisotropically into the air spaces in front of, and behind the coating

- anisotropic secondary scatter patterns that have been observed on coatings

- assessment of light that is scattered anisotropically into the substrate and trapped by total internal reflection

- the measurement of anisotropic scatter distributions during the growth of coatings

- a simple theory of scatter from a columnar microstructure.

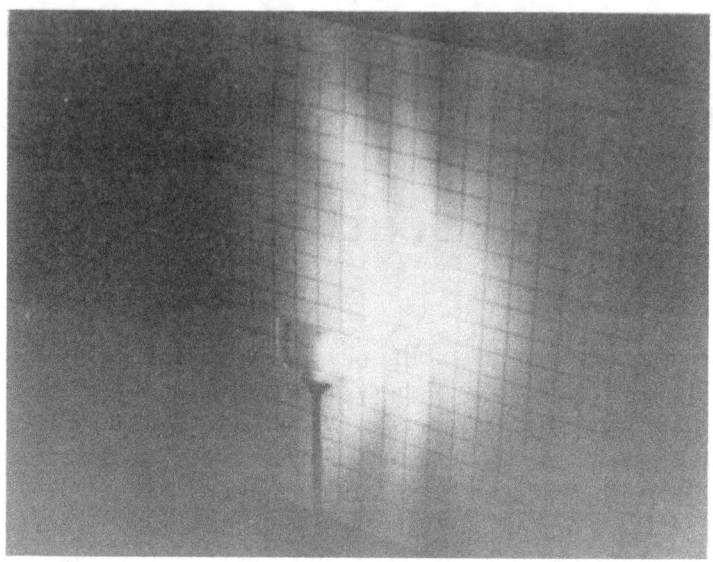

Fig. 10.1. Anisotropic scatter from a titania film photographed on a plane white screen. (Adapted from I.J. Hodgkinson, P.I. Bowmar and Q.H. Wu, *Applied Optics* 34, 163, 1995. Copyright © 1995 Optical Society of America. Reprinted with permission.)

10.1   Scatter into the Air

Anisotropy in the scatter from vacuum deposited birefringent films was discovered without the aid of complicated equipment – the haze from a coated microscope slide illuminated by sunlight coming in a window was observed to change when the slide was rotated by hand.

In some cases the spatial distribution of scatter can be observed directly, although a laser beam is needed for the source. An example is given in Fig. 10.1 for a titania film deposited at 70° and then illuminated at normal incidence with a randomly polarized green He-Ne laser beam. The forward scatter pattern is dominated by a strong principal arc and two weaker secondary arcs. Experiments with polarized sources show that the central arc has, more or less, the polarization of the specularly transmitted beam, and experiments with different wavelengths lead to the conclusion that the central arc is a zero order diffraction peak.

Scatter into a wider angular range can be studied, also by direct observation, when the plane screen in Fig. 10.1 is replaced by a hollow hemisphere painted white on the inside surface. Transmitted and reflected scatter patterns recorded for the same titania film are reproduced in Fig. 10.2. In the upper part of Fig. 10.2 the laser beam is incident from the right and impinges normally on the air/film interface, which is located at the center of the hemisphere. The specularly-transmitted light passes through a small hole in the hemisphere and the transmitted scatter is recorded. In the lower part of the figure the laser beam is incident from behind the hemisphere and impinges normally on the air/film interface, and the reflected scatter pattern is recorded. The pair of bright spots at the top of the photograph identify the incident and specularly-reflected beams.

The hollow hemisphere provides an excellent view of an anisotropic scatter pattern, essentially as a projection from a Lambertian spherical surface on to a plane through the centre of the sphere, but it is not of general use because the scatter from most tilted columnar films is too small for either direct viewing or photography. For this reason photoelectric scanning methods are required.

The scanning apparatus illustrated in Fig. 10.3 can be used to emulate the projection provided by the hollow hemisphere. In the apparatus the sample S and a normally-incident light source L are turned as a unit about a horizontal axis by the rotary stage H. The axis of rotation passes through the intersection point of the incident light beam with the coating. A second stage V carries H and rotates it about a vertical axis passing through the same intersection point.

The highlighted pixel in the square array shown in Fig. 10.4 is the projection of a small area from the surface of the hemisphere. For the case shown the pixel corresponds to scatter at a particular direction in the deposition plane of the film. Scattered light associated with the pixel is obtained directly by driving the stepper-controlled rotary stages H and V to access the required position. Examples of scatter patterns acquired in this way can be viewed by running the script file **vscatter** from the *BTF Toolbox*.

The scanning apparatus can be used to generate polar plots of scatter patterns. Figure 10.5 shows profiles of light scattered into the air space in front of a titania film.

10.2 Scatter From Stress-Related Cracks

Tilted columnar films such as titanium oxide are under anisotropic tensile stress, with greater stress in the deposition plane. Very thick films may fracture as a result of this stress when air is admitted to the vacuum chamber or soon afterwards. A typical fracture pattern is shown in Fig. 10.6. The fracture lines make characteristic angles with the deposition plane, which is indicated by the

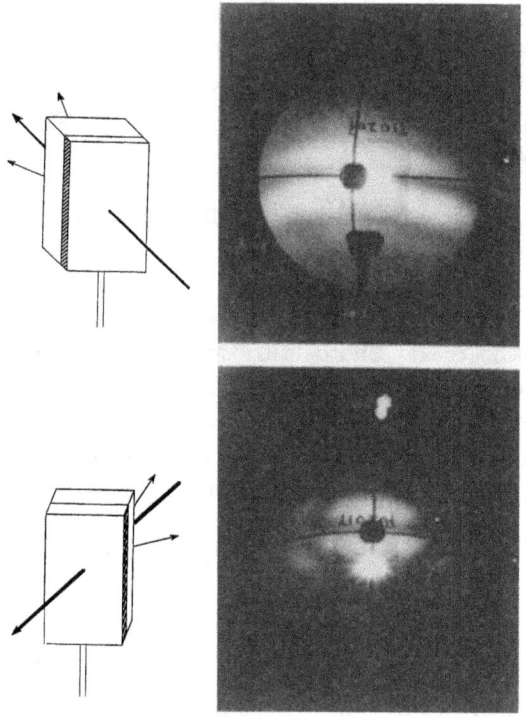

Fig. 10.2. Scatter distributions recorded for a titania film by photographing the light scattered on to the white inside surface of a hollow hemisphere. (Adapted from I.J. Hodgkinson, P.I. Bowmar and Q.H. Wu, *Applied Optics* 34, 163, 1995. Copyright © 1995 Optical Society of America. Reprinted with permission.)

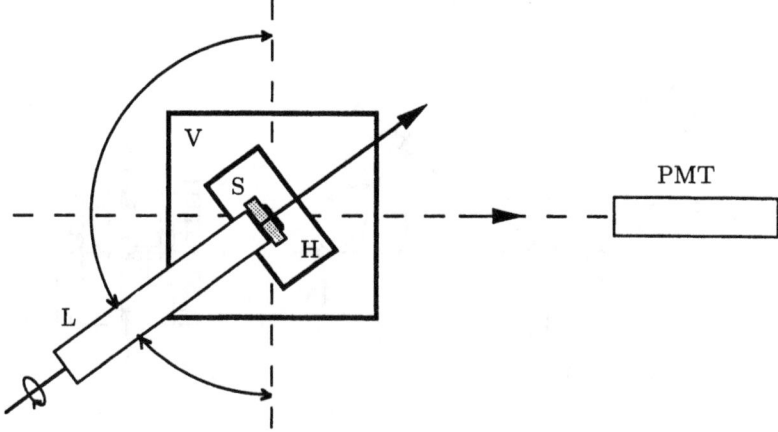

Fig. 10.3. Apparatus for scanning the hemisphere in front of an anisotropic optical coating. (Adapted from I.J. Hodgkinson, S.J. Cloughley, Q.H. Wu and S. Kassam, *Applied Optics* 35, 5563, 1996. Copyright © 1996 Optical Society of America. Reprinted with permission.)

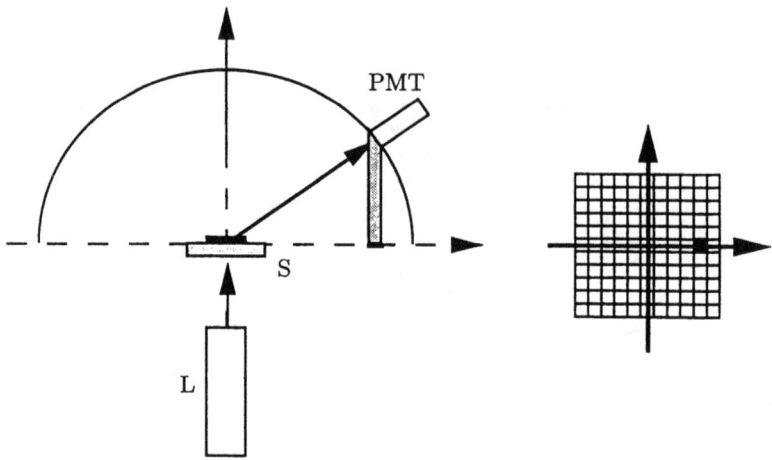

Fig. 10.4. Direct acquisition of a projection of a 3–D distribution of scattered light. (Adapted from I.J. Hodgkinson, S.J. Cloughley, Q.H. Wu and S. Kassam, *Applied Optics* 35, 5563, 1996. Copyright © 1996 Optical Society of America. Reprinted with permission.)

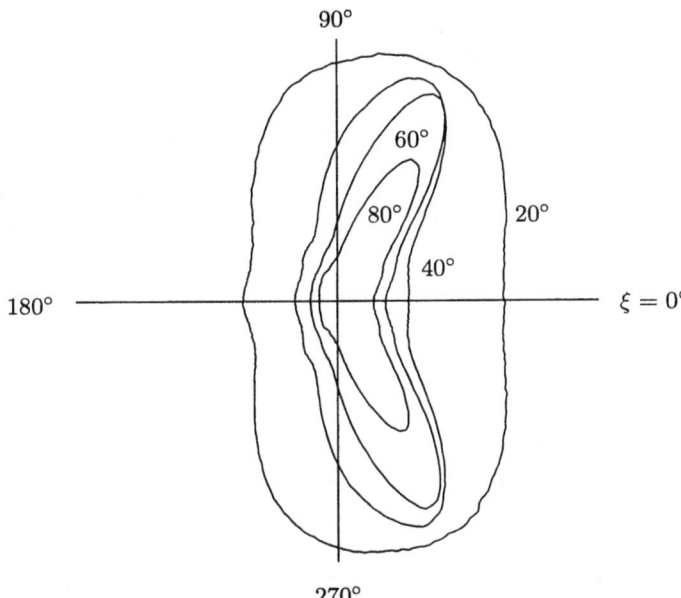

Fig. 10.5. Anisotropic scatter in the air space in front of a titania film illuminated by white light at normal incidence. The labels on the polar plot indicate the angle of scatter. (Adapted from I.J. Hodgkinson and Q.H. Wu, *Optical Interference Coatings*, *Proceedings of the Society of Photo-Optical Instrumentation Engineers* 2253, 882, 1994. Copyright © 1994 SPIE. Reprinted with permission.)

Fig. 10.6. Stress-related cracks in a tilted columnar film. The deposition plane is indicated by the shadow of a solid particle. The cracks diffract light into two planes, perpendicular to the lines of the cracks.

shadow of a solid particle in the figure. Diffraction from the system of cracks leads to a pair of lines in the scatter pattern.

Problem associated with stress in birefringent films can be reduced to an acceptable level by an ion-assisted overcoat of silicon oxide. This provides a compensating compressive stress.

10.3 Scatter Patterns Formed on the Film

Tilted columnar birefringent films (with excessive scatter) exhibit an interesting phenomenon when they are illuminated directly by a laser beam. Brightly illuminated patterns of characteristic shape appear superposed on the film, and are visible on the film at distances of several centimeters from the laser spot.

Typical differences between transmission and reflection patterns formed on tilted-columnar films are highlighted by the patterns recorded for a titania film (Fig. 10.7). The transmission pattern shown in the upper part of the figure consists of two strong lobes pointing away from the laser spot, and the reflection pattern shown in the lower part of the figure contains four weaker lobes in the form of a cross. Tilting the substrate to decrease the angle between the laser beam and the microstructural columns causes the angle between the lobes to decrease.

The patterns are only weakly dependent on wavelength and polarization, and they disappear when a thick block of glass is placed against the back of the substrate, with an index-matching oil to suppress reflections. The latter observation shows that the patterns require substrate/air total internal reflections and are caused by light flux travelling in the substrate, rather than by waveguiding in the film. Additional evidence supporting this conclusion is provided in Fig. 10.7 – in both the transmission pattern and the reflection pattern a bright lobe continues past a scratch that we made in the film.

10.4 Scatter into the Substrate

A considerable fraction of the light scattered by a tilted columnar film may be directed into the substrate and trapped by total internal reflection. A method for measuring the angular distribution of the flux that travels outwards from the spot where the laser beam impinges upon the film is illustrated in Fig. 10.8.

In this method the coated substrate is cut into a disc and the light that emerges from the edge is collected by a fibre and transmitted to a photomultiplier. Values recorded while the substrate is rotated by one turn about its normal, provide data for a polar plot of the substrate flux. As an example, polar plots of substrate fluxes recorded for titania and zirconia are given in Fig. 10.9; these show the same characteristic scatter lobes that are recorded in air space patterns (see Fig. 10.5).

10.5 In Situ Measurement of Scatter

An *in situ* method for measuring scatter during the deposition of a birefringent film is based on collection of scattered light emerging from the substrate or from close to the surface of the film. As shown in Fig. 10.10, a chopped laser beam is incident normally on the film and the emerging light is collected by a shielded glass rod. The signal from the photomultiplier is fed into a lock-in amplifier to increase the signal-to-noise ratio.

The fibre may be positioned to collect from a lobe of a scatter pattern, or the pattern may be scanned by stopping the deposition and rotating the sample by one turn.

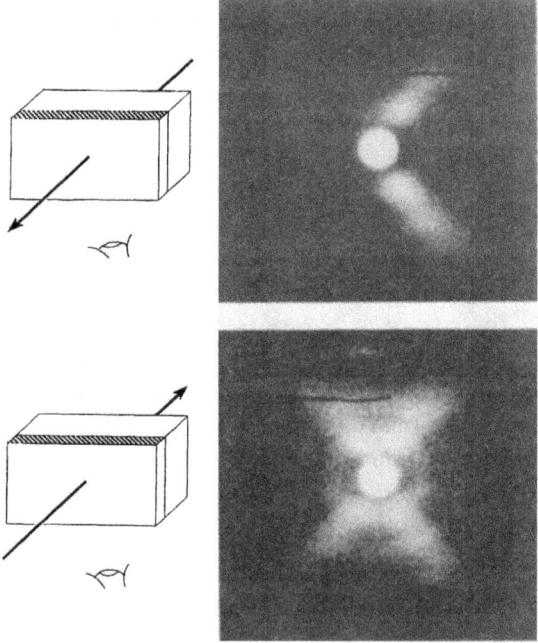

Fig. 10.7. Bright patterns superposed on a titania film illuminated by a laser beam at normal incidence. Upper: bright pattern photographed in transmission. Lower: bright pattern photographed in reflection. (Adapted from I.J. Hodgkinson, P.I. Bowmar and Q.H. Wu, *Applied Optics* 34, 163, 1995. Copyright © 1995 Optical Society of America. Reprinted with permission.)

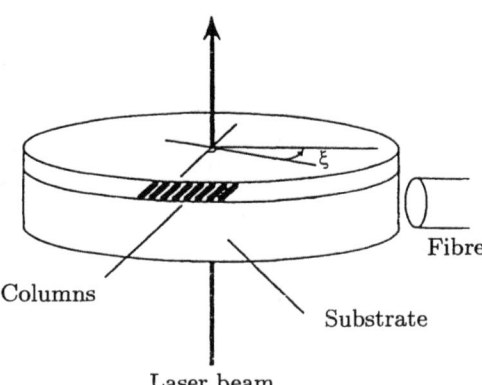

Fig. 10.8. Method used to measure light scattered into the substrate and trapped by total internal reflection. (Adapted from I.J. Hodgkinson, P.I. Bowmar and Q.H. Wu, *Applied Optics* 34, 163, 1995. Copyright © 1995 Optical Society of America. Reprinted with permission.)

10.5.1 Dependence of Haze on Δ

The *in situ* method has been used to study the build-up of anisotropic scatter during the growth of tilted-columnar films. Figure 10.11 shows results obtained for a zirconia film. In this experiment the electron beam gun was powered down at intervals, when the retardation reached 18°, 36°, ... 90°, and the scatter was recorded as the sample was rotated by one turn. The profiles in the figure show that the scatter pattern has a single lobe until a retardation of 36°–54°, when the double peak develops.

10.5.2 Haze from Herring-Bone Stacks

Thickness uniformity of birefringent coatings is improved by depositing, in turn, from opposite sides of the substrate normal. Figure 10.12 shows the effect of multiple-angle depositions on the haze. As in the previous section, the electron beam gun was powered down at specific values of birefringence, and the sample was rotated one turn. In addition, in the current example, the deposition angle was alternated between +65° and −65°.

The profiles in the figure show a ratcheting effect for the scatter lobes - the normal scatter peak is present at $\xi = 0°$ after deposition of the first sublayer, there is a corresponding peak at $\xi = 180°$ after deposition of the second sublayer,

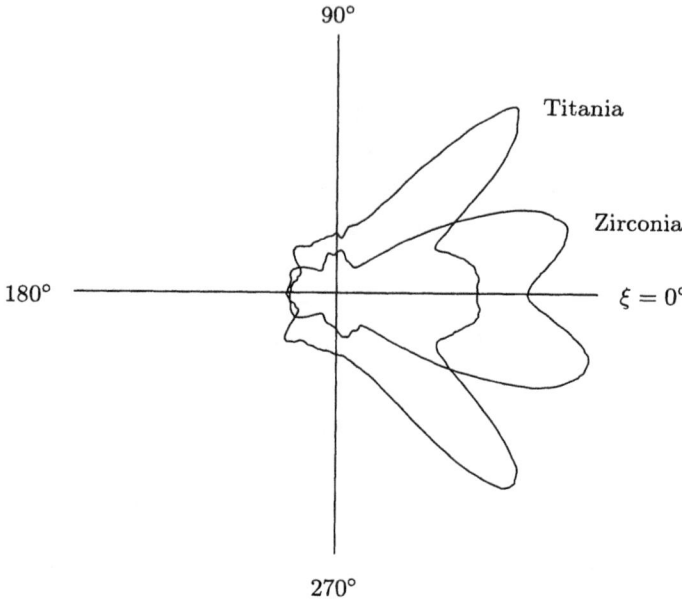

Fig. 10.9. Anisotropic scatter flux trapped in the substrate for titanium oxide and zirconium oxide films illuminated at normal incidence. (Adapted from I.J. Hodgkinson and Q.H. Wu, *Optical Interference Coatings, Proceedings of the Society of Photo-Optical Instrumentation Engineers* 2253, 882, 1994. Copyright © 1994 SPIE. Reprinted with permission.)

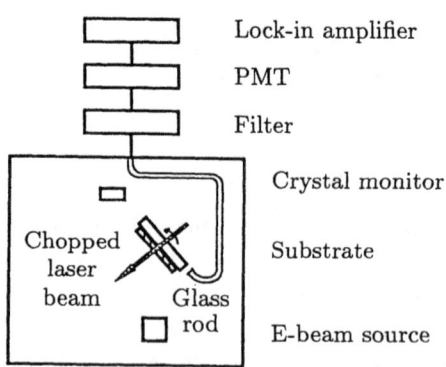

Fig. 10.10. Apparatus for *in situ* measurements of anisotropic scatter. (Adapted from I.J. Hodgkinson and Q.H. Wu, *Optical Interference Coatings, Proceedings of the Society of Photo-Optical Instrumentation Engineers* 2253, 882, 1994. Copyright © 1994 SPIE. Reprinted with permission.)

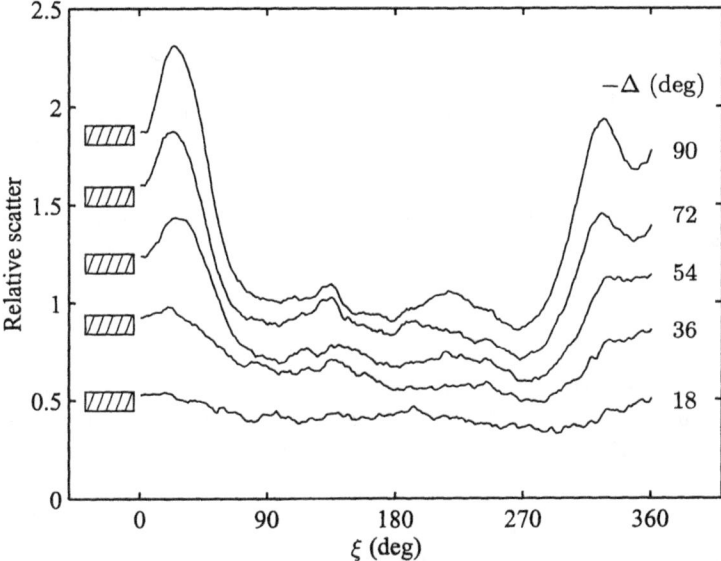

Fig. 10.11. Growth of anisotropic scatter recorded during the deposition of zirconia. The labels show the accumulating columnar structure and phase retardation. (Adapted from I.J. Hodgkinson and Q.H. Wu, *Optical Interference Coatings, Proceedings of the Society of Photo-Optical Instrumentation Engineers* 2253, 882, 1994. Copyright © 1994 SPIE. Reprinted with permission.)

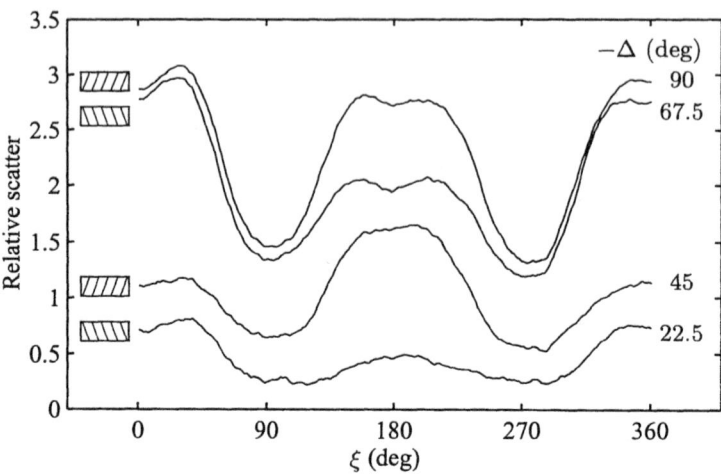

Fig. 10.12.   Anisotropic scatter from a zirconia herring-bone stack.   The labels show the accumulating columnar structure and phase retardation. (Adapted from I.J. Hodgkinson and Q.H. Wu, *Optical Interference Coatings, Proceedings of the Society of Photo-Optical Instrumentation Engineers* 2253, 882, 1994. Copyright © 1994 SPIE. Reprinted with permission.)

the peak at $\xi = 0°$ is enhanced by the third sublayer, etc. The main conclusions reached from this type of experiment are: (i) peak angular scatter is reduced by herring-bone deposition, but (ii) the associated integrated scatter is nearly unchanged.

10.6   Simple Theory of Scatter

In a simple theory of anisotropic scatter from tilted columnar films, the key ingredient is alignment, or correlation of scattering centres on the columns. The scatter pattern is considered to be the diffraction pattern of a column. As shown schematically in Fig. 10.13, the scattered rays are in phase for directions that lie on the surface of a cone. This defines the zero-order interference fringe in the film, but the angles are such that some rays are trapped in the film or in the substrate by total internal reflection.

The fate of a particular ray can be determined by taking a section of the cone parallel to the plane of the film. Such a section is an ellipse, as shown in Fig. 10.14. Two further sections are drawn in Fig. 10.14, a circle corresponding to rays that are travelling in the film at the film/air critical angle, and a second circle representing rays travelling in the film at the film/substrate critical angle.

Intersections of the ellipse with the circles in Fig. 10.14 divide the ellipse into

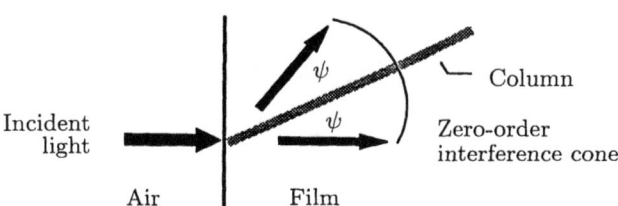

Fig. 10.13. Maximum irradiance of light scattered into the far field occurs on a cone that has the column as axis. (Adapted from I.J. Hodgkinson, P.I. Bowmar and Q.H. Wu, *Applied Optics* 34, 163, 1995. Copyright © 1995 Optical Society of America. Reprinted with permission.)

the segments marked I, II, and III. Segment I represents light that is transmitted into the air and forms the characteristic arcs in transmitted scatter distributions. The pair of segments marked II represent light that is trapped in the substrate and causes the characteristic lobes in substrate flux distributions, as well as the bright patterns observed superposed on the film. Region III represents light trapped in the film.

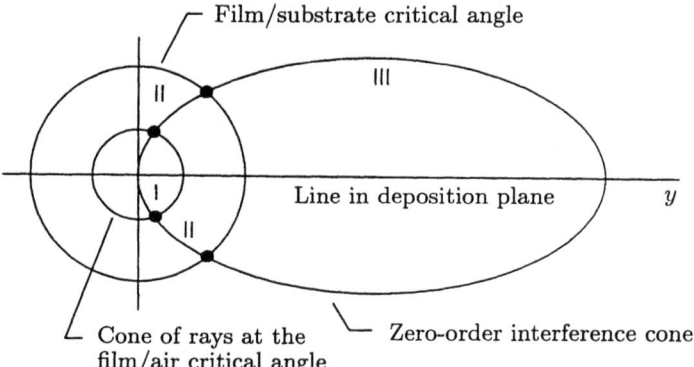

Fig. 10.14. Elliptical section of the zero-order interference cone from a line of scatterers, projected on to a plane parallel to the film. (Adapted from I.J. Hodgkinson, P.I. Bowmar and Q.H. Wu, *Applied Optics* 34, 163, 1995. Copyright © 1995 Optical Society of America. Reprinted with permission.)

Chapter 11

Fluid Transport

The tilted columnar microstructure that endows an evaporated film with unique polarizing properties for light at normal incidence also has the potential to be its Achilles heel. Moisture from the atmosphere is able to enter into the voids between loosely packed columns, and this has the effect of decreasing the difference in refractive index between the crystallites and the voids. Form birefringence depends on this difference in index, and decreases as moisture penetrates into the coating, by an amount such as 30%.

Moisture also has a significant effect on films deposited at normal incidence, provided they have an open microstructure. Indeed, most research on the effects of fluid penetration into optical coatings has related to narrowband interference filters deposited with the vapour incident normally on to the substrate. In such a coating the layers are isotropic for light at normal incidence, and the main concern is an increase in the wavelength of peak transmission of the filter that accompanies moisture penetration.

Problems associated with porosity in isotropic coatings can be eliminated by ion-assisted deposition (IAD). However, IAD decreases birefringence in an oblique film at the same time as it increases packing density. Current techniques require birefringent films to be deposited obliquely without IAD, and then sealed with an isotropic IAD layer.

In this chapter we discuss

- the transport of fluids and the formation of "fluid patches" of characteristic shape in both isotropic and anisotropic dielectric optical coatings

- the associated effect of scatter (or diffraction) from random arrays of fluid patches of circular or elliptical shape

- the effect of fluids on form birefringence.

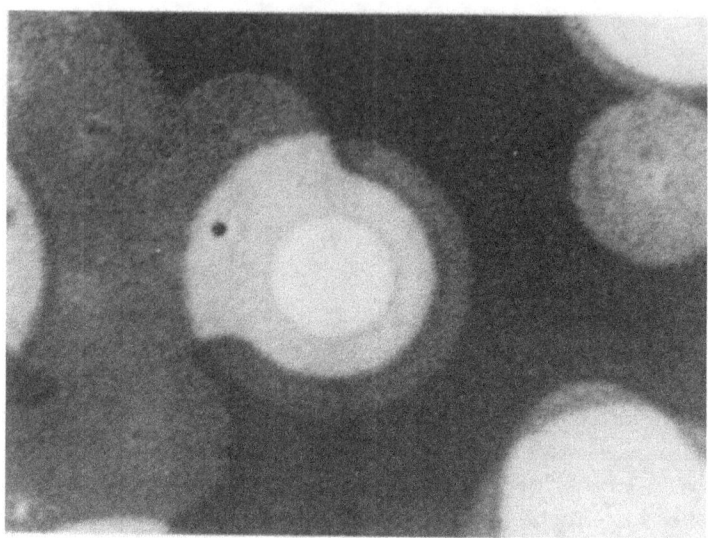

Fig. 11.1.   Circular fluid patches in a 546-nm narrowband interference filter.
(Adapted from I.J. Hodgkinson, M.R. Jacobson, C.C. Lee, H.A. Macleod, R.H. Potoff,
M. Sikkens and R. Sprague, *Thin Solid Films* 138, 289, 1986. Copyright © Elsevier
Sequoia. Reprinted with permission.)

11.1   Fluid Patches

Two different modes of penetration by a fluid can be distinguished. For a single
layer, or a small number of layers, the fluid may enter more or less evenly over
the surface of the film, leading to the entire film having a higher index. A
stack of films may well contain some layers with low porosity, preventing direct
penetration towards the substrate. However, such a coating will usually contain
pinhole shaped pores, that tend to develop on dust particles and the like during
deposition. A fluid is able to enter through such a pore, and spread laterally into
the more loosely packed layers. Sometimes the lateral penetration is remarkably
even in all directions, and "fluid patches" with well defined edges and a circular
shape are formed.[44,45,46] In the example shown in Fig. 11.1, the circular patches
were "locked" in a commercial 546-nm filter when the filter was sealed by the
manufacturer some 35 years ago.

11.1.1 Recording Fluid Patches

A method described by H.A. Macleod and D.A. Richmond[45] was used to produce the map of fluid penetration patches shown in the upper part of Fig. 11.1. In this method the narrowband filter is illuminated normally by a quasi-monochromatic beam of light from a grating monochromator and a low-power microscope is focused on the coating. The contrast of the patches relative to the background illumination can be varied by tuning the monochromator; highest sensitivity is achieved by tuning to one side of the transmittance peak of the filter. A second (and complementary) direct observation technique for viewing the patches, involving the use of fringes of equal chromatic order[47] (FECO) in reflection, displays the peak wavelength of the filter as a function of distance along a line passing through the patches. In some cases the displacement of the fringe relates directly to the increase in refractive index caused by the fluid. An example of the FECO method, for the 546-nm filter, is given in Fig. 11.2.

11.1.2 MDM Narrowband Filters

The use of a metal/dielectric/metal (MDM) narrowband filter allows fluid transport in a single dielectric material to be studied. Fluid penetration fronts are well defined for some combinations of metal and dielectric. Silver is the most suitable metal as it condenses with high packing density, it is stable in a humid atmosphere and its low optical absorption ensures interference fringes with good contrast.

When the spacer layer of a MDM filter is deposited obliquely, the fluid patches that develop have an elliptical shape. Elliptical patches recorded for a silicon oxide film are shown in Fig. 11.3. The upper part of the figure shows the distinct elliptical edges of the patches. The lower part, recorded after the incident light was tuned to a longer wavelength, shows that the quantity of adsorbed fluid (moisture) varies across the patches, and that profiles of constant optical thickness within a patch are also elliptical.

The minor axes of the ellipses are in the direction of the y-axis, in the deposition plane, and the major axes are in the direction of the z-axis, perpendicular to the deposition plane. Denoting the propagation speed of the fluid front in the deposition plane as v_p and the speed perpendicular to the deposition plane as v_s allows the *fluid transport anisotropy* to be defined as

$$A_f = \frac{v_s - v_p}{v_s + v_p}. \tag{11.1}$$

The microscopic parameter A_f can be determined by measuring the axes of the elliptical patches.

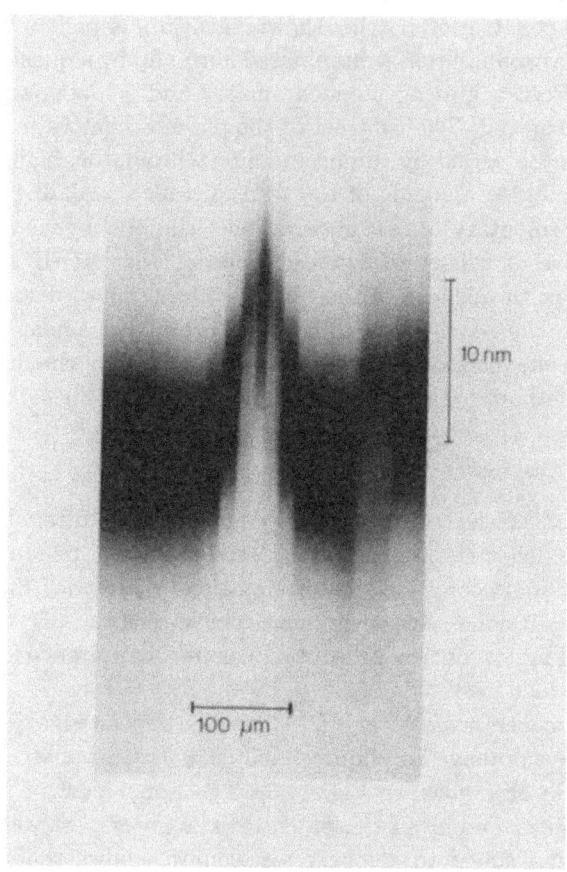

Fig. 11.2. Displacement of the wavelength of peak transmittance along a line through a circular fluid patch. (Adapted from I.J. Hodgkinson, M.R. Jacobson, C.C. Lee, H.A. Macleod, R.H. Potoff, M. Sikkens and R. Sprague, *Thin Solid Films* 138, 289, 1986. Copyright © Elsevier Sequoia. Reprinted with permission.)

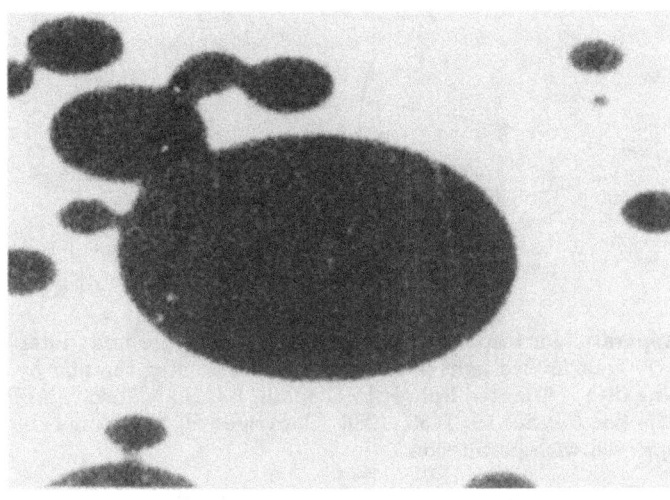

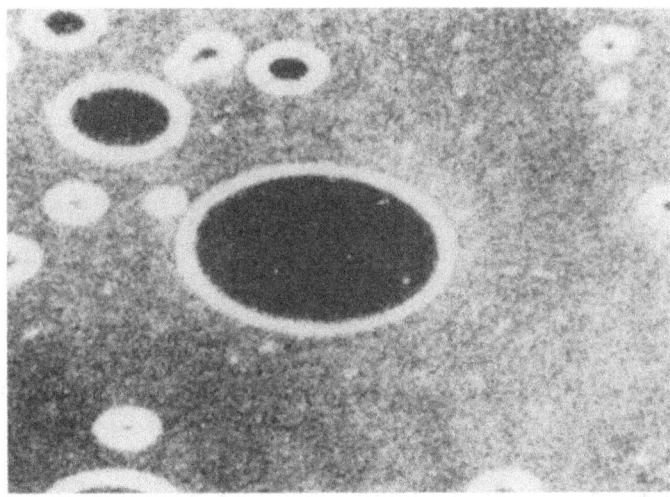

Fig. 11.3. Elliptical water penetration patches in silicon oxide deposited obliquely (upper), and profiles of constant optical thickness within the patches obtained by tuning the wavelength of the incident light to a larger value (lower). (Adapted from I.J. Hodgkinson, M.R. Jacobson, C.C. Lee, H.A. Macleod, R.H. Potoff, M. Sikkens and R. Sprague, *Thin Solid Films* 138, 289, 1986. Copyright © Elsevier Sequoia. Reprinted with permission.)

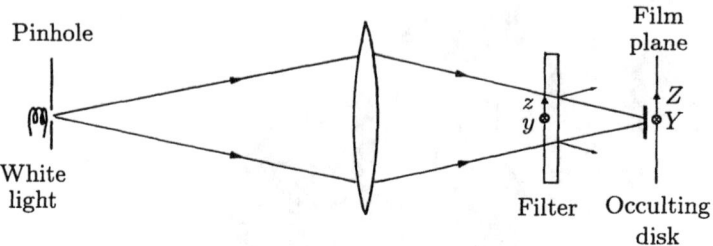

Fig. 11.4. Apparatus for recording scatter from fluid patches in an interference filter. The specularly transmitted light is prevented from reaching the film by a small, circular occulting disk. (Adapted from R.F. Gledhill, I.J. Hodgkinson and P.W. Wilson, *Journal of Applied Physics* 59, 1453, 1986. Copyright © 1986 American Institute of Physics. Reprinted with permission.)

11.2   Scatter from Fluid Patches

Early investigations of water transport in thin films followed observations of speckle or granularity in the light transmitted by interference filters.[48,49] We understand now that the speckle is due to diffraction and scatter from moisture patches, which have different transmittances and reflectances relative to dry areas of the coating.

The apparatus illustrated in Fig. 11.4 can be used to record scatter patterns from fluid patches; the specularly transmitted light is prevented from reaching the film by a small, circular occulting disc.

Experiments confirm that the spatial distribution of the forward scattered light has symmetry determined by the shape of the fluid patches. Thus, the diffraction pattern from circular patches recorded for the 546-nm filter has circular symmetry, as shown in Fig. 11.5.

11.2.1   Scatter Anisotropy

In the case of elliptical patches the diffraction pattern has the same elliptical symmetry as the patches, but with major and minor axes interchanged. This is shown, for an MDM filter with an oblique magnesium fluoride spacer, in Fig. 11.6.

The *scatter anisotropy* can be measured using the expression

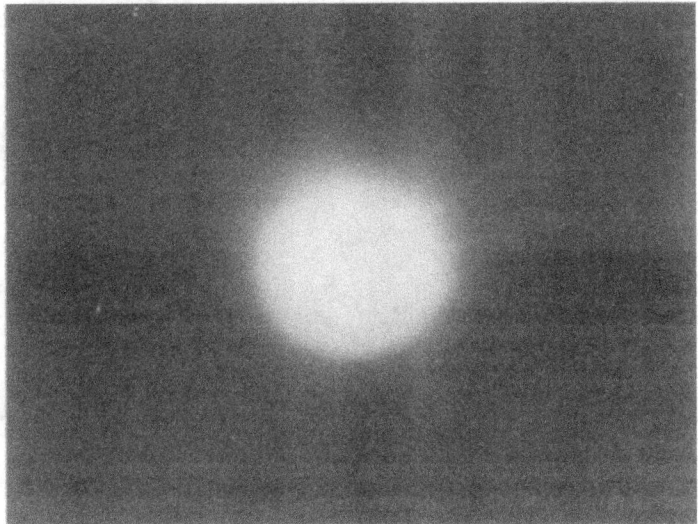

Fig. 11.5. Pattern of light scattered by the circular fluid patches in the 546-nm interference filter. (Adapted from J.R. Gee, I.J. Hodgkinson and P.W. Wilson, *Applied Optics* 25, 2688, 1986. Copyright © 1986 Optical Society of America. Reprinted with permission.)

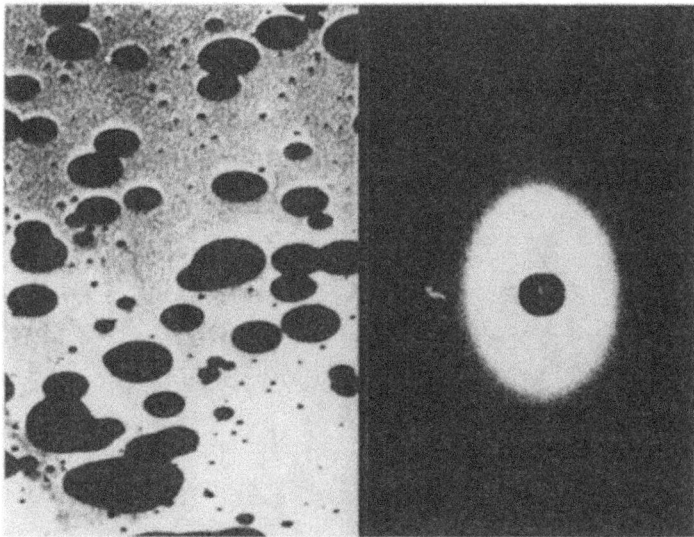

Fig. 11.6. Moisture penetration patches in a magnesium fluoride spacer layer deposited at 45° (left), and spatial distribution of light scattered by the patches (right). (Adapted from R.F. Gledhill, I.J. Hodgkinson and P.W. Wilson, *Journal of Applied Physics* 59, 1453, 1986. Copyright © 1986 American Institute of Physics. Reprinted with permission.)

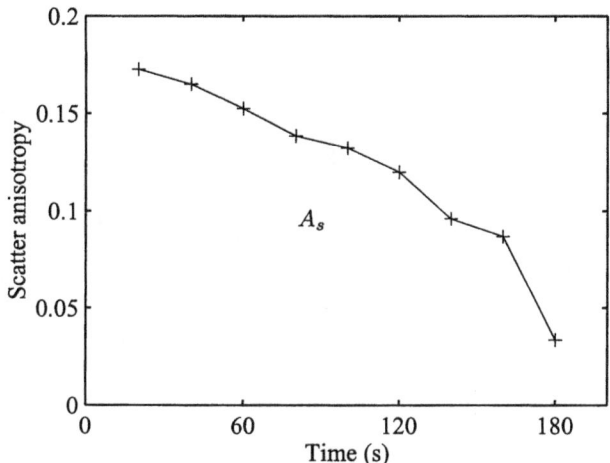

Fig. 11.7. The magnitude of scatter anisotropy A_s decreases with time as moisture patches grow and merge together. (Adapted from R.F. Gledhill, I.J. Hodgkinson and P.W. Wilson, *Journal of Applied Physics* 59, 1453, 1986. Copyright © 1986 American Institute of Physics. Reprinted with permission.)

$$A_s = \frac{k_s - k_p}{k_s + k_p}, \qquad (11.2)$$

where $k_s = k_Z = k \sin \theta_s$ and $k_p = k_Y = k \sin \theta_p$ relate to the angles of scatter θ_s perpendicular to the deposition plane and θ_p in the plane of deposition, measured to a profile of constant irradiance.

In an ideal case, in which the patches are perfect ellipses and do not overlap, the scatter anisotropy is simply related to the fluid transport anisotropy, $A_s = -A_f$. Measurements made on the scatter pattern and the ellipses in Fig. 11.6 yield the values $A_s = -0.18 \pm 0.01$ and $A_f = 0.21 \pm 0.02$.

As the fluid patches grow and merge together the elliptical shape is gradually lost and the scatter anisotropy decreases (Fig. 11.7). Finally, when the entire filter has adsorbed the fluid, and gradients of optical thickness have disappeared, the amount of scatter due to the fluid tends to zero.

11.2.2 Theory of Scatter

An estimate of the magnitude of scatter from fluid patches in optical coatings can be made for situations in which the patches are somewhat larger than the pore or defect at the point of entry, and the patches have well-defined edges

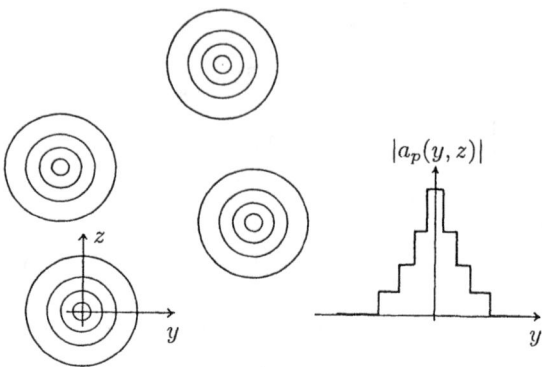

Fig. 11.8.   Random array of moisture patches with identical aperture functions.
(Adapted from J.R. Gee, I.J. Hodgkinson and P.W. Wilson, *Applied Optics* 25, 2688,
1986. Copyright © 1986 Optical Society of America. Reprinted with permission.)

and shape. In this case the scatter pattern at a distance from the coating is
essentially a Fraunhofer (far-field) diffraction pattern.

Suppose that the fluid has formed a set of identical patches distributed
randomly over the surface of the optical coating. A typical number per unit area
is $\eta = 10^8 \, \mathrm{m}^{-2}$. Each patch is characterized by an aperture function $a_p(y, z)$, a
perturbation on the aperture function of the filter. We assume that the incident
field $E_0(y, z)$ is coherent over the area of a patch, but incoherent over several
patches. Then, away from the specularly transmitted or reflected beam, the
diffraction pattern from the array is just the diffraction pattern of a single
aperture, and the total power diffracted incoherently is

$$P = \eta \int \int |a_p(y, z)|^2 dy dz \times \text{incident power.} \qquad (11.3)$$

In the case of power scattered into the reflected beam, $|a_p(y, z)|$ is the mod-
ulus of the change in reflectivity caused by the fluid. Suppose that a thin slab
of material in a coating has phase thickness $d\phi$ before penetration by the fluid
(Fig. 11.9), then we can define a sensitivity

$$s_r = dr/\Delta n d\phi \qquad (11.4)$$

where Δn is the related change in refractive index. Similarly, we define

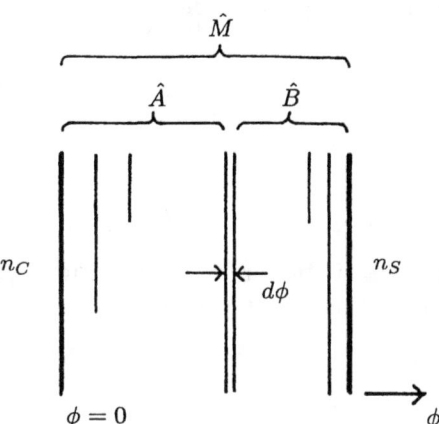

Fig. 11.9. Thin slice of material in an optical coating. (Adapted from J.R. Gee, I.J. Hodgkinson and P.W. Wilson, *Applied Optics* 25, 2688, 1986. Copyright © 1986 Optical Society of America. Reprinted with permission.)

$$s_t = dt/\Delta n d\phi \tag{11.5}$$

as the sensitivity for the transmitted light.

The sensitivities s_r and s_t can be expressed in terms of the elements of the unperturbed 2×2 characteristic matrices $\hat{A}$ and $\hat{B}$ (Fig. 11.9),

$$s_r = it^2(B_{11} + n_S B_{12})^2/n_C \tag{11.6}$$

$$s_t = it^2(A_{22} + n_C A_{12})(B_{11} + n_S B_{12})/n_C, \tag{11.7}$$

or in terms of the ratio of the strength E of the total electric field in the coating to the incident travelling-wave field in the cover,

$$s_r = i(E/E_C^+)^2/n_C \tag{11.8}$$

$$s_t = it(A_{22} + n_C A_{12})(E/E_C^+)/n_C. \tag{11.9}$$

The application of these expressions to three representative optical coatings is discussed next.

11.2.3   General AR Coating

In the case of a general multilayer antireflection coating the expression for s_t simplifies to

$$s_t = it|E/E_C^+|^2/n_C \tag{11.10}$$

and

$$|s_t| = (n_C/n_S)^{1/2}|s_r|. \tag{11.11}$$

As an example, the sensitivities are plotted in Fig. 11.10 for the optimized four-layer quarter-wave coating aLI_1HI_2g with $n_C = 1$, $n_L = 1.35$, $n_{I_1} = 2.15$, $n_{I_2} = 1.78$, $n_H = 2.35$, $n_S = 1.5$.

11.2.4   High Reflectance Coating

The high reflectance coating $a(HL)^N Hg$ has sensitivity

$$s_r = -i4(n_C/n_H^2)(n_L/n_H)^{2(p-1)} \sin^2 \phi, \tag{11.12}$$

at the design wavelength and integration leads to the expressions:-

$$\text{one layer in period p: } \Delta r = -i\pi(n_C/n_H^2)(n_L/n_H)^{2(p-1)}\Delta n,$$
$$\text{all } H \text{ or all } L \text{ layers: } \Delta r = -i\pi n_C(1 - n_L^{2N}/n_H^{2N})(n_H^2 - n_L^2)\Delta n. \tag{11.13}$$

An illustrative example, with $n_C = 1$, $n_L = 1.35$, $n_H = 2.35$, $n_S = 1.52$ and $N = 5$, yields $|\Delta r|^2 = 7 \times 10^{-5}$ for $\Delta n_L = 0.01$ or $\Delta n_H = 0.01$. Hence the scatter from fluid patches in a quarter-wave reflecting coating is small at the design wavelength. However, a perturbation Δn has a much larger effect near the steep sides of the reflection band, as shown in Fig. 11.11. In general, an increase in either n_L or n_H tends to displace the reflection band to a longer wavelength and an increase in the ratio n_H/n_L tends to broaden the band. For this reason, an increase in n_L causes a prominent peak near the low wavelength edge of the band and, similarly, an increase in n_H has most effect near the high wavelength edge.

11.2.5   Narrowband Interference Filter

The narrowband filter $a(HL)^N I^{2m}(LH)^N g$ that is illustrated in Fig. 11.9 has transmission and reflection sensitivities of equal magnitude in the first-order approximation:-

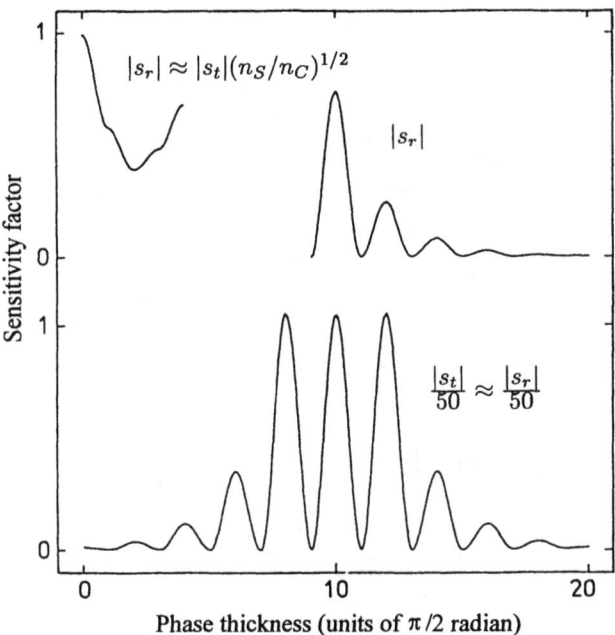

Fig. 11.10. Sensitivity factors for scatter from thin slices with increased refractive indices in an antireflection coating (upper left), in a reflecting coating (upper right), and in a narrowband filter (lower). (Adapted from J.R. Gee, I.J. Hodgkinson and P.W. Wilson, *Applied Optics* 25, 2688, 1986. Copyright © 1986 Optical Society of America. Reprinted with permission.)

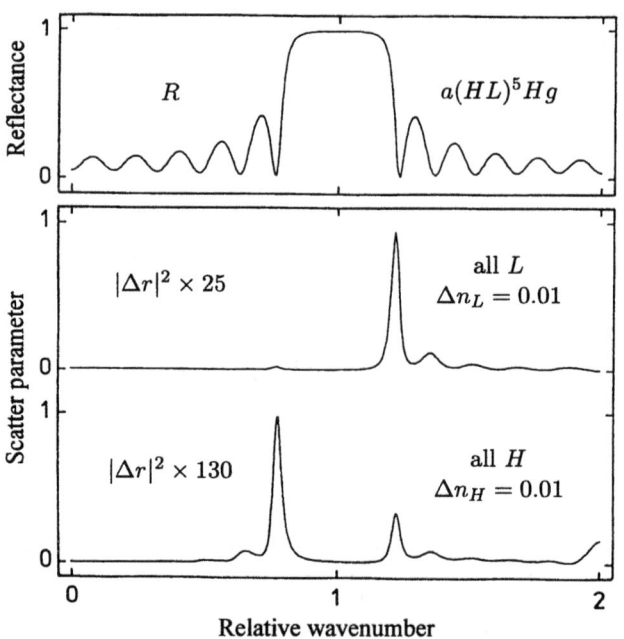

Fig. 11.11. Calculated values of the reflection scatter parameter for a multilayer reflector. (Adapted from J.R. Gee, I.J. Hodgkinson and P.W. Wilson, *Applied Optics* 25, 2688, 1986. Copyright © 1986 Optical Society of America. Reprinted with permission.)

$$s_r = i4n_C(n_C + n_S)^{-2}(n_H/n_L)^{2(N-p)}\cos^2\phi,$$
$$s_t = (-1)^m s_r. \tag{11.14}$$

For homogeneously filled layers the integrated or summed Δr's appropriate to the design wavelength are:-

all I layers: $\Delta r = i2m\pi n_C(n_C + n_S)^{-2}(n_H/n_L)^{2N}\Delta n,$

all H layers: $\Delta r = i2\pi n_C(n_C + n_S)^{-2}(n_H^{2N}/n_L^{2N} - 1)(n_H^2/n_L^2 - 1)^{-1}\Delta n,$

all L layers: $\Delta r = i2\pi n_C(n_C + n_S)^{-2}(1 - n_L^2/n_H^2)^{-1}\Delta n. \tag{11.15}$

The effect of tuning is shown in Fig. 11.12 for the coating $a(HL)^4 4H(LH)^4 g$ with $n_C = 1$, $n_L = 1.35$, $n_H = 2.35$, $n_S = 1.52$. For small values of Δn_L the peak in $|\Delta t|^2$ is similar in shape and width to the filter transmission peak and displaced slightly towards the origin of g. Larger values of Δn_L cause a double peak in the scatter.

The arrangement shown in Fig. 11.13 is suitable for measuring the wavelength dependence of scatter from moisture patches in optical coatings. The results obtained for the commercial 546 nm filter are shown in Fig. 11.14.

11.3   Influence on Birefringence

11.3.1   Change of Birefringence in Fluid Patches

A schematic overview of fluid transport in isotropic and anisotropic layers is given in Fig. 11.15. By adding an analyser (linear polarizer) to the FECO apparatus, small perturbations to the polarizing properties of the coating can be investigated in the vicinity of fluid patches. In the upper part of the figure, a fluid patch is shown displacing an interference fringe towards a longer wavelength and, of course, in this case the displacement is independent of polarization. The lower part of Fig. 11.15 shows that the value of

$$\Delta\lambda = \lambda_s - \lambda_p \tag{11.16}$$

is smaller in a fluid patch, relative to a dry area. Now, we can write

$$A_n = \frac{n_s - n_p}{n_s + n_p} \approx \frac{\lambda_s - \lambda_p}{\lambda_s + \lambda_p}, \tag{11.17}$$

where A_n is the normal incidence refractive anisotropy, and so the smaller value of $\Delta\lambda$ in Fig. 11.15 implies a smaller birefringence.

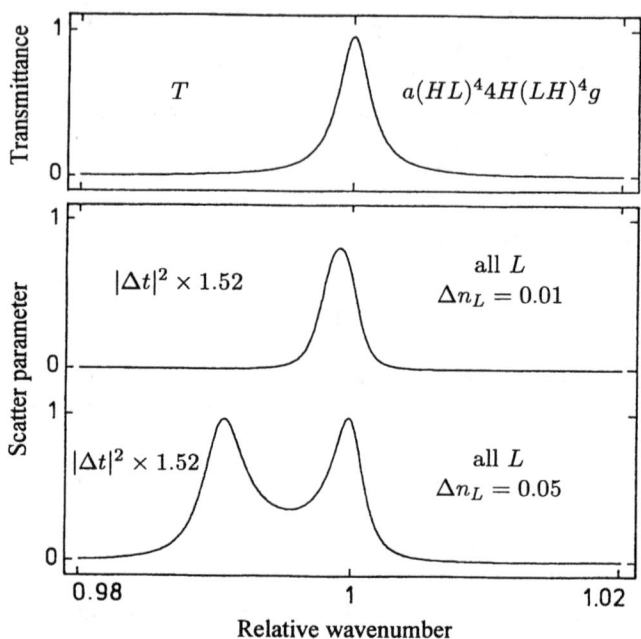

Fig. 11.12. Calculated values of the transmission scatter parameter for a narrowband interference filter. (Adapted from J.R. Gee, I.J. Hodgkinson and P.W. Wilson, *Applied Optics* 25, 2688, 1986. Copyright © 1986 Optical Society of America. Reprinted with permission.)

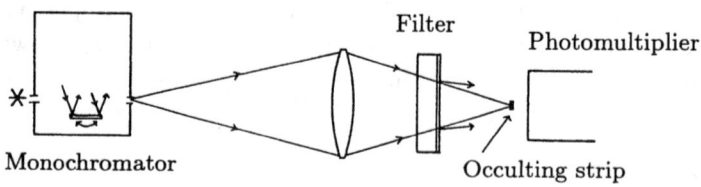

Fig. 11.13. Apparatus for measuring scatter from interference filters during the growth of moisture patches. The monochromator is scanned repeatedly and peak values of scatter are recorded. (Adapted from J.R. Gee, I.J. Hodgkinson and P.W. Wilson, *Applied Optics* 25, 2688, 1986. Copyright © 1986 Optical Society of America. Reprinted with permission.)

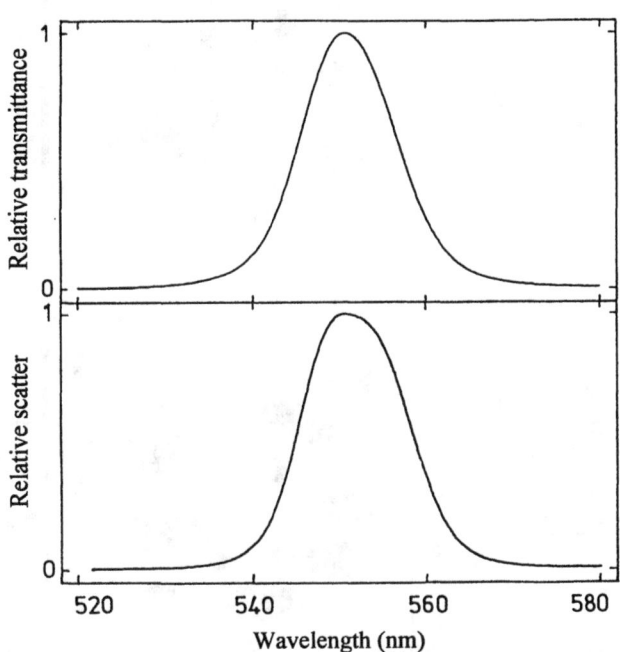

Fig. 11.14. Measured values of transmittance and scatter from fluid patches locked in a commercial 546-nm narrowband filter. (Adapted from J.R. Gee, I.J. Hodgkinson and P.W. Wilson, *Applied Optics* 25, 2688, 1986. Copyright © 1986 Optical Society of America. Reprinted with permission.)

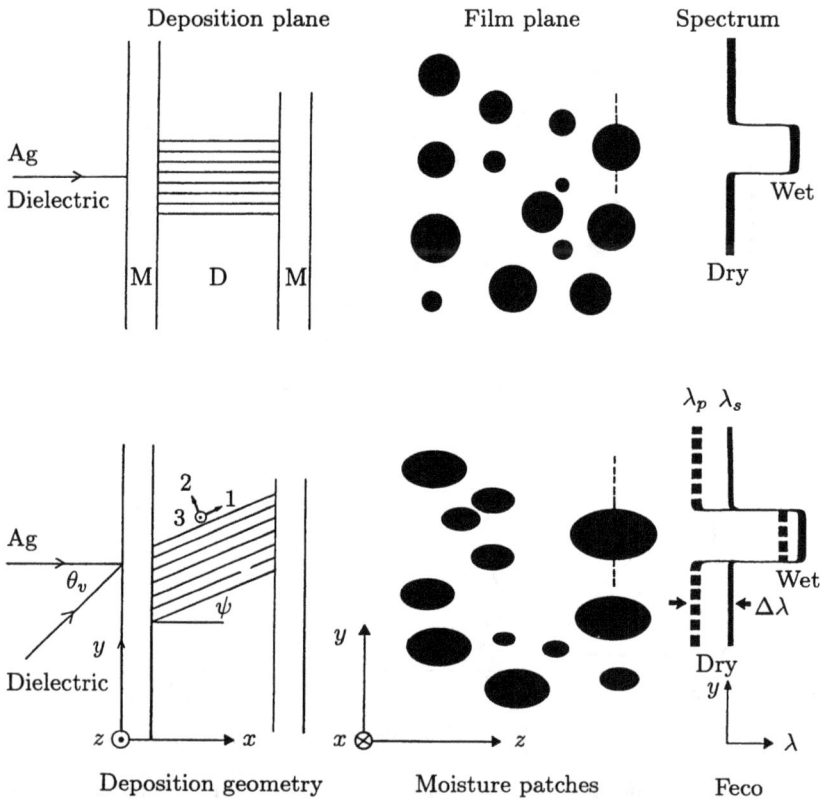

Fig. 11.15. Appearance of fluid patches and fringes of equal chromatic order for a MDM filter deposited at normal incidence (upper), and deposited obliquely (lower). (Adapted from J.R. Gee, I.J. Hodgkinson and H.A. Macleod, *Applied Optics* 24, 3188, 1985. Copyright © 1985 Optical Society of America. Reprinted with permission.)

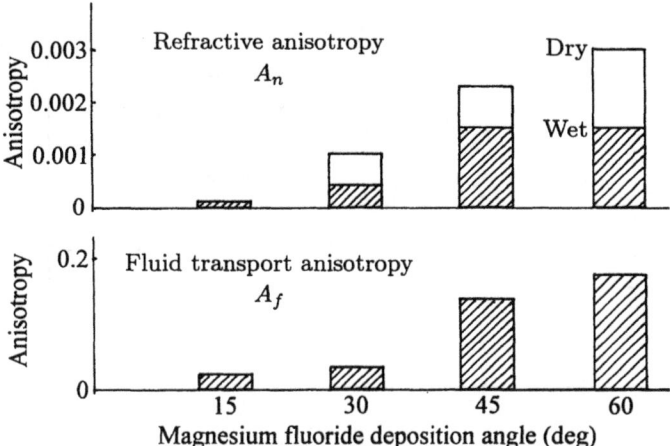

Fig. 11.16. (Upper) Normal incidence refractive anisotropy A_n measured in wet regions (microscopic moisture penetration patches) and in surrounding dry areas of silver-magnesium fluoride-silver interference filters. The value of A_n is reduced by moisture that penetrates into the magnesium fluoride spacer layers. (Lower) Fluid transport anisotropy A_f recorded for the same filters. (Adapted from I.J. Hodgkinson, F. Horowitz, H.A. Macleod, M. Sikkens and J.J. Wharton, *Applied Optics* 24, 1568, 1985. Copyright © 1985 Optical Society of America. Reprinted with permission.)

Figure 11.16 shows experimental results obtained by applying the modified FECO method to silver-magnesium fluoride-silver interference filters. The upper part of the figure shows that A_n is smaller in a fluid patch relative to A_n in the material surrounding the patch, and that the difference increases with increasing deposition angle. The lower part of the figure shows the fluid transport anisotropy, A_f, determined for the same films from measurements of the major and minor diameters of the elliptical fluid patches.

11.3.2 Principal Refractive Indices

The Bragg-Pippard equations (Sect. 8.3.1) can be used to model the influence that an absorbed fluid has on the birefringent properties of tilted-columnar media. Using known values of the principal refractive indices n_1, n_2, n_3 and the void index n_v for the dry material, the *BTF Toolbox* function **bpcdi** (which implements Eqs.(8.24)–(8.26)) can be used to invert the BP crystallite-defined equations. Thus **bpcdi**(n_1, n_2, n_3, n_v) yields the crystallite index n_c, the depolarization factors L_1, L_2, L_3, and the packing density p for the dry material. Now suppose that the voids of the material are completely filled with a fluid of

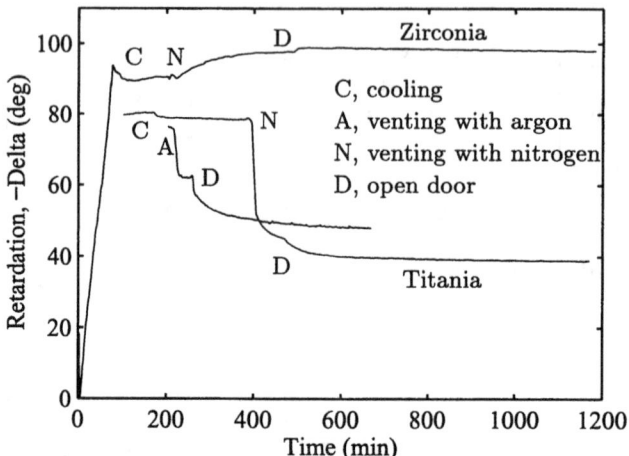

Fig. 11.17. Changes to the phase retardation of thin film wave plates recorded during cooling and venting. (Adapted from I.J. Hodgkinson and Q.H. Wu, *Optical Interference Coatings, Proceedings of the Society of Photo-Optical Instrumentation Engineers* 2253, 882, 1994. Copyright © 1994 SPIE. Reprinted with permission.)

index n_f. Then **bpcd**$(n_c, n_f, L_1, L_2, L_3, p)$, gives the principal refractive indices of the wet material.

Use of the method described above shows that the principal indices are larger for the wet material but the spread of the principal indices is smaller.

11.3.3 Cooling and Venting

The retardation of a thin film wave plate can be monitored by perpendicular incidence ellipsometry (PIE) in transmission during deposition, so that a value such as 90° can be achieved. However, while the wave plate is cooling and while air or some other gas is being admitted to the coating chamber, substantial changes in retardation can occur.

The task of monitoring changes in retardation, while the temperature of the film and some interior parts of the coating chamber change by nearly 300°C, and the pressure changes by one atmosphere, is quite challenging. Apart from mechanical stability of the ellipsometer, the most important requirement is temperature management of windows and mirrors to avoid spurious birefringent effects caused by temperature gradients. Typical results obtained for zirconia and titania during deposition, cooling and venting are plotted in Fig. 11.17.

Chapter 12

Metal Films

It is well known that *in situ* optical measurements can reveal morphological changes that occur during the deposition of thin metal films. For example, the knee feature that is observed in the transmittance versus mass thickness profile during the deposition of silver at normal incidence is due to a refractive index resonance that occurs as the film structure changes from a planar array of globules to a more or less continuous layer. Further, microstructural differences in deposited and sputter-etched gold films, deduced originally from observations of the knee feature and later confirmed by TEM studies, have been used to explain normal incidence optical anisotropies in these films. Thus, the basic morphology of a thin film (particularly a noble metal) may be probed by monitoring the reflectance or transmittance with light at normal incidence as the film is being deposited, and microstructural effects lateral to the deposition plane can be probed by measuring normal incidence optical anisotropies.

In this chapter, we

- discuss the direct experimental acquisition of reflectance and transmittance anisotropy versus thickness profiles for metal films

- consider the intrinsic anisotropic properties of crystallite-defined and void-defined metal composites modelled by the Bragg-Pippard equations

- develop a model to explain the characteristic anisotropic features observed for vacuum-deposited and ion-beam sputter-etched metal films

- include the concept of structural hysteresis loops to allow simulation of the development of anisotropies during the deposition and subsequent ion beam thinning of metal films.

12.1   Growth and Post-Deposition Sputter Etching

Our current understanding of the processes that determine nucleation and growth in thin metal films is due to observations made during and after deposition in an electron microscope.[50] These observations show that there are five characteristic stages in the growth sequence:-

- a distribution of small three-dimensional nuclei forms on the substrate

- the nuclei grow in size, while the number stays the same

- a further increase in size occurs, this time accompanied by a decrease in the number

- a connected network of deposit forms and develops into channels

- a nearly continuous film forms.

The small nuclei which form in the initial stages of deposition on to an isotropic substrate have the shape of an oblate spheroid with axis perpendicular to the substrate. During this period, a spheroid may collect both the atoms that condense on it directly and atoms that condense on bare regions of the substrate and reach it by diffusion.

The spheroids are solid and immobile, and the transport of mass that is necessary to increase the size and reduce the number of the spheroids occurs in a most interesting manner (see Fig. 12.1a). When two spheroids increase in size until they just touch, a neck forms at the point of contact and material is transported rapidly by the process of surface self-diffusion until a new single spheroidal particle is formed. This process, which is sometimes referred to as liquid-like coalescence, occurs rapidly and has the immediate effect of leaving the area around a spheroid, or the channels around an island, free of film material. Secondary nuclei form in the free areas and subsequently coalesce into the islands.

Similar characteristic stages occur during epitaxial growth. In this case dynamic growth through the addition of atoms and very small particles leads to the formation of islands with simple geometrical shapes that are dominated by the influence of the substrate. Liquid-like coalescence of large islands tends to spoil the geometrical shapes of the contributing islands, but subsequent dynamic growth acts to restore it.

When the deposition of a metal film is terminated, several aging processes begin. Coalescence tends to continue after deposition as the film seeks a minimum energy configuration in vacuum. The dynamic process of growth during

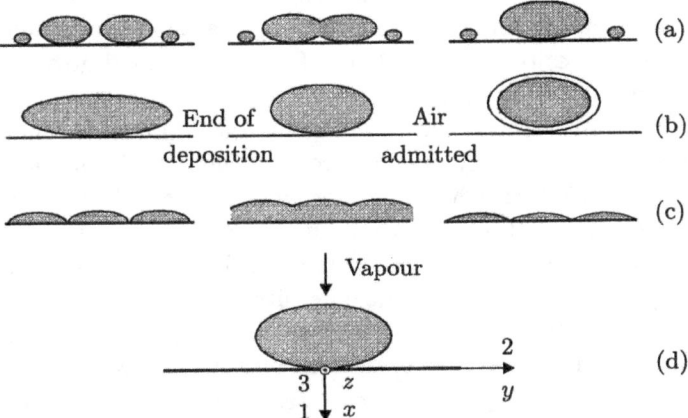

Fig. 12.1. Anisotropy in thin metal films is determined in part by the shape of the globular particles. (a) In the early stages of the growth of a vapour-deposited gold or silver film, high adatom mobility leads to maintenance of an oblate spheroidal shape and coalescence "cleans-up" the substrate. (b) The particles become less oblate as coalescence continues after deposition (left) and the refractive index is influenced by thin pellicle coatings which form after air admittance (right). (c) Possible growth and etching modes for a 100 eV, O_2^+ IAD gold film just before coalescence (left), just after coalescence (centre), and at the point where the film becomes discontinuous during etching (right). (d) Geometric axes. (Adapted from I.J. Hodgkinson and P.W. Wilson, *CRC Critical Reviews in Solid State and Materials Science* 15, 27, 1988. Copyright © 1988 CRC Press. Reprinted with permission.)

deposition favours a relatively flat spheroidal shape, but the equilibrium shape of an isolated globule of the metal is a sphere. The interaction between the substrate and the globule has a flattening effect on the shape, and in practice, an equilibrium value of about 2 is obtained for the mean axial ratio. Further aging processes occur when air is admitted, and these are consistent with the formation of dielectric pellicle coatings due to the adsorption of gas and moisture or chemical changes such as sulphuration (see Fig. 12.1b).

Both the optical properties and the adhesion of gold films can be improved by O_2^+ IAD.[51] Oxygen ion-bombardment seems to promote wetting of the substrate, and films that are subsequently sputtered by the ion beam remain continuous down to thicknesses of less than 1 nm. Possible idealized growth and etching modes are shown in Fig. 12.1c.

When a metal film (deposited without ion assistance) is thinned by argon ion etching the morphology does not retrace the deposition states. This can be seen in Fig. 12.2, in which the morphology of a gold film is recorded at mass thicknesses of $1, 2, 3, 4$ and 5 nm during deposition and at the same mass thicknesses during post-deposition etching. In this case etching thins the existing islands, and initially, the particle density on the substrate remains nearly constant. When a fully coalesced film is etched, the morphological changes are similar to those described in the previous paragraph. Early on in the process the fully coalesced state persists, then holes appear in the film, and eventually a low density of disconnected islands remains. Fig. 12.3 provides a schematic overview of the changes in shape and density that occur as globular, island, and fully coalesced gold films are deposited and then etched.

Liquid-like coalescence and secondary nucleation are retarded when reactive metals such as aluminium are deposited in the presence of even small amounts of oxygen. During such a deposition the supply of oxygen is uniform over the surface of the metal, but the supply of metal atoms is uneven due to shadowing. Oxidation retards the process of surface diffusion and a columnar microstructure develops.[52]

12.2 Direct Recording of Optical Anisotropies

The apparatus shown in Fig. 12.4 can be used for recording anisotropy in reflectance and transmittance during the deposition of metal films and while post-deposition processes take place. Note that the vapour is incident obliquely on to the substrate, but the light is incident normally. Values of the reflectances R_s, R_p and the transmittances T_s, T_p are recorded as the linear polarizer is moved from one position to the other. Here the subscripts s and p refer to the polarization direction of the light with respect to the plane of deposition. Absorptances

Deposit $\rightarrow$

100 nm

$\leftarrow$ Etch

Fig. 12.2. Changes in morphology during the growth of a 5-nm gold film (frames 1–5), and during post-deposition argon ion etching (frames 6–9). substrate: mica coated with carbon. Deposition parameters: angle = 2°, rate = 0.3 nm/s. Etching parameters: 350 eV, 2 A/m² Ar⁺ beam incident at 15°. Optical measurements: transmittance measured at wavelength 700 nm, angle of incidence 0°. (Adapted from I.J. Hodgkinson and J. Lemmon, *Journal of Applied Physics* 67, 6876, 1990. Copyright © 1990 American Institute of Physics. Reprinted with permission.)

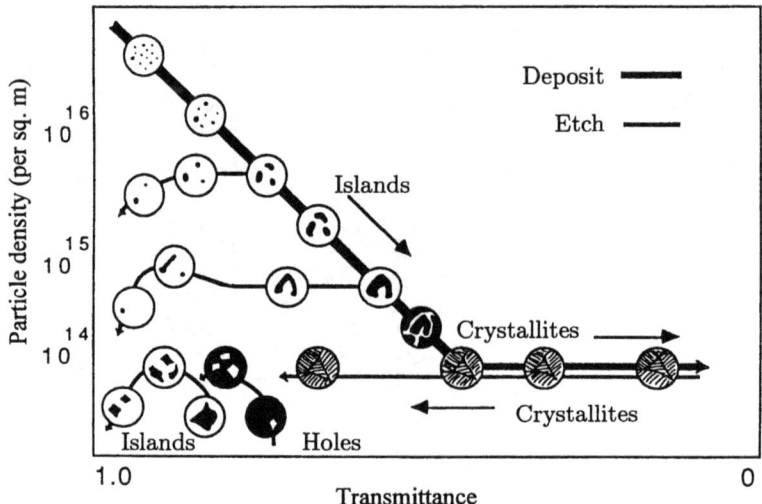

Fig. 12.3. Schematic overview of the changes in particle shape and density that take place as gold films are deposited and then etched. (Adapted from I.J. Hodgkinson and J. Lemmon, *Journal of Applied Physics* 67, 6876, 1990. Copyright © 1990 American Institute of Physics. Reprinted with permission.)

are calculated using the equations $A_s = 1 - R_s - T_s$ and $A_p = 1 - R_p - T_p$.

12.2.1 Silver and Gold

Figs. 12.5 and 12.6 show that both silver and gold deposited obliquely exhibit anisotropy in reflectance, transmittance and absorptance during the early stages of growth. However, when the thickness exceeds 30 nm or so, and the films are fully coalesced, the anisotropies are negligible. The double reversal of sign shown for gold in Fig. 12.6 is also observed for silver deposited at low rates. For both metals the monitor wavelength was 600 nm, and additional experiments show that the relative heights and the positions of the first and second peaks depend on wavelength.

12.2.2 Aluminium

Reflectance, transmittance and absorptance anisotropies recorded during the deposition of aluminium in the absence of oxygen are shown in Fig. 12.7. Modelling described in the following sections indicates that these anisotropy profiles are basically similar to the curves for silver and gold, and that the second peak in the transmittance profile is suppressed, as it tends to occur at the foot of the transmittance curve where coalescence is well advanced. Hence the anisotropies observed when metals are evaporated obliquely in the absence of oxygen may be described as *intrinsic* anisotropies.

The presence of residual oxygen in the evaporation chamber has negligible effect on noble metals such as silver and gold, but leads to the development of columnar microstructure and *residual* anisotropy in reactive metals such as aluminium. As an illustration, the transmittance anisotropy recorded as aluminium was deposited as above, but with a background pressure of 2×10^{-4} mbar O_2, is plotted in Fig. 12.8. This anisotropy profile is consistent with the establishment of a columnar structure at a thickness of about 4 nm.

Coatings that exhibit anisotropic reflection (and transmittance) are said to be *angularly-selective*. Potential applications for angularly-selective coatings include controlling the entry of light through windows into buildings and into automobiles.

The *reflection anisotropy* and the *transmission anisotropy* of an angularly-selective coating can be defined as

$$A_r = (R_s - R_p)/(R_s + R_p) \tag{12.1}$$

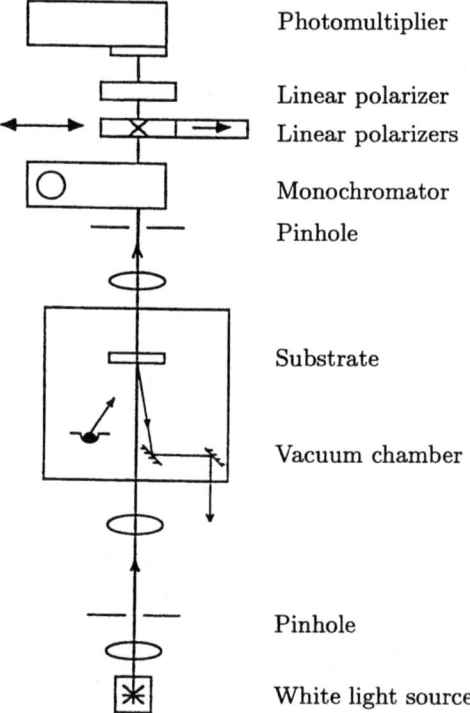

Photomultiplier

Linear polarizer

Linear polarizers

Monochromator

Pinhole

Substrate

Vacuum chamber

Pinhole

White light source

Fig. 12.4. Apparatus for monitoring optical anisotropy in transmittance during the deposition of thin films. An equivalent detection system (not shown) records anisotropy in reflection. (Adapted from I.J. Hodgkinson, *Applied Optics* 30, 1303, 1991. Copyright © 1991 Optical Society of America. Reprinted with permission.)

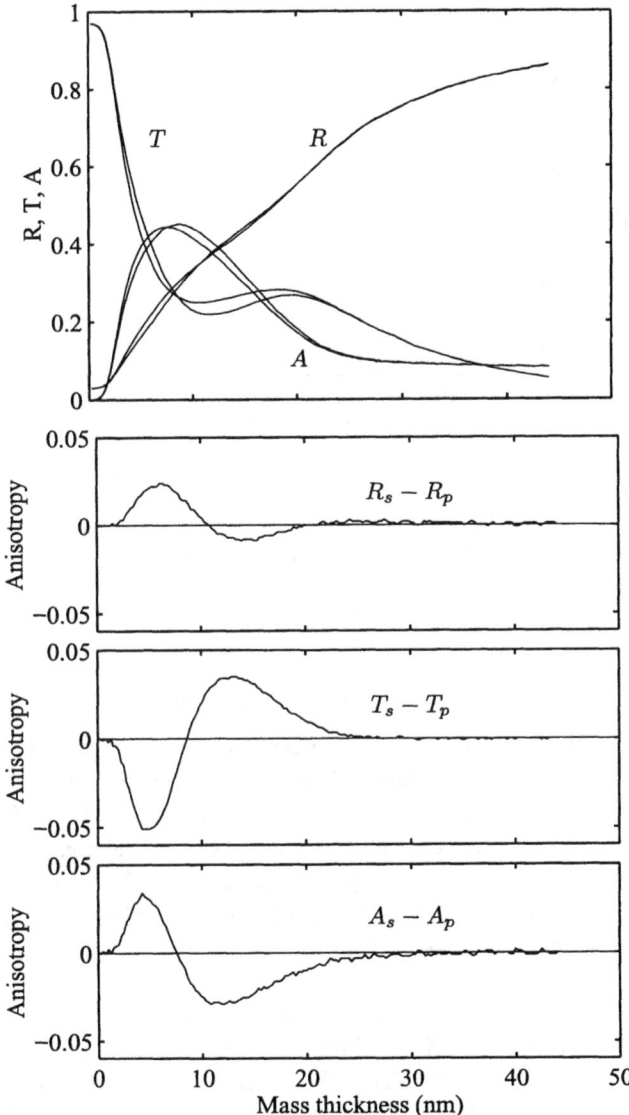

Fig. 12.5. (Upper) Reflectance, transmittance, and absorptance recorded with light of wavelength 600 nm during the deposition of silver at angle 45° and rate 0.5 nm/s. (Lower) Optical anisotropies measured for the silver film. (Adapted from I.J. Hodgkinson, *Applied Optics* 30, 1303, 1991. Copyright © 1991 Optical Society of America. Reprinted with permission.)

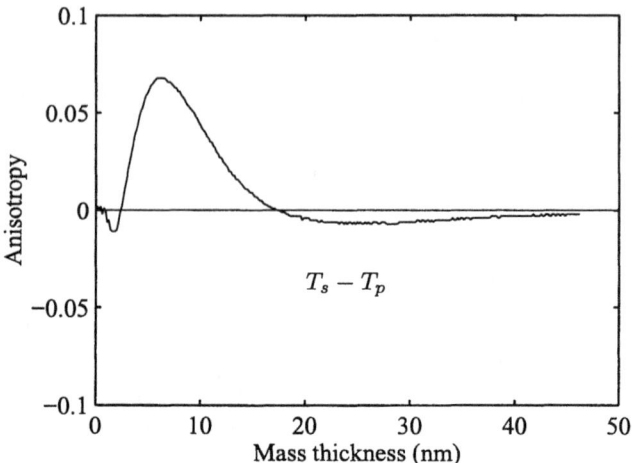

Fig. 12.6. Anisotropy profile $T_s - T_p$ v. mass thickness, measured during the deposition of gold, exhibits a positive peak and two negative peaks: angle of deposition, 45°; deposition rate, 0.3 nm/s; monitor wavelength, 600 nm. (Adapted from I.J. Hodgkinson, *Applied Optics* 30, 1303, 1991. Copyright © 1991 Optical Society of America. Reprinted with permission.)

and

$$A_t = (T_s - T_p)/(T_s + T_p). \tag{12.2}$$

Experimentally, the normal incidence reflection anisotropy for aluminium deposited in oxygen at a given pressure is found to increase both with increasing vapour angle and decreasing deposition rate, as shown in Fig. 12.9.

Increasing the pressure of oxygen also causes a rapid increase in A_r at normal incidence for aluminium coatings. At oblique angles of incidence A_r behaves as shown in Fig. 12.10. The anisotropy of the transmitted beam is negative ($T_s < T_p$) and substantially larger than the reflection anisotropy.

Normal incidence reflection anisotropy in aluminium has implications for the coating of telescope mirrors with polarization-insensitive coatings. In a typical arrangement for coating a large mirror the aluminium is evaporated serially from a set of sources which are located on a circle, as shown in Fig. 12.11. Both the diameter of the mirror and the distance between the source and the substrate are made approximately equal to the mirror diameter, so that the coating has nearly uniform thickness and minimal effect on the optical figure of the mirror.

Each element of the surface of a mirror coated in this way receives a unique set of thin layers, with deposition angles in the range 0°–45°. If the residual

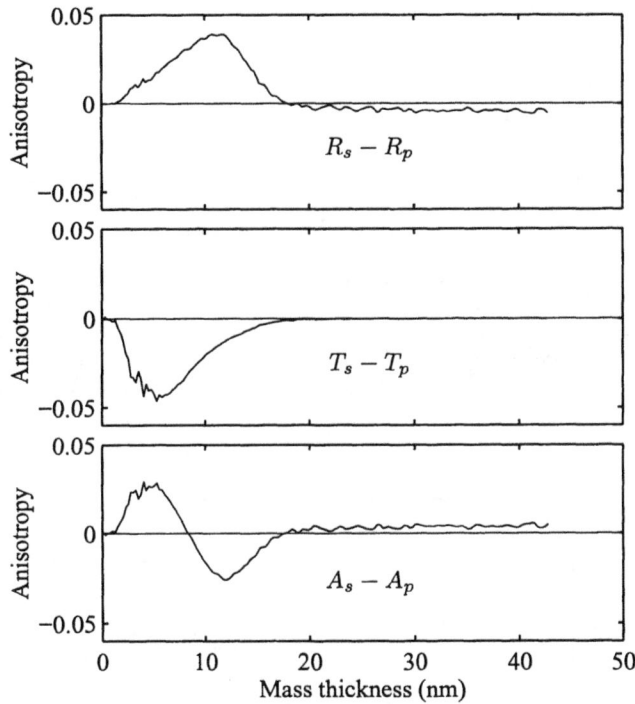

Fig. 12.7. Aluminium deposited in the absence of oxygen exhibits intrinsic anisotropy similar to that of gold and silver. Angle of deposition, 45°; deposition rate, 0.3 nm/s; monitor wavelength, 600 nm. (Adapted from I.J. Hodgkinson, *Applied Optics* 30, 1303, 1991. Copyright © 1991 Optical Society of America. Reprinted with permission.)

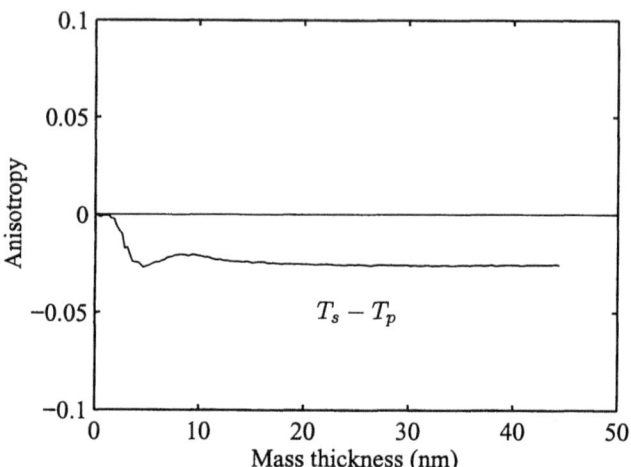

Fig. 12.8. When oxygen is present in the residual gases in the coating chamber, aluminium shows residual anisotropy due to the establishment of columnar microstructural growth. Angle of deposition, 45°; deposition rate, 0.3 nm/s; partial pressure of O_2, 2×10^{-4} mbar; monitor wavelength, 600 nm. (Adapted from I.J. Hodgkinson, *Applied Optics* 30, 1303, 1991. Copyright © 1991 Optical Society of America. Reprinted with permission.)

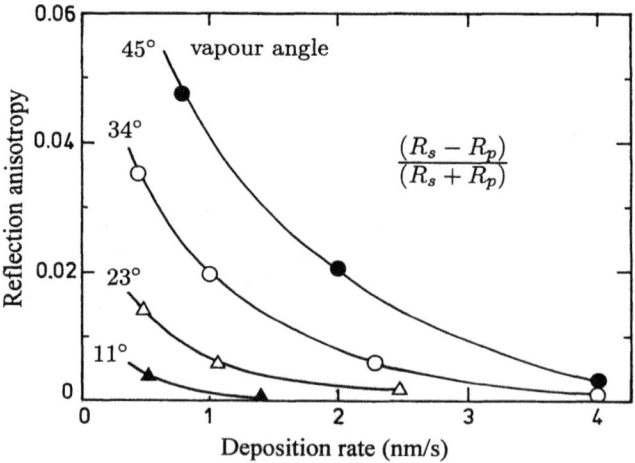

Fig. 12.9. Normal incidence reflection anisotropy from aluminium films deposited in O_2 at 10^{-5} Torr. (Adapted from J.R. Gee, I.J. Hodgkinson and P.W. Wilson, *Journal of Vacuum Science and Technology A* 4, 1875, 1986. Copyright © 1986 American Vacuum Society. Reprinted with permission.)

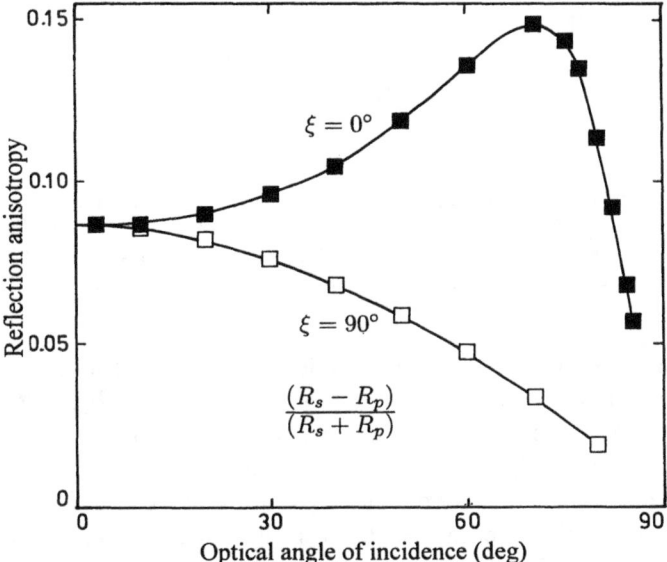

Fig. 12.10. Dependence of reflection anisotropy on optical angle of incidence (Adapted from J.R. Gee, I.J. Hodgkinson and P.W. Wilson, *Journal of Vacuum Science and Technology A* 4, 1875, 1986. Copyright © 1986 American Vacuum Society. Reprinted with permission.)

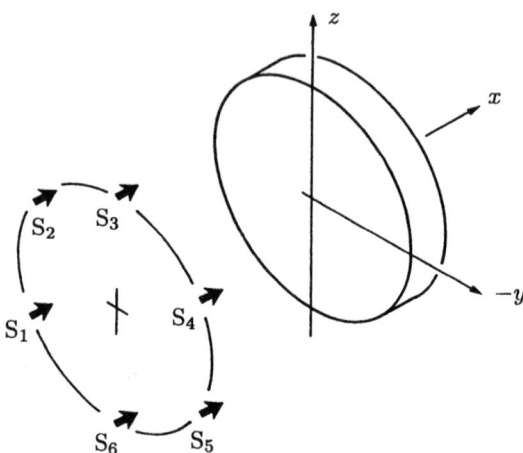

Fig. 12.11.   Approximate arrangement of sources for coating a large surface with aluminium.   (Adapted from J.R. Gee, I.J. Hodgkinson and P.W. Wilson, *Journal of Vacuum Science and Technology A* 4, 1875, 1986. Copyright © 1986 American Vacuum Society. Reprinted with permission.)

gases contain oxygen, then reflection anisotropy may occur, and the mirror surface will contain polarization-sensitive regions with $R_s > R_p$. The anisotropy depends most on the last layer deposited, and is not cancelled by sequential depositions at angles $\pm\theta_v$ (Fig. 12.12). When linearly polarized light is reflected from such a mirror, the reflection anisotropy effectively causes a rotation of the vibration direction in the sense away from the deposition plane. Rotation angles of about $\pm 3°$ are possible in coatings that have a normal visual appearance.

The requirements for small anisotropy are just those for small absorption and scattering losses. The partial pressure of oxygen should be kept small. Deposition at low rates, which can occur unintentionally when most of the charge in a filament has been evaporated but material has migrated to the cooler ends of the source, should be avoided. If a choice exists, filaments that give small deposition angles should be fired last.

12.2.3   Aging

When the deposition of a metal film is terminated while the layer is in a globular microstructural state, structural relaxation processes lead to slow temporal changes in the optical properties of the film. The changes recorded in the transmittance anisotropy during the aging of a 10-nm silver film in vacuum are

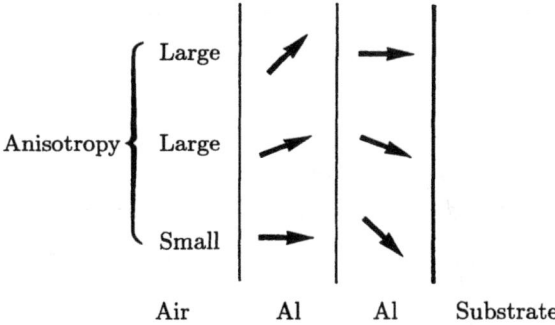

Fig. 12.12. Reflection anisotropy depends mainly on the last film deposited. (Adapted from J.R. Gee, I.J. Hodgkinson and P.W. Wilson, *Journal of Vacuum Science and Technology A* 4, 1875, 1986. Copyright © 1986 American Vacuum Society. Reprinted with permission.)

illustrated in Fig. 12.13. Smaller and slower changes were recorded for gold films aging in vacuum. To a first approximation, the changes that take place as a globular metal film ages in vacuum can be described as back-tracking at a slowing logarithmic rate through the deposition values. For aluminium, changes of anisotropy associated with oxidation occur when the chamber is vented.

12.2.4 Argon Ion Sputter Etching

Three examples of anisotropy that occur during sputter etching are considered here. Fig. 12.14 shows the knee feature recorded during the deposition of a gold film at normal incidence, and again during argon ion sputter etching. These curves indicate that structural hysteresis occurs in a deposition/etch cycle.

The effect of the initial thickness (and hence structure) of gold films on transmittance anisotropy induced by sputter etching is illustrated in Fig. 12.15.

Fig. 12.16 shows the anisotropy recorded during a deposition/etching cycle for a "thick" gold film. Note that the anisotropy remains close to zero during the etching path, until the film breaks up at a transmittance of about 0.7.

12.3 Computer Modelling of Anisotropy in Metals

For the purposes of modelling, we consider a vacuum deposited "metal" film to be equivalent to a mixture of bulk metal and a dielectric medium such as air. Two limiting cases of anisotropic, aggregated media[53] are illustrated in

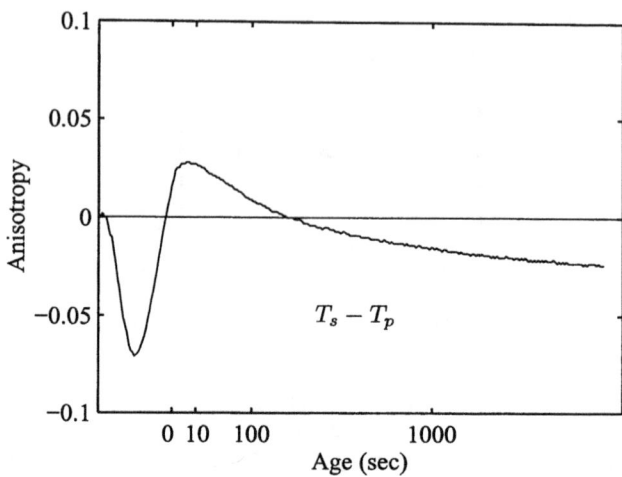

Fig. 12.13. Transmittances measured during the deposition and aging of a silver film in vacuum. Angle of deposition, 45°; deposition rate, 0.5 nm/s; mass thickness, 10 nm; monitor wavelength, 600 nm. (Adapted from I.J. Hodgkinson, *Applied Optics* 30, 1303, 1991. Copyright © 1991 Optical Society of America. Reprinted with permission.)

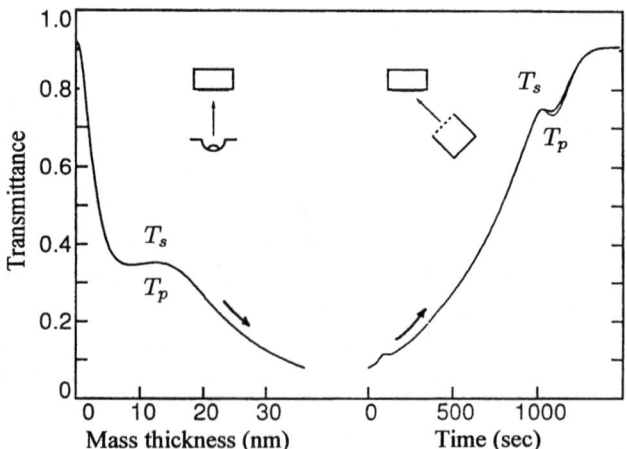

Fig. 12.14. Knee-shaped features observed in the transmittance of a gold film during deposition onto a glass substrate (right) and during post-deposition sputter etching (left). (Adapted from I.J. Hodgkinson, *Journal of Applied Physics* 68, 768, 1990. Copyright © 1990 American Institute of Physics. Reprinted with permission.)

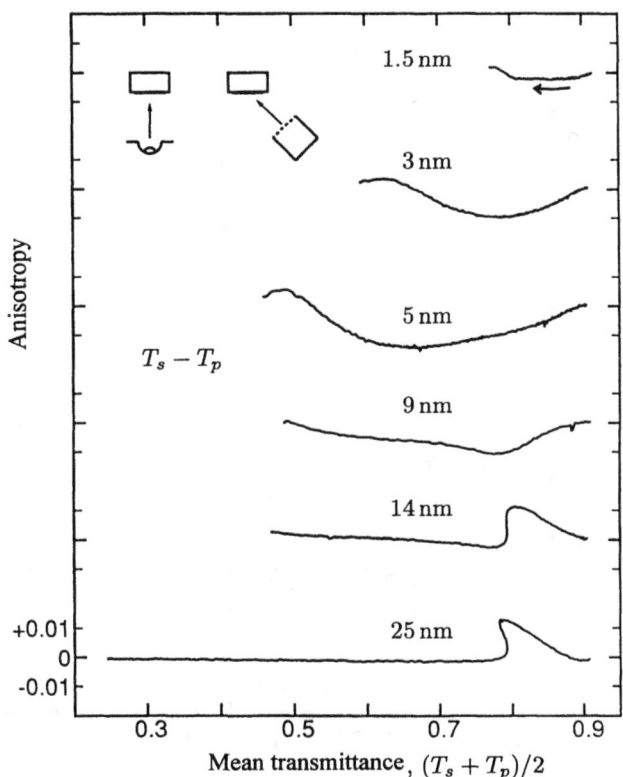

Fig. 12.15. Dependence of anisotropy of gold films on initial mass thickness. (Adapted from I.J. Hodgkinson, *Journal of Applied Physics* 68, 768, 1990. Copyright © 1990 American Institute of Physics. Reprinted with permission.)

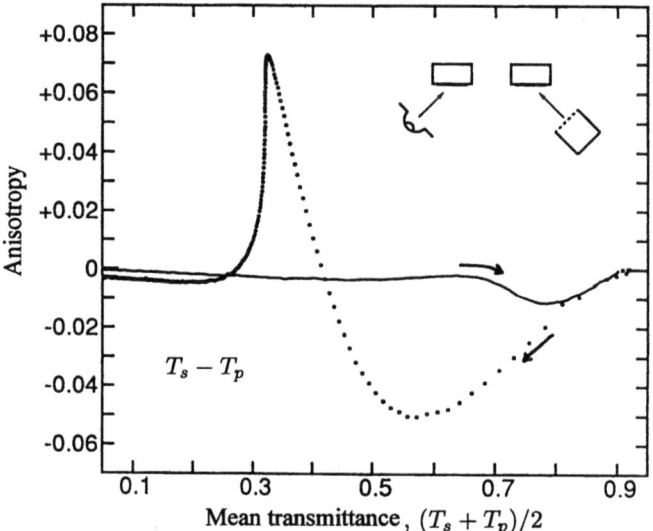

Fig. 12.16. Anisotropy recorded during the deposition of a gold film (dotted line) and during post-deposition etching. (Adapted from I.J. Hodgkinson, *Journal of Applied Physics* 68, 768, 1990. Copyright © 1990 American Institute of Physics. Reprinted with permission.)

Fig. 12.17. In the upper part of the figure ellipsoidal metal particles of dielectric constant ε_c and similar shape and orientation are distributed, rather sparsely, throughout a lossless medium of dielectric constant ε_v. This is referred to as a crystallite-defined (cd) medium. In the second limiting case, which corresponds to a void-defined (vd) medium, a metallic material of dielectric constant ε_c has similarly oriented ellipsoidal particles or voids of dielectric constant ε_v distributed throughout its volume.

Immediately after the start of the deposition of a metal film, the dominant morphology is cd, but by the time the film is fully coalesced, the dominant form has changed to vd. Hence a model for deposition needs to cater for a changing mixture of cd and vd material. Similarly, when a metal film is thinned by sputter etching, the ratio of the two materials changes in favour of the cd form.

We proceed by considering the bulk metal, list the effective media equations for ε_c and ε_v, and then show that the resonance in the absorption occurs in cd material.

12.3.1 Bulk Metals

A bulk metal is characterized by a complex refractive index[54]

$$n = n_R + in_I \tag{12.3}$$

or, equivalently, by a complex dielectric constant

$$\varepsilon = \varepsilon_R + i\varepsilon_I. \tag{12.4}$$

Values of the optical constants n_R and n_I are listed in Table 12.1 for bulk aluminium, silver and gold. (See **alumin**, **gold** and **silver** in the *BTF Toolbox*.) The positive sign of the imaginary part of n ensures that plane waves attenuate as they propagate into a metal. The attenuated wave is described by $\exp(i2\pi nx/\lambda) = \exp(i2\pi n_R/\lambda)\exp(-2\pi n_I x/\lambda)$. As an example, at wavelength 700 nm the refractive index of gold is $n = 0.17 + 3.97i$, and the wave amplitudes attenuate by 50% in a distance of 18 nm.

The effective media equations that are listed in the next section provide values for dielectric constants, rather than refractive indices. At $\lambda = 700$ nm the dielectric constant for the bulk gold, determined by squaring n, is $\varepsilon = -15.73 + 1.35i$. In general it is useful to have equations that relate ε, n and the real and imaginary parts of these quantities:-

$$\varepsilon_R = n_R^2 - n_I^2,$$

Table 12.1. Optical constants of aluminium, gold and silver.[55]

Wavelength	Aluminum		Gold		Silver	
$\lambda/$nm	$(n_a)_R$	$(n_a)_I$	$(n_g)_R$	$(n_g)_I$	$(n_s)_R$	$(n_s)_I$
550	0.76	5.32	0.34	2.37	0.055	3.32
600	0.97	6.00	0.23	2.97	0.060	3.75
650	1.24	6.60	0.19	3.50	0.070	4.20
700	1.55	7.00	0.17	3.97	0.075	4.62
750	1.80	7.12	0.16	4.42	0.080	5.05

$$\varepsilon_I = 2n_R n_I, \tag{12.5}$$

$$n_R = \left[\frac{\varepsilon_R + (\varepsilon_R^2 + \varepsilon_I^2)^{1/2}}{2} \right]^{1/2},$$

$$n_I = \left[\frac{-\varepsilon_R + (\varepsilon_R^2 + \varepsilon_I^2)^{1/2}}{2} \right]^{1/2}. \tag{12.6}$$

12.3.2 Depolarization Factors

The effect of ellipsoidal particles or voids on the principal refractive indices of a cd or vd aggregated metal can be described in terms of the packing density p (volume fraction of metal in each case), and three depolarization factors, L_j, that add to 1. Here the subscript j refers to the axes labelled in Fig. 12.17. For modelling appropriate to normally incident light the radius r_2 of the spheroids needs to be less than the radius r_3, in order to give anisotropies with the correct sign. It is sufficient to use prolate spheroids, with $r_1 = r_2 < r_3$ and eccentricity $e = (r_1^2/r_3^2 - 1)^{1/2}$. The depolarization factors for prolate spheroids are

$$L_3 = \frac{1 - e^2}{2e^3} (\log \frac{1 + e}{1 - e} - 2e),$$

$$L_2 = \frac{1}{2}(1 - L_3),$$

$$L_1 = L_2. \tag{12.7}$$

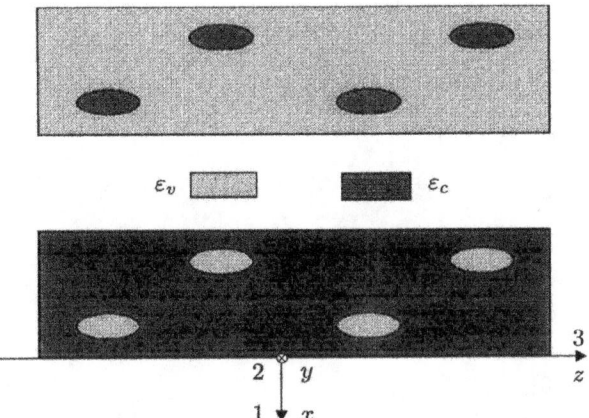

Fig. 12.17. Aggregated media with anisotropy defined by ellipsoidal crystallites of a metal (upper) and by ellipsoidal voids or particles of a dielectric material (lower). The metal is represented by the dark areas. (Adapted from I.J. Hodgkinson, *Applied Optics* 30, 1303, 1991. Copyright © 1991 Optical Society of America. Reprinted with permission.)

12.3.3 Isotropic Resonance

As a preliminary step, before considering the cause of anisotropy in noble metals, we discuss resonance in an isotropic aggregated metal film. The cd and vd materials are modelled by the Bragg-Pippard equations, Eqs.(8.19) and (8.20). As shown in Fig. 12.18, a resonance peak occurs in the modulus of the cd refractive index, $|n_{cd}|$, at a particular value p_0 of the packing density. Figure 12.18 is plotted for spherical crystallites and voids ($L = L_1 = L_2 = L_3 = 1/3$) and using optical constants for bulk gold from Table 12.1. To indicate that the cd case is unrealizable physically when $p \approx 1$ and that the vd case is unrealizable when $p \approx 0$, the appropriate ends of the curves in Fig. 12.18 are shown as broken lines. As well, violations of the requirement that the particle density should be sparse are most severe in these regions.

Approximate expressions relevant to the resonance in the cd material can be derived readily when Eq.(8.19) is rewritten so that it has the denominator $[\varepsilon_v + (1-p)(\varepsilon_R - \varepsilon_v)L]^2 + [(1-p)\varepsilon_I L]^2$, where ε_R and ε_I are the real and imaginary parts of ε_c. The resonant peak occurs in the effective dielectric constant when the left-hand term in the new denominator is equal to zero. Putting $\varepsilon_v = 1$ for voids of air and assuming that $\varepsilon_R \approx -n_I^2 >> \varepsilon_v$ gives the expression

$$p_0 \approx 1 - 1/n_I^2 L \qquad (12.8)$$

for the location of the resonance on the p-axis. At resonance, when $p = p_0$, the

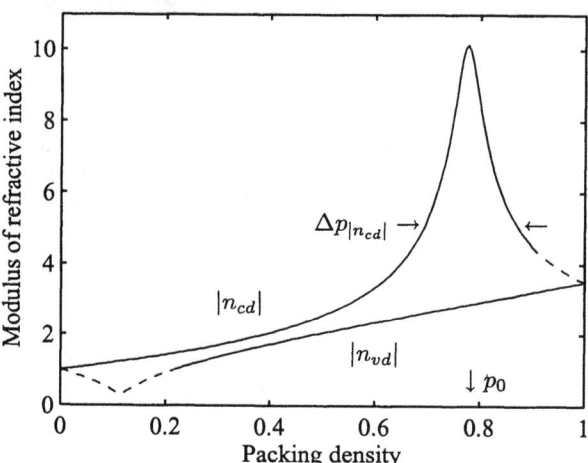

Fig. 12.18. A resonance peak occurs in the modulus of the refractive index of the crystallite-defined material at a specific value (p_0) of the packing density, but not in the void-defined material. The optical constants of bulk gold at a wavelength of 650 nm (Table 12.1) were used in the calculations and the particles have a spherical shape, i.e. $L_1 = L_2 = L_3 = 1/3$. (Adapted from I.J. Hodgkinson, *Applied Optics* 30, 1303, 1991. Copyright © 1991 Optical Society of America. Reprinted with permission.)

value of the effective dielectric constant is given by

$$\varepsilon_{cd} \approx i p_0 n_I^3 / 2 n_R, \tag{12.9}$$

and the effective refractive index is

$$n_{cd} \approx (p_0 n_I^3 / 4 n_R)^{1/2} (1 + i), \tag{12.10}$$

The width (FWHM) of the resonance peak can be expressed as

$$\Delta p_{|\varepsilon_{cd}|} \approx 4\sqrt{3}\, n_R / L n_I^3, \tag{12.11}$$

or as

$$\Delta p_{|n_{cd}|} \approx 4\sqrt{15}\, n_R / L n_I^3. \tag{12.12}$$

The refractive-index-based parameters that define the resonance are shown as labels on Fig. 12.18.

12.3.4   Anisotropic Resonance

To explain the origin of anisotropy in nobel metals we consider aggregated media in which the particles or voids are prolate spheroids with $L_1 = L_2 = 0.35$ and $L_3 = 0.30$. The upper part of Fig. 12.19 shows that, whereas the location (p_0) and the height of the resonant peak both increase with increasing wavelength, the opposite is true of both the width ($\Delta p_{|n_{cd}|}$) and the separation (Δp_0) of the curves with the same wavelength. These trends can be predicted using the values for the optical constants of gold listed in Table 12.1 together with Eqs. (12.8), (12.10), (12.12), and the approximation

$$\Delta p_0 \approx \Delta L / (n_I^2 L^2), \tag{12.13}$$

that follows from Eq. (12.8); $\Delta L = L_2 - L_3$ is a measure of the particle shape anisotropy.

The remaining parts of Fig. 12.19 show the characteristic shapes of the graphs $|n_{cd\,3}| - |n_{cd\,2}|$ v. p for the resonant cd-material (middle) and for the vd-material (lower). Thus, as an increase in refractive index usually leads to a decrease in the transmittance of the metallic film, we can conclude that the large negative → large positive → small negative excursions in transmittance anisotropy observed during the deposition of gold films may be deduced from the intrinsic anisotropic properties of a composite BP material that changes from the cd form to vd form as the packing density increases towards unity.

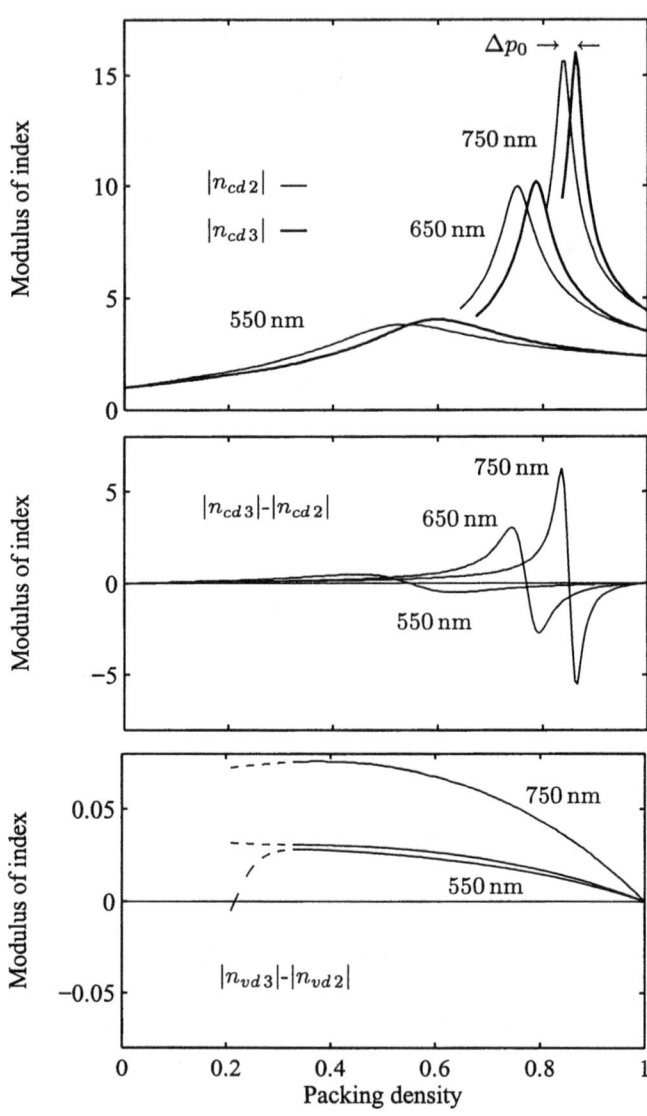

Fig. 12.19. Dependence of the resonance (upper) and the anisotropy (middle) of crystallite-defined material on wavelength and polarization, and the very much smaller anisotropy of the void-defined material (lower). The optical constants used in the calculations are those of bulk gold (Table 12.1), and the particles are prolate ellipsoids with $L_1 = L_2 = 0.35$ and $L_3 = 0.3$. (Adapted from I.J. Hodgkinson, *Applied Optics* 30, 1303, 1991. Copyright © 1991 Optical Society of America. Reprinted with permission.)

12.4 Modelling Deposition and Etching

Non-trivial problems are encountered when an attempt is made to model the anisotropy profiles recorded *in situ* during the growth and etching of metal films. An important difficulty is that the parameter p, fundamental to the BP equations, is not accessible during experimental measurements of anisotropy; during deposition the reflectances and transmittances are monitored as functions of the mass thickness d recorded by a quartz crystal monitor. Thus, one problem is to obtain a suitable relationship linking p and d. Experiments with gold indicate that the curve of p v. d has a step feature associated with the resonance. But, rather than "write in" anisotropy via the p v. d function, we use the smooth exponential function

$$p(d) = p(0) + [p(\infty) - p(0)] \left[1 - \exp(-d/\Delta d_p)\right], \qquad (12.14)$$

to model the packing density.

Complicated changes in morphology occur as a film evolves from the globular to the nearly bulk form, and these are represented, very approximately, by considering the film to be a mixture of the ideal cd and vd structures. The effective dielectric constant of the mixture is assumed to be given by the equation

$$\varepsilon = f\varepsilon_{cd} + (1 - f)\varepsilon_{vd}, \qquad (12.15)$$

where the structure fraction f is a function that changes smoothly between the levels $f(-\infty)$ and $f(\infty)$,

$$f(d) = \frac{f(-\infty)}{1 + \exp[(d - d_f)/\Delta d_f]} + \frac{f(\infty)}{1 + \exp[-(d - d_f)/\Delta d_f]}. \qquad (12.16)$$

At $d = d_f$, the structure fraction f is equal to the mean of $f(-\infty)$ and $f(\infty)$. In practice $f(0)$, $f(\infty)$, d_f and Δd_f are used as parameters for the structure fraction, and the value of $f(-\infty)$ required for Eq. (12.16) is determined using the equation

$$f(-\infty) = \{f(\infty) - f(0)/[1 + \exp(d/\Delta d_f)]\} \times [1 + \exp(-d/\Delta d_f)]. \qquad (12.17)$$

The values $f(0) = 1$, $f(\infty) = 0$ and $p(\infty) = 1$ were used in all of the simulations discussed here. In this case at $d = d_f$ the value of f is 0.5, and the structure is 50% cd, 50% vd.

Another problem is that the knee feature in the T v. d curve, as calculated using the refractive index for a bulk metal, is too sharp and too large relative to the shape of the experimental curve. In practice several factors act to smooth the resonance in the refractive index, and hence the shape of the knee. In a real isotropic metal film, such as gold deposited at normal incidence, the particles are not identical spheres, but are approximately ellipsoids with a range of sizes and eccentricities and with the orientations of the horizontal axes distributed randomly. The asymmetry in particle shape that causes the anisotropy in a film deposited obliquely is a small perturbation superposed on this distribution. This is referred to as the *first anisotropy*[19] and, as it may be caused by the bunching of particles or columns as well as by particle shape, it is perhaps best described in general by the appropriate depolarization factors. Even when the metal film is very thin, the islands touch and coalesce, so the structure fraction f is probably always less than 1. Lowering the f v. d profile softens the resonance. The restriction imposed on the mean free path (mfp) of electrons in very small metal islands increases n_R but leaves n_I nearly constant. Thus, according to Eqs (12.8), (12.10) and (12.12), the restricted mfp does not shift the location of the resonance but acts to decrease the height and increase the width. For simplicity, only the latter effect (restricted mfp) is used to influence the shape of the resonant peak in the model. The electronic mfp is assumed to be proportional to the mass thickness,

$$l_{mfp} = k\,d, \tag{12.18}$$

and the method outlined by S. Norrman et al[56] is used to calculate ε_c and hence n_R and n_I.

12.4.1 Simulated Deposition of Gold

The curves for p and f illustrated in Fig. 12.20 provide a reasonable representation of the normal incidence transmittances for gold at wavelength 700 nm. Considerable control has been exercised over the shape of the resonance, mainly through assignment of the value 0.2 to the parameter k in Eq. (12.18). The other parameters were set at $\Delta L = 0.08$, $p(0) = 0.6$, $\Delta d_p = 8$ nm, $d_f = 5$ nm, and $\Delta d_f = 4$ nm. In practice the anisotropy occurs over a larger mass thickness range, indicating that p actually changes more slowly in the vicinity of the resonance.

As an additional check on the validity of the model, characteristic ellipsometric curves were recorded for a gold film as it was deposited at 45° and monitored at 630 nm. These are reproduced in Fig. 12.21; note that isotropic gold would yield dot Δ v. Ψ curves at $\Psi = 45°$, $\Delta = 0$ in (a) and (c), and horizontal line Δ v. d curves through $\Delta = 0$ in (b) and (d).

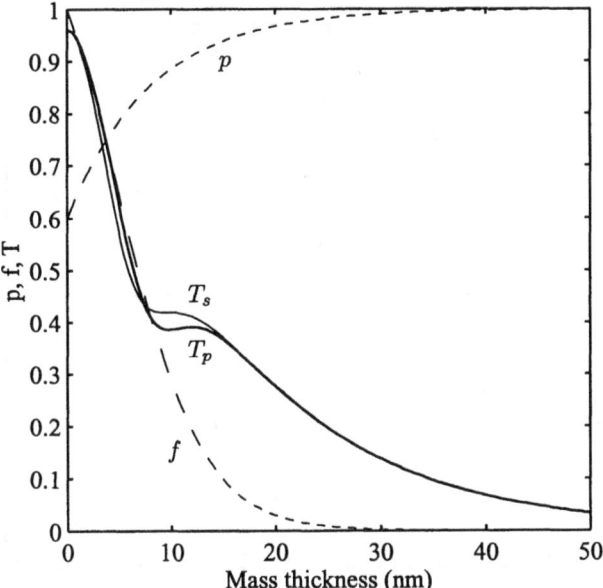

Fig. 12.20. Profiles of transmittance at normal incidence, T_s v. d and T_p v. d, simulated by the BP model for a gold film deposited at 45°. As the mass thickness of the film increases, the packing density p increases and the fraction f of the crystallite-defined structure decreases. (Adapted from I.J. Hodgkinson, *Applied Optics* 30, 1303, 1991. Copyright © 1991 Optical Society of America. Reprinted with permission.)

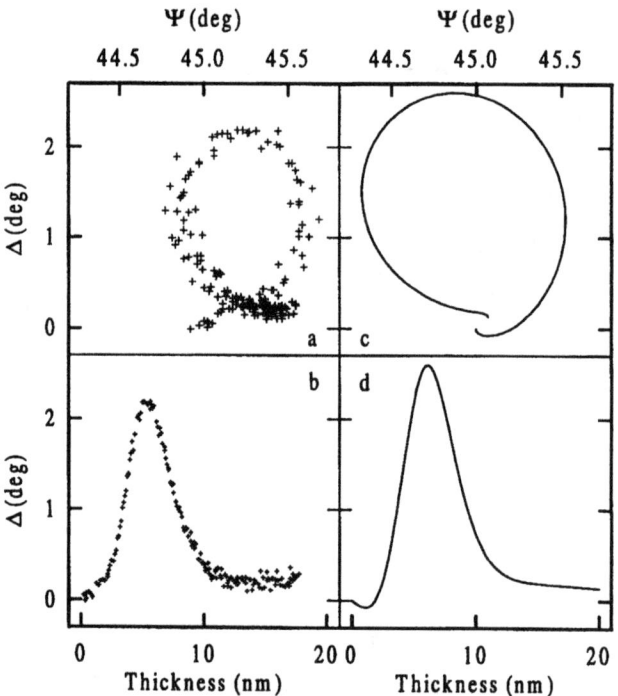

Fig. 12.21. Characteristic ellipsometric profiles recorded during the deposition of a gold film (a and b), and simulated using a computer model (c and d). Adapted from Q.H. Wu and I.J. Hodgkinson, *Journal of Optics: Nouvelle Revue D'Optique* 25, 43, 1994. Copyright © Masson, Paris, 1994. Reprinted with permission.)

12.4.2 Simulated Deposition of Silver

The simulated R, T, and A curves for silver shown in the upper part of Fig. 12.22, and the corresponding anisotropy profiles drawn to a larger scale in the lower part of the figure, provide an overview that is appropriate to both silver and gold, and that compares well with the experimental profiles (Fig. 12.5). For these simulations the parameters were set at $\Delta L = 0.05$, $p(0) = 0.3$, $\Delta d_p = 6\,\mathrm{nm}$, $d_f = 10\,\mathrm{nm}$, $\Delta d_f = 4\,\mathrm{nm}$ and $k = 0.125$. Further, by using the model, it can be deduced that decreases in p and f associated with relaxation of particle shape towards a more spherical form cause aging effects similar to those observed experimentally (Fig. 12.13).

12.4.3 Simulated Deposition of Aluminium

The model was used to simulate the intrinsic anisotropies of aluminium, i.e. the anisotropies that are recorded when the residual pressure of oxygen is kept very low (Fig. 12.7). The simulation in Fig. 12.23, for $\lambda = 600\,\mathrm{nm}$, was made using the parameter values $\Delta L = 0.03$, $p(0) = 0.5$, $\Delta d_p = 5\,\mathrm{nm}$, $d_f = 6\,\mathrm{nm}$, $\Delta d_f = 3\,\mathrm{nm}$ and $k = 0.125$.

12.4.4 Simulated Deposition / Etch Paths

During post-deposition sputter etching of gold films, p and f initially change more slowly with change in d, relative to the deposition path, because the basic island or continuous film structure tends to remain constant initially and does not devolve along the deposition path. Structural hysteresis loops, such as those shown in Fig. 12.24, enable the features observed experimentally in the deposition / etch anisotropy profiles of gold to be simulated. In Fig. 12.25 the normal incidence transmittance anisotropy is simulated for two extreme cases of film structure and plotted as a function of the mean transmittance, $(T_s + T_p)/2$, because d is inaccessible during etching experiments.

In the upper part of Fig. 12.25 the simulation for a 5-nm gold film deposited at normal incidence and then etched with an argon ion-beam incident at $45°$ has the basic form of the experimental results (Fig. 12.15). Such a gold film has an island structure throughout. In the simulation the anisotropy during etching was assumed to increase exponentially with removal of mass, from an initial value of 0 towards a limit of 0.05 with a mass thickness constant of 1 nm; the other parameters were set at $p(0) = 0$, $\Delta d_p = 2\,\mathrm{nm}$, $d_f = 2\,\mathrm{nm}$, $\Delta d_f = 1\,\mathrm{nm}$ and $k = 0.2$.

The lower part of Fig. 12.25 shows the simulation of anisotropy in a 40-nm gold film deposited at $45°$ and then sputter etched with an argon ion beam

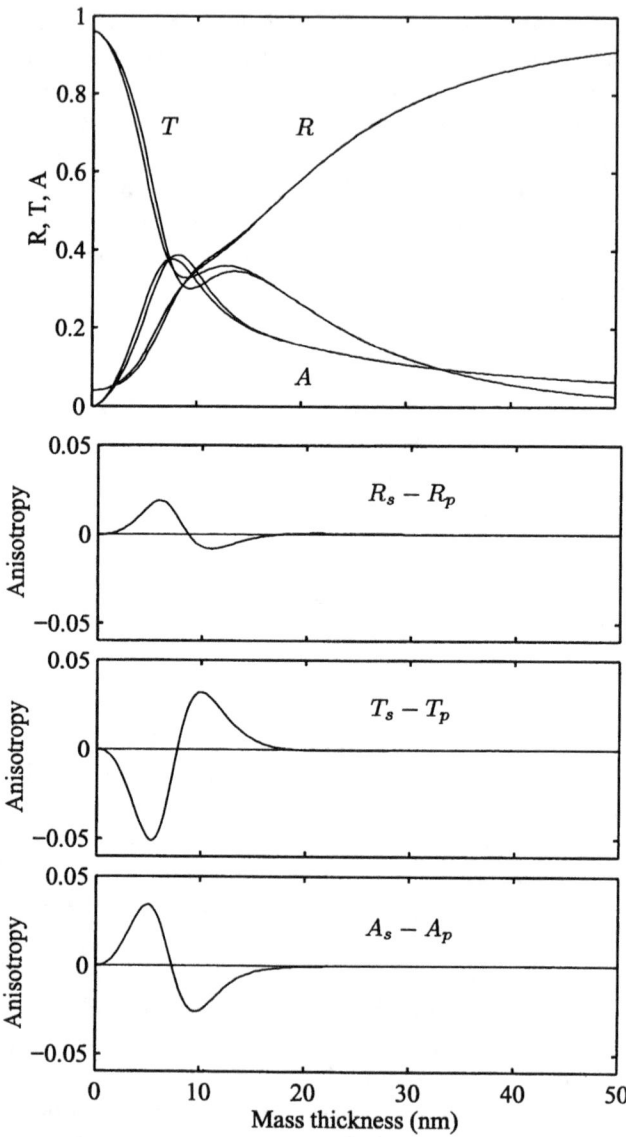

Fig. 12.22. Simulations of the normal incidence reflectances, transmittances, and absorptances at 600 nm for a silver film deposited at 45°. Experimental results for this film are plotted in Fig. 12.5. (Adapted from I.J. Hodgkinson, *Applied Optics* 30, 1303, 1991. Copyright © 1991 Optical Society of America. Reprinted with permission.)

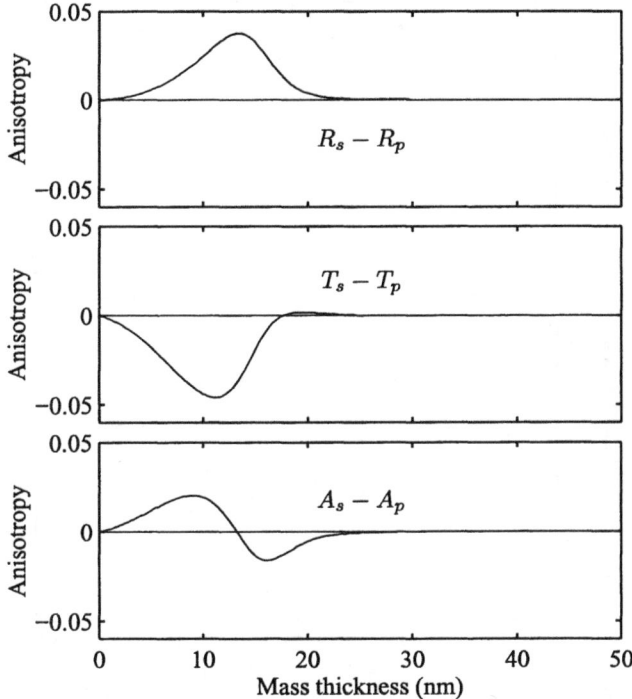

Fig. 12.23. Simulated normal incidence anisotropy curves for aluminium deposited at 45° show the intrinsic anisotropic effects observed experimentally at wavelength 600 nm when aluminium is deposited in the absence of oxygen. See Fig. 5 for the experimental values. (Adapted from I.J. Hodgkinson, *Applied Optics* 30, 1303, 1991. Copyright © 1991 Optical Society of America. Reprinted with permission.)

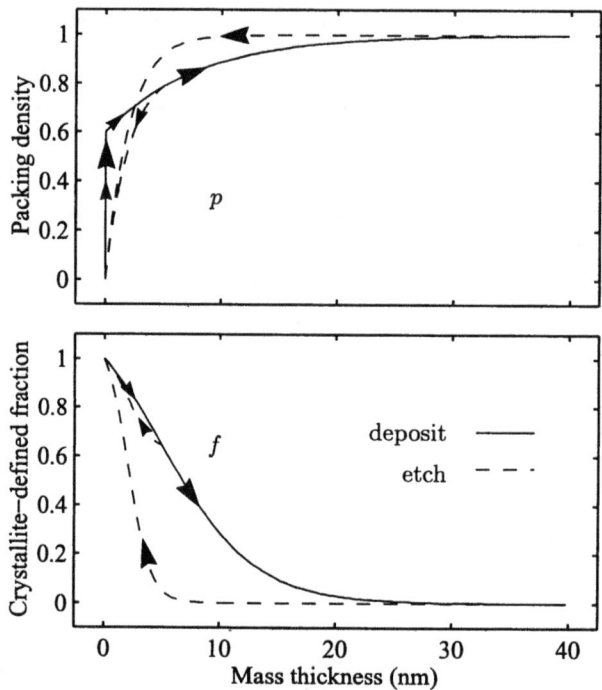

Fig. 12.24. Structural hysteresis loops, for the packing density p and the crystallite-defined fraction f, used to simulate the development of normal incidence transmittance anisotropy as 5-nm and 40-nm gold films are deposited and subsequently thinned by argon ion-beam sputter etching. (Adapted from I.J. Hodgkinson, *Applied Optics* 30, 1303, 1991. Copyright © 1991 Optical Society of America. Reprinted with permission.)

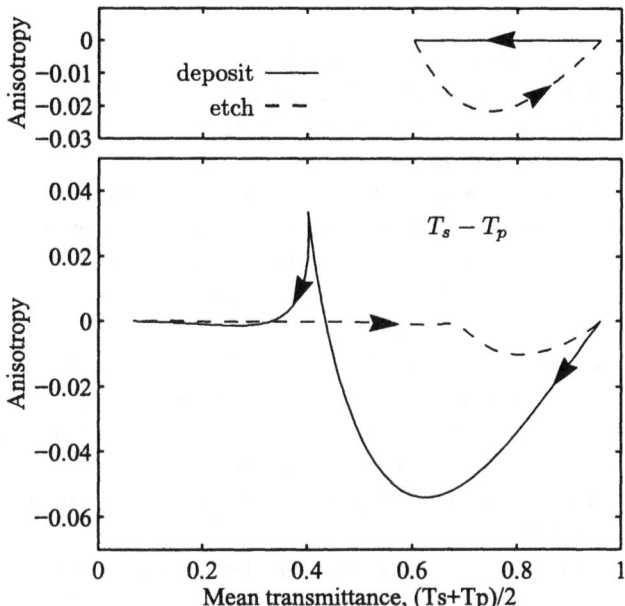

Fig. 12.25. Simulations of the development of anisotropy as 5-nm (upper) and 40-nm (lower) gold films are deposited and then thinned by argon ion sputter etching at 45°. The 5-nm film is deposited at normal incidence and so has zero anisotropy due to the deposition (see Fig. 12.15). The transmittance anisotropy of the 40-nm film reaches a large value during deposition at 45° but has nearly vanished when the film is fully coalesced. During etching the anisotropy of this film is nearly zero until it breaks up into islands (see Fig. 12.16 for the experimental profiles). (Adapted from I.J. Hodgkinson, *Applied Optics* 30, 1303, 1991. Copyright © 1991 Optical Society of America. Reprinted with permission.)

at 45° in the same plane of incidence (Fig. 12.16). In this case the deposited film is fully coalesced and the anisotropy during etching remains close to zero, until the film breaks up when the mean transmittance is about 0.7. As in the previous case, the main features of the experimental observations are simulated correctly.

12.5   Summary

When a metal is deposited obliquely, particle shape anisotropy causes the resonance for the s-polarization to occur before the resonance for the p-polarization. This is the cause of relatively large intrinsic optical anisotropies that occur during the growth of metal films. For a noble metal the anisotropies reduce to zero as the thickness increases, and a fully coalesced film is formed. For a reactive metal, such as aluminium deposited in the presence of oxygen, a columnar growth is established and is the cause of residual anisotropies. Effects such as particle shape relaxation and oxidation cause slow changes in the anisotropies immediately after deposition or venting.

Crystallite-defined and void-defined media (modelled by the BP equations) together contain the intrinsic anisotropic properties exhibited by thin metal films. By regarding a thin metal film as a thickness-dependent mixture of the two structures, with a thickness-dependent packing density, the characteristic anisotropic features observed experimentally during the deposition of metal films can be simulated.

The main features of the model that has been applied to the deposition of metals are:-

1. The particle shape anisotropy is constant during deposition.

2. The packing density increases smoothly with mass thickness.

3. The fraction of cd-material (the structure fraction) decreases smoothly with mass thickness.

4. The knee in the transmittance curve is smoothed by a restricted electronic mfp.

5. The first and second peaks in the anisotropy curves are caused by displaced refractive index resonance peaks in the cd-material.

6. The third peak is caused by form birefringence in the vd-material.

Part III

Applications of Birefringent Media

Chapter 13

Linear polarizers

An ideal linear polarizer can be regarded as a device that produces linearly polarized light from any input light. Thus a mixture of unpolarized light and elliptically polarized light, for example, should lead to a transmitted beam of linearly polarized light. Several different types of linear polarizer are available,[57] and it is possible to classify them according to the physical principles that are utilized. Already we have seen that any polarization state can be resolved into two superposed linearly polarized beams vibrating, say, along the y and z-axes; and the action required of a linear polarizer can be considered to be isolation of one of the component beams.

▨ In *dichroic polarizers* one of the component beams is absorbed and the other is transmitted.

▨ In *tilted plate polarizers* the p-polarized beam utilizes transmission without loss at the Brewster angle, whereas the s-polarized beam suffers high reflectance. In cases, (such as *thin film polarizers*), where both beams are available for use, the devices are strictly *polarizing beam splitters*, but we will consider them as linear polarizers in this chapter.

▨ In *crystalline prism polarizers* the two beams are separated, in angle at first, and then spatially as they diverge.

Linear polarizers are used extensively as internal components in lasers and for modifying the polarization state of laser beams. For some applications there is a need for polarizers that will withstand high laser flux densities, and in other cases the polarization purity of the transmitted beam is an important parameter. We begin this chapter by considering the specification of real, rather than ideal polarizers, and then discuss examples of the three types highlighted above.

13.1   Real Polarizers

In Sect. 3.2 we saw that coherent polarized beams of light, such as from a laser, can be described using Jones vectors; and in Sect. 4.2 the action of various polarizing devices on coherent light was described using 2×2 Jones matrices. However, the concept of an ideal linear polarizer, represented for the horizontal orientation of the transmission axis by the Jones matrix

$$\hat{J} = \begin{bmatrix} 1 & 0 \\ 0 & 0 \end{bmatrix}, \tag{13.1}$$

is found wanting in practice. A real polarizer transmits less than 100% of the light that vibrates in the 'parallel' direction and transmits some of the beam vibrating in the 'perpendicular' direction. As well, the optical path lengths for transmission through the linear polarizer may be different for the two polarizations.

In most polarizers the p polarization is transmitted and the s polarization is rejected. For this reason we label the parallel and perpendicular vibration directions as p and s respectively. Then by writing the relevant transmission coefficients as $T_p^{1/2} \exp(i\phi_p)$ and $T_s^{1/2} \exp(i\phi_s)$, we can define a more practical Jones matrix as

$$\hat{J} = \begin{bmatrix} T_p^{1/2} \exp(i\phi_p) & 0 \\ 0 & T_s^{1/2} \exp(i\phi_s) \end{bmatrix}. \tag{13.2}$$

For linearly polarized incident light, shown vibrating at angle θ to the y-axis in Fig. 13.1, the Jones vector of the transmitted beam is given by

$$\begin{bmatrix} E_{yt} \\ E_{zt} \end{bmatrix} = \begin{bmatrix} T_p^{1/2} \exp(i\phi_p) & 0 \\ 0 & T_s^{1/2} \exp(i\phi_s) \end{bmatrix} \begin{bmatrix} E \cos \theta \\ E \sin \theta \end{bmatrix}. \tag{13.3}$$

Thus the total transmittance is given by the equation

$$T = T_p \cos^2 \theta + T_s \sin^2 \theta. \tag{13.4}$$

Alignment of E with the y-axis by putting $\theta = 0°$ shows that the principal transmittance is T_p, and alignment of E with the z-axis by putting $\theta = 90°$ shows that the minor transmittance is T_s. Other performance characteristics can be derived from T_p and T_s. A list including names is given below, but note that some of the definitions differ between suppliers of polarizing products.

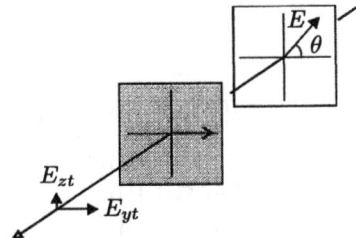

Fig. 13.1. Light transmission by a real polarizer.

Transmission efficiency T_p is the transmission of the polarizer to linearly polarized incident light.

Degree of polarization $(T_p - T_s)/(T_p + T_s)$ provides a value between 0 for unpolarized transmitted light and 1 for $T_s = 0$, corresponding to linearly polarized transmitted light.

Extinction ratio T_p/T_s is a large number for a high quality polarizer. Note that some manufacturers define extinction ratio as the small number T_s/T_p.

Total transmittance $T_T = (T_p + T_s)/2$ is the total light transmitted by a single polarizer for unpolarized incident light.

Open transmittance $H_0 = (T_p^2 + T_s^2)/2$ is the transmittance of two aligned polarizers for unpolarized incident light.

Closed transmittance $H_{90} = T_p T_s$ is the transmittance of two crossed polarizers for unpolarized incident light.

As well it is common practice to specify the range of acceptance angles, and the laser damage threshold (LDT).

13.2   Dichroic Polarizers

Some crystalline materials, such as tourmaline, exhibit different absorption coefficients for different polarizations of the incident light. Historically, when this phenomenon was investigated it was found that an incident beam of white light produces two distinctive colours and hence the name *dichroism* was coined. Today dichroism can be understood in terms of the proximity of the light frequency to atomic absorption bands. In a typical case for visible light, a transparent material has absorption bands in the ultraviolet. If the material is uniaxial

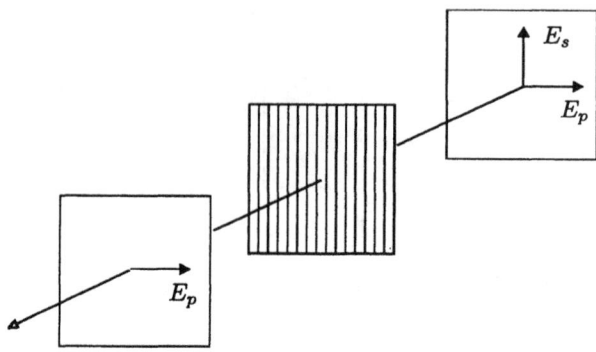

Fig. 13.2. Wire grid polarizer.

(for simplicity), then for a visible wavelength both the ordinary and extraordinary refractive indices may be real, and the material is said to be birefringent. For a smaller wavelength, one refractive index may be real and the other may be complex corresponding to high absorption. Such a material is said to be dichroic.

Tourmaline is a naturally occurring polarizer, but in practice, specially engineered dichroic materials are used. In dichroic *sheet polarizer* the unwanted polarization is absorbed as heat, and as a consequence, sheet polarizer cannot be used in an intense laser beam. A century ago Hertz used a grid of aligned wires to polarize radio waves. The principle of operation of the *wire grid polarizer* is shown in Fig. 13.2. The electric field component E_s vibrating parallel to the wires causes ac currents to flow along the wires, and the energy of the incident electromagnetic wave is converted to heat. On the other hand, the strength of current flow perpendicular to the wires is limited by higher electrical resistance – the spacing between the wires is less than the wavelength of the waves – and the electric field component E_p is transmitted.

Modern dichroic sheet polarizer is a molecular analogue of the wire grid polarizer. A sheet of polymeric plastic is stretched so that long-chain molecules in the material align in the stretching direction and pigment molecules are attached to the chains. Preferential absorption of one polarization occurs in a direction determined by the orientation of the chemical bonds at the attachment sites. Typical performance values are $H_0 \approx 0.2$ and $H_{90} \approx 10^{-4}$ for visible light. The large absorption that is implied here prohibits the use of sheet polarizer in high energy applications. The advantages of this form of polarizer relate to cost and ease of use – it can be obtained in large sheets and cut by the user into specific shapes, etc.

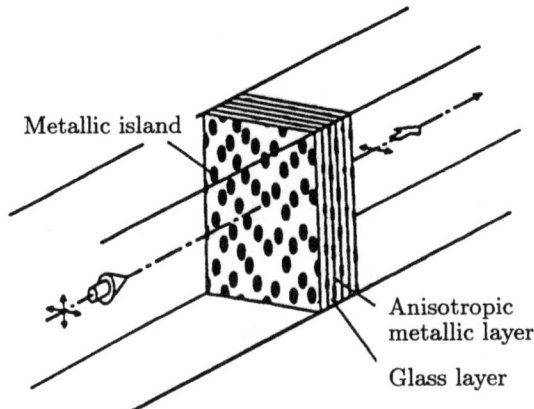

Metallic island

Anisotropic
metallic layer

Glass layer

Fig. 13.3. Stretched metallic island polarizer. Adapted from K. Baba and M. Miyagi, *Optics Letters* 16, 964, 1991. Copyright © 1991 Optical Society of America. Reprinted with permission.)

Two variations of the common sheet polarizer are in use. In the first, an array of parallel stripes of metal formed by vacuum deposition on to glass operates as a linear polarizer in the near-infrared, for example, in the wavelength range 1250–1550 nm that spans two optical communication windows.

In the second variation, called the *dichroic glass polarizer*, aligned, elongated silver crystals embedded in a glass disk preferentially absorb light vibrating in the direction of a common axis of the crystals. The surfaces of the glass can be worked smooth and coated with anti-reflection coatings. In this way a high value of T_p, of the order of 95%, is achieved; other typical values are $T_p/T_s \approx 10^4$, $\pm 15°$ acceptance angle, laser damage threshold $5\,\mathrm{J/cm^2}$ for transmission and $0.1\,\mathrm{J/cm^2}$ for blocking. In a recent development spherical-globular metallic-island films are deposited on to glass, overcoated with a sputtered glass film, heated and stretched to elongate the metallic islands. Several layers are required to achieve a satisfactory extinction ratio, as illustrated in Fig. 13.3.

13.3 Tilted Plate and Thin Film Polarizers

13.3.1 Plate Polarizers

Figure 13.4 (upper) shows a glass plate oriented at the Brewster angle in a laser beam. From the elements of the Jones matrix for the plate (given in Sect. 4.2), we see that the transmission efficiency $T_p = 1$ and $T_s = q^4$ where

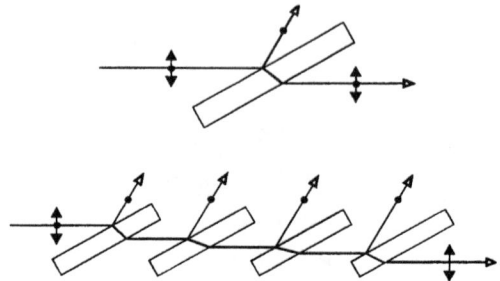

Fig. 13.4. Single-plate polarizer (upper) and multiple-plate polarizer (lower).

$q = 2n/(1 + n^2)$. For a plate of refractive index $n = 1.5$, q is approximately 0.923 and the extinction ratio T_p/T_s is approximately 1.4.

One method of increasing the extinction ratio, while retaining the transmission ratio of unity, is to use several plates in series as shown in Fig.13.4 (lower). This device is known as a *pile-of-plates polarizer*. From the form of the Jones matrix for a single plate, or from the ray path, it can be seen that the extinction ratio of a stack of N plates is $1/q^{4N}$. The extinction ratio increases slowly with N, because $1/q$ is just slightly greater than 1. For a stack of 10 glass plates, for example, the extinction ratio is only 24.

13.3.2 Coated-Plate Polarizers

In practice the nearly equivalent arrangement of a set of H and L layers deposited on to a single plate is used as a thin film analogue of the multiple-plate polarizer. First, though, consider a plate with both surfaces coated with an H layer. By applying Eq. (5.22) to an air/film/glass sequence we can determine the condition for r_p to equal zero as $\gamma_{Hp}^2 = \gamma_{Cp}\gamma_{Sp}$, and reorganize it as a quadratic in β^2

$$(n_H^8 - n^4)\beta^4 + [2n_H^2 n^4 - n_H^8(1 + n^2)]\beta^2 + n_H^8 n^2 - n_H^4 n^4 = 0. \tag{13.5}$$

For a film with index $n_H = 2.3$ on a substrate with $n = 1.5$, solution of Eq. (13.5) gives $\beta = 0.908$, corresponding to an angle of incidence of 72.3°. An extinction ratio of 13 is obtained for the plate with coatings on both sides.

Next consider a glass plate coated on one surface with a quarter-wave reflecting stack, as illustrated in Fig. 13.5 for the system $ag[HL]^N Ha$. Figure 13.6 shows the p and s transmittances for this arrangement with $n = 1.5$, $n_L = 1.38$,

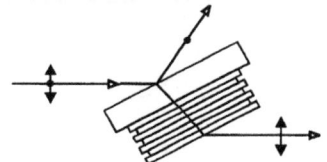

Fig. 13.5. Coated-plate polarizer.

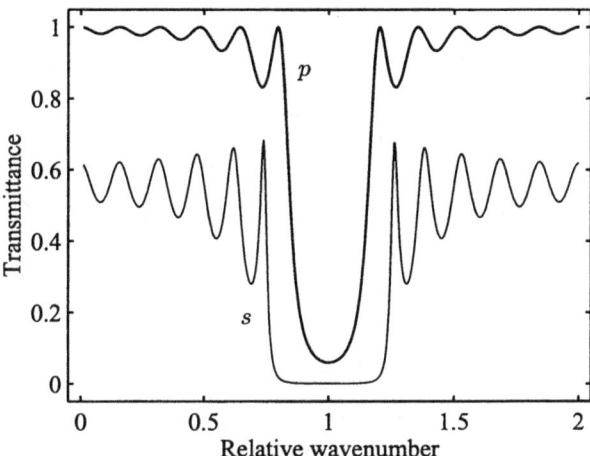

Fig. 13.6. Calculated values of T_p and T_s for a reflecting stack on a glass plate.

$n_H = 2.3$ and $N = 5$. The angle of incidence is the Brewster angle for the air-glass interface, 56.3°. Polarizers of this type, i.e. thin film versions of the pile of plates polarizer, are usually operated on the long wavelength edge of the transmission band of the reflecting stack, in the narrow wavelength zone between the p and s labels in Fig. 13.6, where T_p is close to unity and T_s is close to zero.

In practice the polarizers are used at defined laser wavelengths, and the restriction on operating bandwidth is not a problem. However, it is difficult to manufacture the coating so that the narrow peak near the p label in Fig. 13.6 occurs exactly at the required wavelength. An improved impedance and phase matched design that flattens the ripple has been reported.[58]

Figure 13.7 shows that the extinction coefficient, for the plate coated with the standard reflecting coating, is larger than 100 over a much greater wavelength band. Note that $\log(T_p/T_s)$ is plotted, and that T_p is unacceptably small

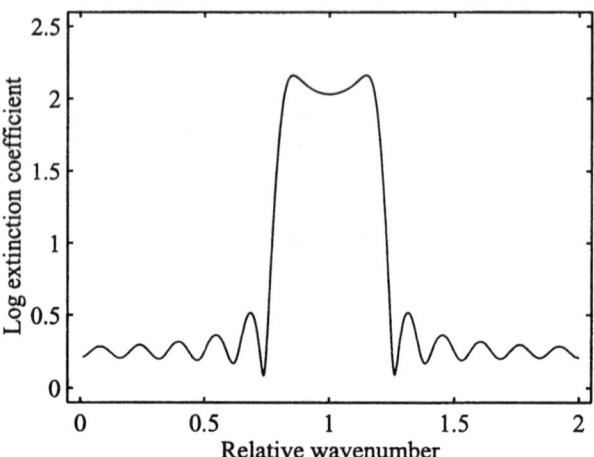

Fig. 13.7. Extinction coefficient for a reflecting stack on a glass plate.

over much of the wavelength band.

13.3.3 Embedded Thin Film Polarizers

The reason why the broad bandwidth of the extinction coefficient of the coated plate polarizer cannot be accessed is due to the asymmetry imposed by the air and glass bounding media. This disadvantage can be overcome, at the cost of a smaller laser damage threshold, by embedding the optical coating in glass. In the case of a plate polarizer a second glass plate is cemented over the coated surface, and it is the use of cement that lowers the LDT.

The most common form of embedded polarizer, though, is the polarizing cube beam-splitter. In the form that is illustrated in Fig. 13.8 the hypotenuse-face of a prism is coated and a second prism is attached with cement to complete the cube. An advantage of this type of polarizer is that the optical beam is not deviated laterally on transmission, as it would be in a plate polarizer.

An embedded coating can be designed to have $T_p = 1$, or $T_p \approx 1$, for all wavelengths. We discuss the principles that are involved by considering the design $g[HL]^N Hg$. First of all we note that the Brewster angle condition is satisfied in each L and H layer if

$$\beta = n_p \sin \theta_p = \frac{n_L n_H}{(n_L^2 + n_H^2)^{1/2}}. \tag{13.6}$$

Suppose that the above condition is satisfied by the prism index n_p and the angle of incidence on the coating in the prism θ_p. Then the only remaining

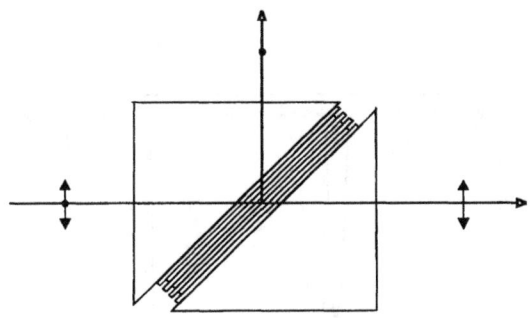

Fig. 13.8. Polarizing cube beam-splitter.

Fresnel reflections are from the two prism–H interfaces, the p reflectance from the entire structure is small and $T_p \approx 1$. If the prism index $n_p = n_L$ then these glass–H Fresnel reflections disappear as well and $T_p = 1$. An example illustrating the case $T_p \approx 1$ for a polarizing-cube beam-splitter is given in Figs. 13.9 and 13.10.

13.3.4 Birefringent Fabry-Perot Polarizing Filter

The theory of the Fabry-Perot interferometer and the related narrowband interference filter, constructed from isotropic materials, is well known and discussed in standard optics textbooks.[59] For this reason we shall confine our attention in this section to the adaptation of the theory to the birefringent polarizing filter.

Consider the basic arrangement of a birefringent spacer layer of thickness d sandwiched between isotropic reflecting films, as illustrated in Fig. 13.11 (a substrate is not shown). The birefringent spacer is defined, as in other sections of this book, by the principal indices n_1, n_2, n_3, and the angles η, ψ and ξ. In this section η, ξ and the angle of incidence θ are all assumed to be zero. This means that the p-polarized light (parallel to the deposition plane) "sees" refractive index n_p (defined by Eq.(3.34)) and the s-polarized light encounters index $n_s = n_3$.

The reflecting coatings could be metallic or dielectric, in either case we assume that each coating is characterized by reflectance R, transmittance T and absorptance $A = 1 - R - T$. Phase changes on reflection are neglected, as they are not major contributors to the principles that we wish to discuss. This is also our justification for another assumption, that R, T and hence A are all independent of polarization.

Multiple reflections in an isotropic plate lead to an equation for the trans-

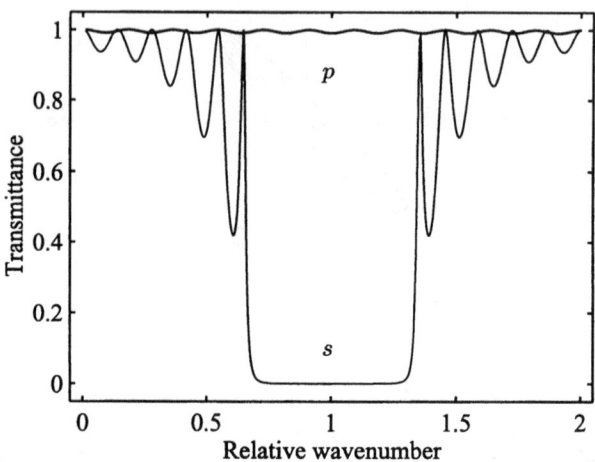

Fig. 13.9. Calculated p and s transmittances for a polarizing beam splitting cube.

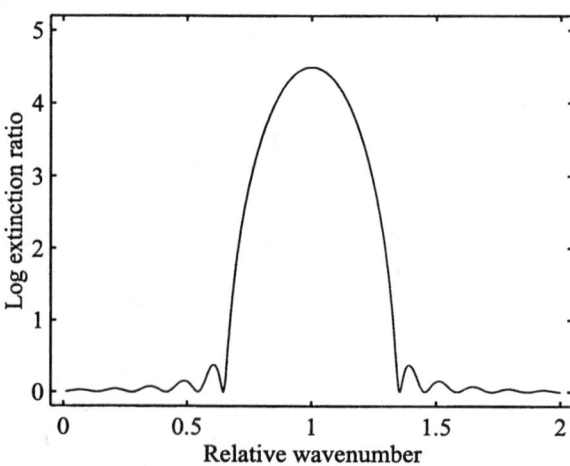

Fig. 13.10. Extinction coefficient for a polarizing beam splitting cube.

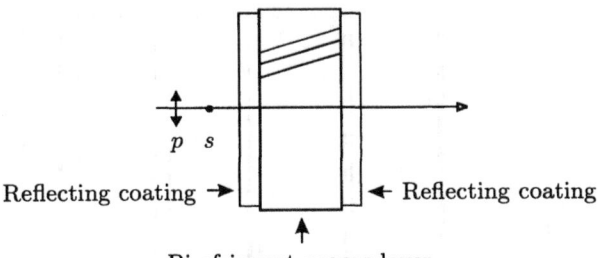

Fig. 13.11. Fabry-Perot polarizing filter.

mittance that is known as the Airy function. For the birefringent etalon we can write individual Airy functions for the p and s-polarizations

$$T_p(\delta_p) = \frac{1}{(1+A/T)^2} \times \frac{1}{1+\frac{4R}{(1-R)^2}\sin^2 \delta_p/2},$$

$$T_s(\delta_s) = \frac{1}{(1+A/T)^2} \times \frac{1}{1+\frac{4R}{(1-R)^2}\sin^2 \delta_s/2}, \tag{13.7}$$

where

$$\delta_p = \frac{2\pi}{\lambda}n_p d,$$

$$\delta_s = \frac{2\pi}{\lambda}n_s d. \tag{13.8}$$

A typical plot of the two transmittances (Fig. 13.12) shows two peaked functions (fringes) with slightly different periods. Note, from Eqs.(13.7), that the effect of absorption is to scale, but not change, the shape of the Airy functions. The p and s peaks occur whenever the appropriate condition for constructive interference is satisfied

$$\delta_p = 2m\pi,$$

$$\delta_s = 2m\pi. \tag{13.9}$$

Here the integer m is called the *order of interference*.

From Eqs.(13.7) it is a straightforward task to derive expressions for the maximum transmittance (at the tip of each peak) and the minimum transmittance (in the trough midway between peaks). Thus the maximum transmittance is given by

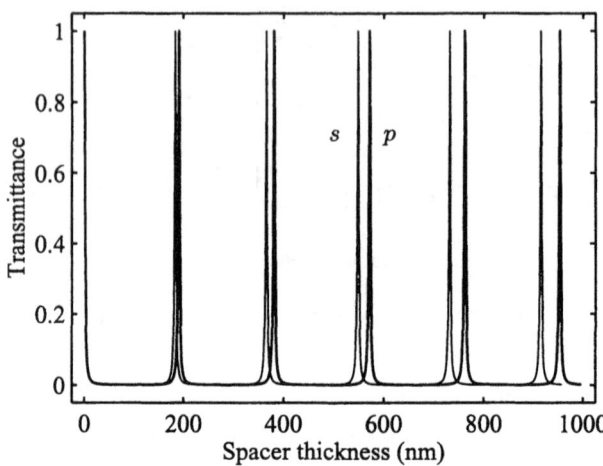

Fig. 13.12. Calculated transmittance curves for a Fabry-Perot polarizing filter.

$$(T_p)_{max} = (T_s)_{max} = \frac{1}{(1 + A/T)^2},\tag{13.10}$$

and hence in the absence of absorption both polarizations give maximum transmittances of 1. The minimum transmittances are given by

$$(T_p)_{min} = (T_s)_{min} = \frac{1}{(1 + A/T)^2} \times \frac{1}{1 + \frac{4R}{(1-R)^2}}.\tag{13.11}$$

The *fringe finesse* $\mathcal{F}$, defined as the fringe spacing divided by the FWHM of the peak, is an inverse measure of the fringe width. In "phase units" the fringe spacing is 2π, in order of interference or "m units" it is 1. In "wavelength units" the fringe spacing is called the *free spectral range* or the *spectral range without overlap*, and is given by the equation

$$\Delta\lambda_{fsr} = \frac{\lambda^2}{2nd}.\tag{13.12}$$

An expression for the finesse,

$$\mathcal{F} = \frac{\pi R^{1/2}}{1 - R},\tag{13.13}$$

can be derived from the form of the Airy function.

As the range of d in Fig. 13.12 is increased, the peaks of T_p and T_s periodically move apart and back together. A pair of p and s fringes, such as those shown in Fig. 13.12, will cut at half-height if the equations

$$2n_p d = m\lambda,$$

$$2n_s d = (m + \frac{1}{\mathcal{F}})\lambda, \tag{13.14}$$

are satisfied simultaneously. That occurs at

$$d_{1/2} = \frac{\lambda}{2|n_p - n_s|\mathcal{F}}. \tag{13.15}$$

An ideal value of thickness of the birefringent spacer, the smallest value required for a quarter-wave plate,

$$d = \frac{\lambda}{4|n_p - n_s|}, \tag{13.16}$$

would position the s fringes mid-way between the p fringes. However in practice, a smaller value, say 3 or 4 times the value of $d_{1/2}$, is more likely to minimize losses in a thin film spacer.

Finally, let us suppose that any value of birefringence is available. We can determine general expressions for the transmission efficiency and the extinction ratio of the Fabry-Perot polarizer. In particular, the transmission efficiency is just the maximum transmittance which we write as

$$T_p = \frac{1}{(1 + A/T)^2}, \tag{13.17}$$

and the extinction ratio, at a p-peak midway between a pair of s-peaks, is

$$T_p/T_s = 1 + \frac{4R}{(1 - R)^2}. \tag{13.18}$$

Further birefringent thin film polarizer designs are discussed in Chapter 16.

13.4 Crystalline Prism Polarizers

The highest quality polarizers, with the largest transmission efficiency, extinction ratio and LDT, are constructed from crystalline materials. We consider a representative range of crystalline polarizers that are available from commercial suppliers.

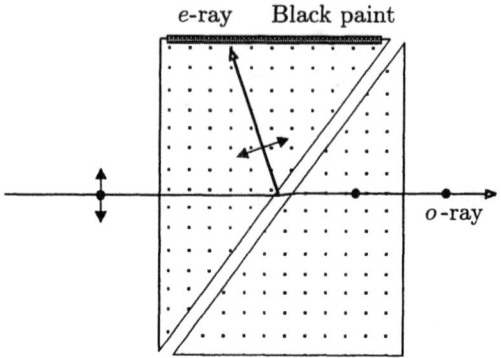

Fig. 13.13. Glan-Foucault prism polarizer.

13.4.1 Glan-Foucault Prism

The *Glan-Foucault polarizer*[57] consists of two prisms of calcite arranged with optic axes parallel and to make a rectangular block, as shown in Fig. 13.13. In one version, called the *Glan-Thompson polarizer*[57] the two prisms are either cemented together or the space between the prisms is filled with a liquid. With an air space between the prisms, as shown in the illustration, the device is called a *Glan-air polarizer*.[59] In both cases the operating principle is angular separation of the *o* and *e* rays at the calcite/air interface, where the *o*-ray suffers total internal reflection and the *e*-ray is transmitted.

The *o* and *e* rays arrive at the oblique interface travelling in the same direction, so that they have a common angle of incidence θ. For the Glan-air polarizer the condition for the *o*-ray to be totally reflected and the *e*-ray to be partially transmitted can be stated as

$$n_o \sin \theta > 1 > n_e \sin \theta. \tag{13.19}$$

Thus, θ is required to be larger than $\sin^{-1}(1/n_o) = 37.1°$, but smaller than $\sin^{-1}(1/n_e) = 38.6°$.

We can use the matrix method to calculate the transmittance of an *e*-ray through the Glan-air polarizer, including or excluding the effects of interference in the air gap. With the assumption that interference can be neglected, the total transmittance through the air gap is the square of the calcite-to-air transmittance. For $\lambda = 633$ nm the transmittance is 0.977, and so 2.3% of the *e*-ray is reflected at the oblique air layer. As well, 7.6% in total is reflected at the entrance and exit faces. In the *Glan-laser polarizer*[60] a higher transmission efficiency is achieved by modifying the interface angle to minimize reflection

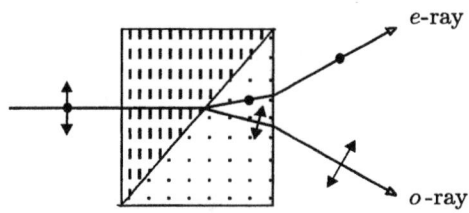

Fig. 13.14. Wollaston prism polarizer.

losses, and the entrance and exit faces are usually coated with AR coatings. A typical value for the transmission efficiency of a coated Glan-laser polarizer is 96%.

Some applications, including Q-switching, require a polarizer to be placed within the cavity of a laser. The highest available transmission efficiencies are needed for intracavity use, and values of about 98% are achieved by using Brewster angle entrance and exit windows. Another factor that needs to taken into account here, the effect of multiple-reflections within the prism, has been considered in detail by Z. Knittl.[61]

13.4.2 Wollaston Prism

In the *Wollaston prism*[57] two crystal prisms are positioned with perpendicular optic axes. In Fig. 13.14 two separated rays are shown passing through a Wollaston prism. The two rays have different character in the two prisms. Thus the o-ray in the left side is an e-ray in the right side. Two linearly polarized rays diverge as they pass through the oblique interface, and the angle of divergence increases as the rays pass out of the prism. Note that the angles of emergence of the e-ray and the o-ray are not exactly equal. As well, the two angles of emergence are wavelength-dependent.

13.4.3 Rochon Prism

The *Rochon prism*[57] illustrated in Fig. 13.15 is constructed by coupling a calcite prism with a glass prism; the refractive index and dispersion of the glass matches the appropriate index of calcite. The result is achromatic transmission of one ray without deviation.

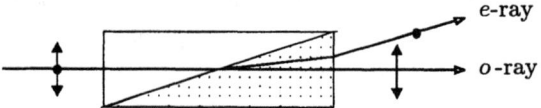

Fig. 13.15. Rochon prism polarizer.

Chapter 14

Phase Retarders

A polarized beam of light can be represented as the superposition of two linearly polarized beams. As in other sections of this book, we assume that the beam is travelling along the x-axis, that one of the components vibrates in the y-direction and the other vibrates along z. If the electric fields along y and z are in phase then the polarization of the beam is linear. If the field strengths are unequal and E_y lags E_z by $\pi/2$, then the polarization state of the beam is right elliptical.

From the above it follows that the way to change the state of polarization of a beam of light is to change the relative phase of the y and z components. In practice any optical component, in which off-axis reflections or transmissions occur, will have an effect on the phase difference. Several devices have been designed to produce a controlled change of phase, and these are usually referred to as retarders. Some make use of different p and s phase changes on total internal reflection, and others utilize the different propagation speeds of basis waves in a birefringent medium.

The material in this chapter is organized to allow comparisons to be made between

▨ established macroscopic material designs that utilize crystals and glasses, and

▨ birefringent thin film designs.

The latter are in an early development stage. As this chapter is being written (somewhat later than its deadline of course), working values of the three principal refractive indices of birefringent thin films are becoming available for the first time.

14.1   Crystalline Wave Plates

14.1.1   Quartz and Magnesium Fluoride

Crystal quartz grown artificially (using a hydrothermal process), is the preferred material for crystalline wave plates.[60] Quartz has a transparency range ($> 90\%$) extending from $0.25\,\mu$m to $2.5\,\mu$m. It is a positive uniaxial material, $n_e > n_o$, and the refractive indices n_e and n_o can be calculated using the Laurent equation[60,65]

$$n^2 = B_1 + B_2\lambda^2 + \frac{B_3}{\lambda^2} + \frac{C_1}{\lambda^4} + \frac{C_2}{\lambda^6} + \frac{C_3}{\lambda^8}. \tag{14.1}$$

The dispersion of each refractive index, and hence the dispersion of the birefringence of quartz, can be obtained by differentiating Eq.(14.1) with respect to λ,

$$\frac{dn}{d\lambda} = \left[B_2\lambda - \frac{B_3}{\lambda^3} - \frac{2C_1}{\lambda^5} - \frac{3C_2}{\lambda^7} - \frac{4C_3}{\lambda^9} \right] /n. \tag{14.2}$$

The function **quartz** in the *BTF Toolbox* returns values of n_e, n_o, $dn_e/d\lambda$ and $dn_o/d\lambda$ for a wavelength entered in μm units.

Some wave plate designs require a second birefringent material, and for this, magnesium fluoride is a popular choice. MgF_2, is a tetragonal crystal also grown artificially (using the vacuum Stockbarger technique).[60] The optical classification of magnesium fluoride is positive uniaxial, as for quartz, but the transparency range of 0.2–$0.6\,\mu$m is broader. The refractive indices n_e and n_o of MgF_2 may be calculated using the Sellmeier equation[60,65]

$$n^2 = 1 + \frac{B_1\lambda^2}{\lambda^2 - C_1} + \frac{B_2\lambda^2}{\lambda^2 - C_2} + \frac{B_3\lambda^2}{\lambda^2 - C_3}. \tag{14.3}$$

Differentiation with respect to λ gives an equation for the dispersion of the refractive indices,

$$\frac{dn}{d\lambda} = -\lambda \left[\frac{B_1 C_1}{(\lambda^2 - C_1)^2} + \frac{B_2 C_2}{(\lambda^2 - C_2)^2} + \frac{B_3 C_3}{(\lambda^2 - C_3)^2} \right] /n. \tag{14.4}$$

The function **mgf2** in the *BTF Toolbox* returns values of refractive indices and dispersions, in a similar way to the function **quartz**.

Values of n_e, n_o, $n_e - n_o$ and $d(n_e - n_o)/d\lambda$ calculated for a range of wavelengths using the functions **mgf2** and **quartz** are listed in Table 14.1. Reference to the table shows that, at all wavelengths in the range 0.25–2.5 μm, quartz has a smaller birefringence than MgF$_2$. In some applications, such as achromatic wave plates, quartz and MgF$_2$ plates are used in tandem. Here the critical requirement is different values of $d(n_e - n_o)/d\lambda$ divided by $n_e - n_o$.

14.1.2   Multiple-Order Wave Plates

Wave plates are normally constructed as a disc with the optic axis of the crystal parallel to the surface of the disc. For the positive materials quartz and MgF$_2$, light travels at slowest speed when it is polarized parallel to the optic axis. Thus, for these materials the optic axis is the slow axis.

The retardance R of a plate of thickness d is usually expressed in waves and, as it is normal practice to neglect interference effects, we have

$$R = (n_e - n_o)d/\lambda. \tag{14.5}$$

Hence the physical thickness of a *multiple-order quarter-wave plate* is

$$d = (m + 0.25)\lambda/(n_e - n_o), \tag{14.6}$$

where m is the order. Similarly, for a multiple-order half-wave plate,

$$d = (m + 0.5)\lambda/(n_e - n_o). \tag{14.7}$$

Equation (14.6), with $m = 0$, shows that at $\lambda = 0.85\,\mu$m the thinnest quarter-wave plate made from quartz would have a thickness of just 24 μm. Such a thin plate would be difficult to cut and polish, and for this reason, thicker plates with larger values of m are used in practice. For example, for $m = 10$ the required value of d is close to 1 mm.

A series of wavelengths satisfies the quarter-wave condition for a multiple-order wave plate. The closest smaller wavelength for the 1 mm plate is

$$\begin{aligned}
\lambda_{m+1} &= \frac{10 + 0.25}{11 + 0.25}\lambda_m \\
&= 0.774\,\mu\text{m}. \tag{14.8}
\end{aligned}$$

Some designs take advantage of the multiplicity of wavelengths that satisfy the wave plate condition. As an example, a *dual-wavelength wave plate*, that is full-wave at the Nd:YAG fundamental wavelength of 1.064 μm and half-wave

Table 14.1. Refractive indices of MgF$_2$ and quartz.

	Magnesium Fluoride				Quartz			
λ	n_e	n_o	$n_e - n_o$	$\frac{d(n_e-n_o)}{d\lambda}$ $\times 100$	n_e	n_o	$n_e - n_o$	$\frac{d(n_e-n_o)}{d\lambda}$ $\times 100$
0.25	1.4156	1.4028	0.0128	-0.9698	1.6116	1.6005	0.0111	-2.1821
0.30	1.4054	1.3930	0.0124	-0.5647	1.5883	1.5780	0.0103	-1.1414
0.35	1.3996	1.3874	0.0122	-0.3555	1.5754	1.5655	0.0099	-0.6857
0.40	1.3959	1.3839	0.0121	-0.2394	1.5673	1.5577	0.0096	-0.4400
0.45	1.3935	1.3815	0.0120	-0.1704	1.5619	1.5525	0.0094	-0.2986
0.50	1.3917	1.3798	0.0119	-0.1271	1.5580	1.5488	0.0093	-0.2140
0.55	1.3904	1.3785	0.0118	-0.0987	1.5552	1.5460	0.0092	-0.1616
0.60	1.3893	1.3775	0.0118	-0.0794	1.5529	1.5438	0.0091	-0.1283
0.65	1.3885	1.3767	0.0118	-0.0660	1.5511	1.5421	0.0090	-0.1067
0.70	1.3878	1.3761	0.0117	-0.0565	1.5497	1.5407	0.0090	-0.0925
0.75	1.3872	1.3755	0.0117	-0.0497	1.5484	1.5394	0.0090	-0.0831
0.80	1.3867	1.3751	0.0117	-0.0447	1.5473	1.5384	0.0089	-0.0770
0.85	1.3863	1.3746	0.0117	-0.0411	1.5463	1.5374	0.0089	-0.0731
0.90	1.3859	1.3743	0.0116	-0.0384	1.5454	1.5366	0.0088	-0.0708
0.95	1.3855	1.3739	0.0116	-0.0365	1.5446	1.5358	0.0088	-0.0696
1.00	1.3852	1.3736	0.0116	-0.0352	1.5438	1.5350	0.0088	-0.0693
1.10	1.3846	1.3730	0.0116	-0.0338	1.5423	1.5336	0.0087	-0.0703
1.20	1.3840	1.3724	0.0115	-0.0334	1.5409	1.5323	0.0086	-0.0728
1.30	1.3834	1.3719	0.0115	-0.0338	1.5395	1.5309	0.0086	-0.0763
1.40	1.3828	1.3713	0.0115	-0.0347	1.5381	1.5296	0.0085	-0.0803
1.50	1.3822	1.3708	0.0114	-0.0359	1.5367	1.5283	0.0084	-0.0848
1.60	1.3816	1.3702	0.0114	-0.0373	1.5352	1.5269	0.0083	-0.0895
1.70	1.3810	1.3697	0.0114	-0.0389	1.5337	1.5255	0.0082	-0.0945
1.80	1.3804	1.3691	0.0113	-0.0406	1.5321	1.5240	0.0081	-0.0996
1.90	1.3798	1.3685	0.0113	-0.0425	1.5305	1.5225	0.0080	-0.1048
2.00	1.3791	1.3678	0.0112	-0.0444	1.5288	1.5209	0.0079	-0.1102
2.10	1.3784	1.3672	0.0112	-0.0464	1.5270	1.5192	0.0078	-0.1156
2.20	1.3776	1.3665	0.0111	-0.0485	1.5252	1.5175	0.0077	-0.1211
2.30	1.3769	1.3658	0.0111	-0.0506	1.5233	1.5157	0.0076	-0.1267
2.40	1.3761	1.3650	0.0110	-0.0527	1.5213	1.5138	0.0074	-0.1323
2.50	1.3753	1.3643	0.0110	-0.0549	1.5192	1.5119	0.0073	-0.1381

at the harmonic at $0.355\,\mu$m, can be used to transmit the fundamental without change in polarization state, but rotate the plane of polarization of the harmonic. Such a plate is called a *harmonic wave plate*.

14.1.3 Zero-Order Wave Plate

Some advantages and disadvantages of multiple-order wave plates are discussed in the previous section. Now focusing on the disadvantages, multiple-order plates are sensitive to effects such as changes in temperature and small angular misalignments. These effects depend on the total thickness of the plate and can be reduced by constructing a double plate of quartz, with optic axes at 90° to each other, so that one plate cancels the retardation of the other apart from the required 0.25-wave or 0.5-wave retardation. An example is the combination of a 20-wave plate and a 20.25-wave plate with fast and slow axes aligned. For such a *zero-order wave plate* temperature and angular effects are associated with the thickness corresponding to 0.25-wave retardation.

14.1.4 Achromatic Wave Plates

The zero-order quarter-wave plate, constructed from two thicker plates of the same material as described in the previous section, has the chromatic properties of a single quarter-wave plate of thickness d given by Eq.(14.6) with $m = 0$. The rate at which the retardance of the plate changes with wavelength can be obtained by differentiating Eq.(14.5) and substituting 0.25 for $(n_e - n_o)d/\lambda$,

$$\frac{dR}{d\lambda} = \frac{0.25}{n_e - n_o}\frac{d(n_e - n_o)}{d\lambda} - \frac{0.25}{\lambda}. \tag{14.9}$$

The right-hand side of Eq.(14.9) is dominated by the second term, and direct cancellation of the two terms is not possible using a single material. A plot of retardance versus wavelength for the zero-order quarter-wave plate is given in the upper part of Fig. 14.1; the ideal retardation is 0.25-waves. The lower part of the same figure shows the transmittance $T(\lambda)$ of the wave plate followed by an ideal linear polarizer and illuminated with unpolarized light, as illustrated in Fig. 14.2. The transmittance of a wave plate in this configuration is given by the equation

$$T(\lambda) = 0.5 + 0.5\cos[2\pi w(\lambda)]. \tag{14.10}$$

With $w = 0$ (zero retardation) the ideal transmittance is $T = 1$. The ideal value for a quarter-wave plate, with $w = 0.25$, is $T = 0.5$, and the ideal value for a half-wave plate is $T = 0$.

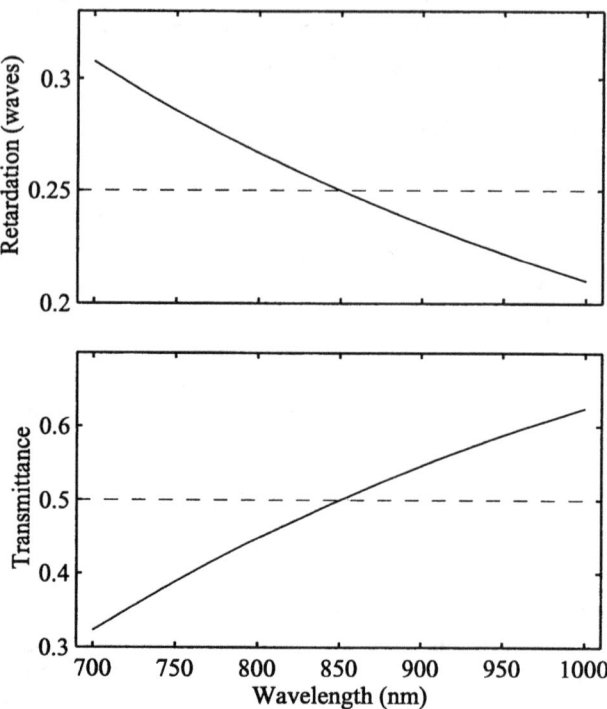

Fig. 14.1. Variation of retardance and transmittance with wavelength calculated for a zero-order quarter-wave plate.

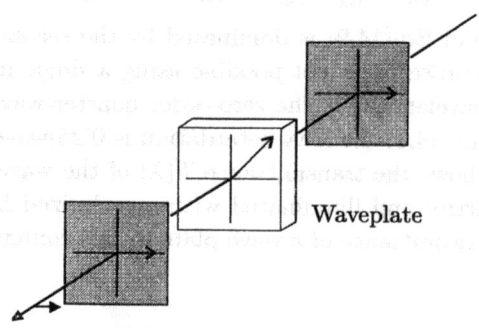

Fig. 14.2. Arrangement for defining the transmittance $T(\lambda)$ of a wave plate. The fast axis of the wave plate makes an angle of 45° with respect to the transmission axes of the aligned linear polarizers.

Now consider a wave plate constructed from two materials, MgF$_2$ (M) and quartz (Q). A series of plates may be used, and we assume that the axes of the individual plates are aligned. Considering the M-plates alone, and possible cancellation of retardances when fast and slow axes are aligned, we define the equivalent physical thickness of the M-plates as d_M and the uncompensated retardance as w_M. Similarly, the equivalent thickness of the Q-plates alone is d_Q and the uncompensated retardance is w_Q. Then the retardance of the complete wave plate in waves is given by the equation

$$R = \frac{(n_e - n_o)_M d_M}{\lambda} - \frac{(n_e - n_o)_Q d_Q}{\lambda}. \tag{14.11}$$

Differentiating Eq.(14.11) with respect to λ, equating $dR/d\lambda$ to zero and rearranging gives

$$\frac{w_M}{w_Q} = -\frac{\left[\dfrac{1}{n_e - n_o}\dfrac{d(n_e - n_o)}{d\lambda} - \dfrac{1}{\lambda}\right]_Q}{\left[\dfrac{1}{n_e - n_o}\dfrac{d(n_e - n_o)}{d\lambda} - \dfrac{1}{\lambda}\right]_M}. \tag{14.12}$$

Now consider the design of an achromatic wave plate for the wavelength range 0.7–1.1 μm. For a two-material, two-plate, zero-order, quarter-wave plate we have to satisfy an additional equation,

$$w_M + w_Q = 0.25, \tag{14.13}$$

and there is some advantage in using a central wavelength that requires w_M and w_Q to be integral numbers of quarter-waves. By inspection these conditions are found to be satisfied with $\lambda = 0.828\,\mu$m, $w_M = 7$ waves, $w_Q = -6.75$ waves.

The critical parameter for achromatization is the ratio of uncompensated retardances of magnesium fluoride and quartz. Thickness can be added to the two-plate design, provided that the ratio of uncompensated retardances given by Eq.(14.12) is maintained. Thus the quarter-wave plate design 27M -26.75Q 20Q -20M has the same spectral characteristic as 7M -6.75Q.

Figure 14.3 shows the performance of such an *achromatic quarter-wave plate*. Note that, when $\Delta w = w(\lambda) - 0.25$ is small, Eq.(14.10) can be replaced by

$$T(\lambda) \approx 0.5 - \pi \Delta w. \tag{14.14}$$

The performance of an *achromatic half-wave plate* is shown in Figure 14.4. In this case Eq.(14.10) can be approximated by

$$T(\lambda) \approx (\pi^2 \Delta w)^2. \tag{14.15}$$

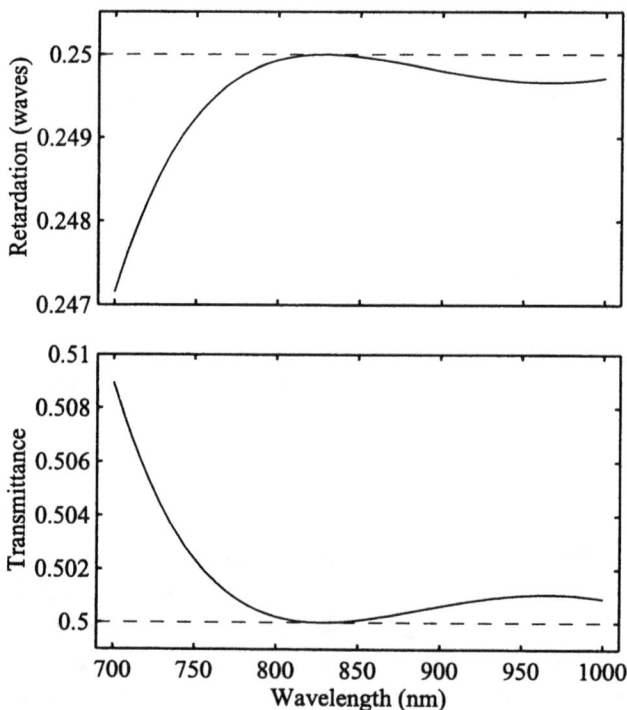

Fig. 14.3. Retardance and transmittance calculated for an achromatic quarter-wave plate.

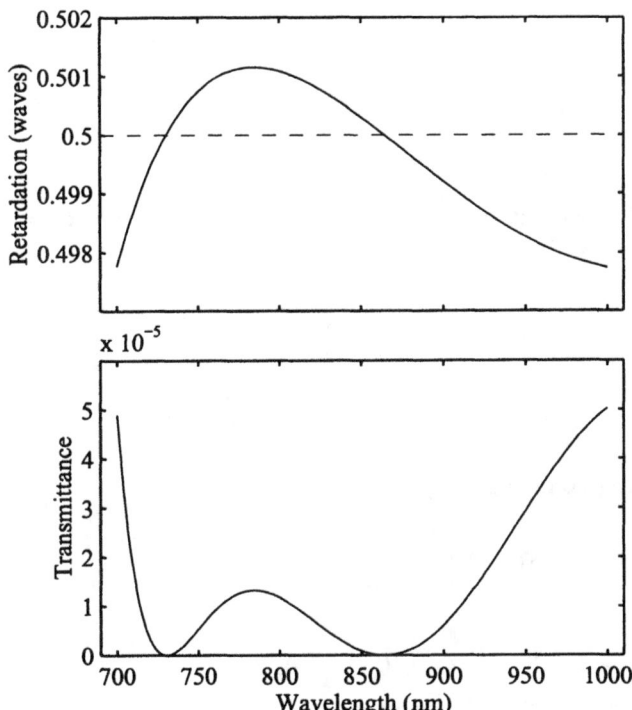

Fig. 14.4. Retardance and transmittance calculated for an achromatic half-wave plate.

14.1.5   Wide-Field Elements

The phase retardation of a wave plate is equal to the difference in the phase thicknesses seen by waves that travel as e-waves and o-waves,

$$\Delta = k(\alpha_e^+ - \alpha_o^+)d. \tag{14.16}$$

Here α_e^+ and α_o^+ are given by the quadratics in Eq.(3.38),

$$\alpha_e^2 = n_e^2 \left[1 - \beta^2 \left(\frac{\cos^2 \xi}{n_o^2} + \frac{\sin^2 \xi}{n_e^2} \right) \right], \tag{14.17}$$

and

$$\alpha_o^2 = n_o^2 \left(1 - \frac{\beta^2}{n_o^2} \right). \tag{14.18}$$

For small angles of incidence, $\beta^2 \ll n_e^2, n_o^2$, we can write approximate solutions for α_e^+ and α_o^+ as

$$\alpha_e^+ = n_e \left[1 - \frac{1}{2}\beta^2 \left(\frac{\cos^2 \xi}{n_o^2} + \frac{\sin^2 \xi}{n_e^2} \right) \right], \tag{14.19}$$

$$\alpha_o^+ = n_o \left(1 - \frac{1}{2}\frac{\beta^2}{n_o^2} \right), \tag{14.20}$$

and hence (after some algebra)

$$\Delta(\beta, \xi) = \Delta(0,0) \left\{ 1 - \frac{\beta^2}{2n_o^2} \left[1 - \frac{(n_e + n_o)\sin^2 \xi}{n_e} \right] \right\}. \tag{14.21}$$

where

$$\Delta(0,0) = kd(n_e - n_o). \tag{14.22}$$

Next, we suppose that the birefringence is small, $n_e - n_o \ll n_e, n_o$, and that the plate is immersed in air ($n = 1$), so that $\beta \approx \theta$. With these simplifications, substitution of Eqs.(14.19) and (14.20) into Eq.(14.16) yields

$$\Delta(\theta, \xi) = kd(n_e - n_o) \left[1 - \frac{\theta^2}{2n_o^2}(1 - 2\sin^2 \xi) \right]. \tag{14.23}$$

Equation (14.23) shows that angular effects are most significant when the plate is oriented with $\xi = 0°$ or $\xi = 90°$, and least significant when $\xi = \pm 45°$. The *wide-field element* [62] devised by Lyot and illustrated here in Fig. 14.5 makes use of these results. The split arrangement on the right-hand side of the figure has the same retardation as the single plate shown on the left-hand side, as can

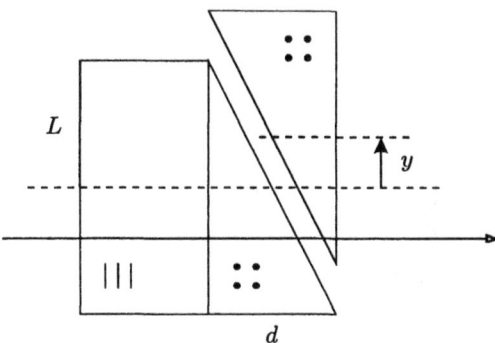

Fig. 14.7. Soleil-Babinet compensator.

gives the full-aperture phase retardation as a function of the displacement of the moveable wedge.

Berek Compensator

The phase retardation of a wave plate depends on the angle of incidence, θ, of the light beam. The *Berek compensator*,[2] a wave plate with optic axis perpendicular to the surface, makes use of this angular effect. When $\theta = 0$ the phase retardation is equal to zero, and when the plate is tilted, the phase retardation is given by $k(\alpha_1^+ - \alpha_2^+)d$. Using Eq.(3.39) for α_1^+ and α_2^+, and assuming that $\Delta n = n_e - n_o \ll n_o$ and that θ is a small angle, leads to the approximation

$$\Delta(\theta) = \frac{\Delta n}{2n_o}\theta^2. \tag{14.29}$$

14.2 Birefringent Thin Film Analogues

14.2.1 Thin Film Wave Plates

A single biaxial film $[n_1\ n_2\ n_3\ 0\ \psi\ 0\ d/\lambda]$ deposited on to a glass substrate, as illustrated in Fig. 14.8, acts as a thin film wave plate.[31] The retardance of the plate is sensitive to angular rotation, particularly to rotation about the z-axis (perpendicular to the plane of the diagram). Thus, the single film wave plate can be coarse-tuned by rotation about z and fine-tuned by rotation about y.

If the plate is required to have a large aperture, then thickness (and hence phase) variations associated with the deposition geometry can be overcome by depositing the double layer $[n_1\ n_2\ n_3\ 0\ \psi\ \pi\ 0.5d/\lambda]$, $[n_1\ n_2\ n_3\ 0\ \psi\ 0\ 0.5d/\lambda]$ shown in Fig. 14.9.[31] The herring-bone microstructure serves a double purpose,

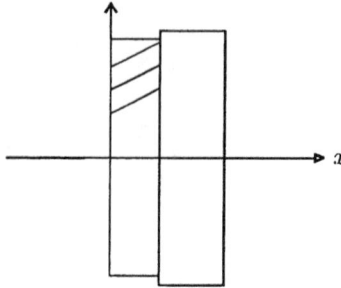

Fig. 14.8. Single film wave plate.

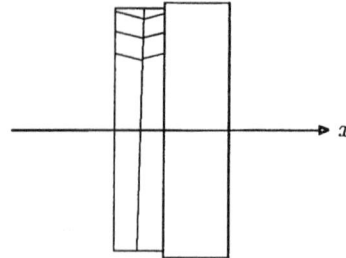

Fig. 14.9. Double film wave plate.

as it reduces the sensitivity of retardance to rotation of the plate about the
z-axis. Thus the herring-bone thin film wave plate is a wide-angle wave plate.
For small angles of rotation the retardance varies quadratically with angle.

In practice, an additional ion-assisted isotropic layer is required to seal the
columnar microstructure of a biaxial thin film. Further, it is difficult to deposit
a film with an exact value of phase retardation, and for this reason, a means of
making post-deposition adjustments is needed. The composite structure shown
in Fig. 14.10 allows adjustment through lateral displacement of the two halves
before they are cemented or fixed together. It is equivalent to the single film
plate described above.

An alternative placement of the components, shown in Fig. 14.11, leads to a
composite plate equivalent to the the herring-bone or wide-angle plate described
in the previous paragraph.

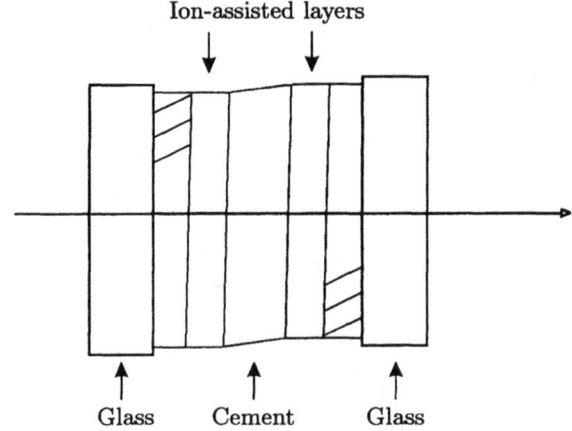

Fig. 14.10. Composite single film wave plate.

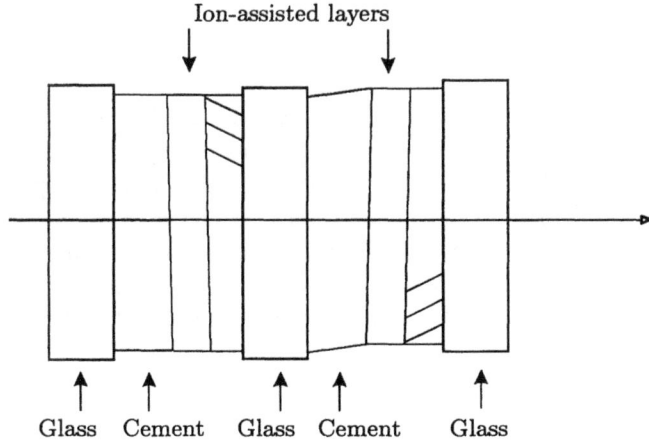

Fig. 14.11. Composite double film wave plate.

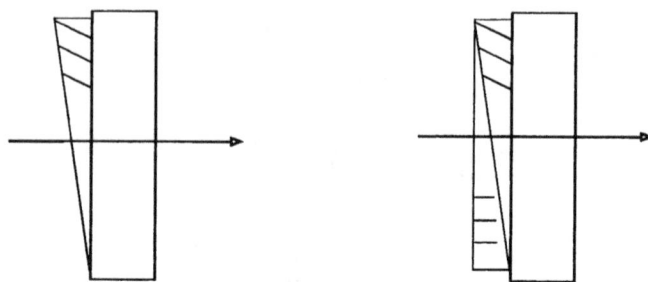

Fig. 14.12. Wedged biaxial film (left) and thin film Babinet compensator (right).

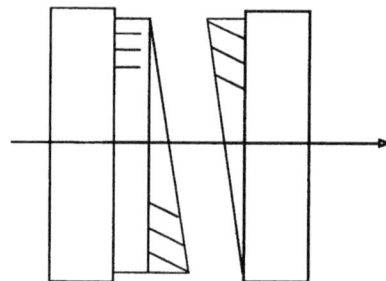

Fig. 14.13. Thin film Soleil-Babinet compensator.

14.2.2   Thin Film Babinet Compensator

The wedged biaxial film shown in the left-hand side of Fig. 14.12 produces a variable phase rotation, but always with the same sign. We have produced such a coating by moving a substrate mask steadily during deposition. The pair of films shown in the right-hand side of the figure forms a thin film analogue of the Babinet compensator. Here the materials are $[n_1 \; n_2 \; n_3 \; 0 \; \psi \; 0]$, $[n_1 \; n_2 \; n_3 \; 0 \; \psi \; \pi/2]$ and both films are wedged along the y-direction. Positive and negative retardations can be selected for a narrow beam of light.

14.2.3   Thin Film Soleil-Babinet Compensator

The set of three films shown in Fig. 14.13, a plane parallel layer of material $[n_1 \; n_2 \; n_3 \; 0 \; \psi \; \pi/2]$ and a wedged layer of material $[n_1 \; n_2 \; n_3 \; 0 \; \psi \; 0]$ on one substrate and a wedged layer of material $[n_1 \; n_2 \; n_3 \; 0 \; \psi \; 0]$ on a second substrate, makes a thin film analogue of the Soleil-Babinet compensator.

14.2.4   Thin Film Berek Compensator

The uniaxial film of material $[n_1\ n_2\ n_2\ 0\ 0\ 0]$, deposited at normal incidence, is an angular analogue of the Berek compensator.

Chapter 15

Birefringent Filters

Birefringent filters are used in a variety of applications and scientific disciplines. As examples, the Lyot-Ohman filter is used in astronomy and another series-of-plates filter is used as an intracavity tuning device for dye lasers. Birefringent filters have specific properties that cannot be matched by conventional high-resolution devices such as the Fabry-Perot interferometer. Thus, the Lyot-Ohman filter constructed with wide-field elements provides both a wide field of view and a narrow band of wavelengths.

In the above examples the primary function of the birefringent filter is to produce a narrow band of wavelengths. A secondary function is to filter the polarization of the transmitted beam. In the case of tuning a dye laser, with a stack of birefringent plates tilted at the Brewster angle, the secondary function ensures that the output beam from the laser is linearly polarized. Hence it is convenient to broaden the scope of devices classed as birefringent filters, to include elements that filter or change the state of polarization of a beam of light. Within this broadened scheme the primary function of a birefringent filter may be to isolate or eliminate a particular *state of polarization*, or to isolate or eliminate a particular *wavelength* or narrow band of wavelengths.

In this penultimate chapter we discuss

▨ polarization state filters

▨ wavelength filters.

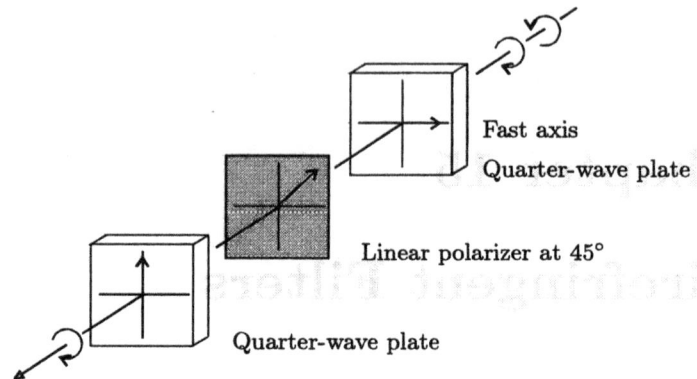

Fig. 15.1. Circular polarizer.

15.1 Polarization State Filters

15.1.1 Linear Polarizer

The linear polarizer can be regarded as a polarization state filter. That is, the action performed by a linear polarizer is to pass light polarized parallel to the transmission direction and reject light polarized in the perpendicular direction.

15.1.2 Circular Polarizer

The *circular polarizer* illustrated in Fig. 15.1 transmits the $\mathcal{R}$-state and rejects the $\mathcal{L}$-state. If the incident light is unpolarized then, as before, the transmitted light is in the $\mathcal{R}$-state.

15.1.3 Rotator

A half-wave plate acts as a *rotator* for linearly polarized light. As illustrated in Fig. 15.2, incident light polarized at angle ξ to the fast axis of the half-wave plate emerges polarized at angle $-\xi$; that is, the plane of polarization is rotated by 2ξ.

The half-wave plate rotator is sensitive to alignment, which is an advantage in that any angle of rotation of the vibration direction of the light can be attained, but a disadvantage if the requirement is for a fixed rotation, such as $90°$. A rotator that is alignment-insensitive can be made from a plate of quartz, cut so that the light passes down the optic axis (Fig.15.3). The rotation of the beam is due to optical activity in the crystal. Fresnel provided an explanation of the effect. The linearly polarized state is equivalent to the superposition

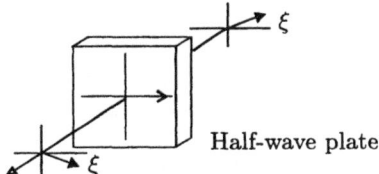

Fig. 15.2. Half-wave plate rotator.

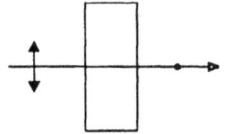

Fig. 15.3. Optically-active rotator.

of an $\mathcal{R}$-state and an $\mathcal{L}$-state. The two circularly polarized states propagate at different speeds – the material is said to exhibit *circular birefringence* – and the direction of the resultant linear polarization rotates with distance of propagation.

15.1.4 Depolarizer

The element illustrated in Fig. 15.4 is known as a *depolarizer* or as a *scrambler*. It can be constructed from glass and quartz and is used to depolarize a beam of light. An incident beam, parallel and linearly polarized, emerges as a beam that is still parallel but effectively depolarized; the state of polarization is elliptical at any point in the transmitted beam but the shape of the ellipse changes rapidly with distance across the beam.

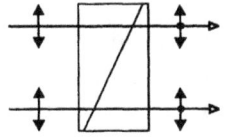

Fig. 15.4. Depolarizer.

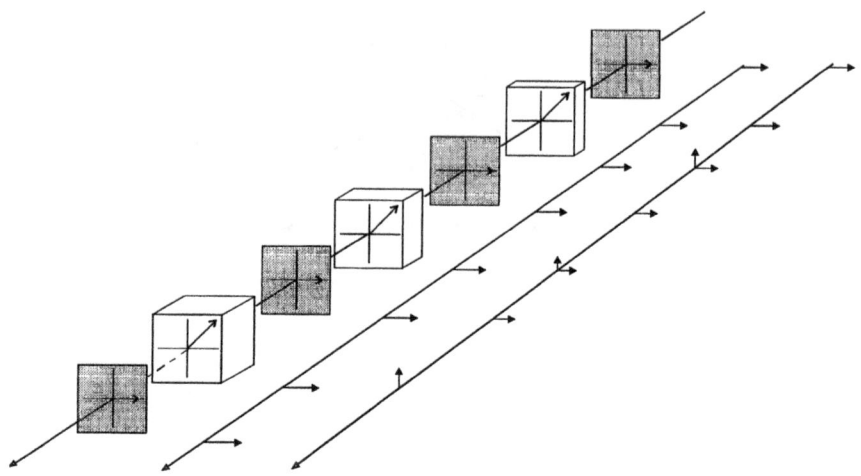

Fig. 15.5. Lyot-Ohman birefringent filter.

15.2 Wavelength Filters

15.2.1 Lyot-Ohman Filter

The Lyot-Ohman filter[62] is illustrated in Fig. 15.5. It consists of a series of
birefringent plates sandwiched between polarizers. The transmission axes of
the polarizers are all aligned, with the y-axis in our illustration. In general
the birefringent plates have thicknesses given by the geometric progression, d,
$2d$, $4d$, $8d$, ..., and the axes of the birefringent plates are aligned at 45° to
the transmission axes of the polarizers. For simplicity we assume that the
birefringent material is uniaxial and that the plates are cut so that the optic
axis is in the surface. Dispersion of birefringence is neglected, and the polarizers
are assumed to be ideal.

The basic unit of the Lyot-Ohman filter is a simple wave plate oriented at 45°
between a pair of polarizers in transmission mode, as shown in Fig. 15.6. The
transmittance of the first unit, when it is illuminated by light linearly polarized
along the y-axis, is given by

$$T_1 = \cos^2 \frac{\delta}{2}, \tag{15.1}$$

where

$$\delta = k(n_e - n_o)d. \tag{15.2}$$

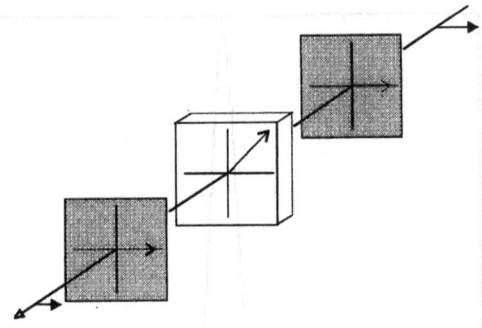

Fig. 15.6. Basic unit of Lyot-Ohman filter.

Suppose that the complete filter has N stages. Then the overall transmittance of the N stages (which are in series) can be found as the product of transmittance terms that are similar in form to T_1. This gives the equation

$$T = \frac{1}{2}\cos^2(\frac{\delta}{2})\cos^2(\frac{2\delta}{2})\cos^2(\frac{4\delta}{2})\ldots\cos^2(\frac{2^{N-1}\delta}{2}) \qquad (15.3)$$

for the transmittance. The transmittance curve for 4 plates ($N = 4$) is plotted in Fig. 15.7.

The Lyot-Ohman filter can be described as a multiple-beam interference filter, and it is instructive to consider an alternative derivation of the transmittance function, which highlights the way that the Lyot-Ohman filter generates a set of beams with different phases. We begin by considering the action of the first wave plate, and suppose that light polarized along the direction marked on the plate is transmitted without change of phase whereas light polarized perpendicular to the marked direction has a phase retardation represented by $e^{i\delta}$.

Linearly polarized light of unit amplitude that is incident on the first plate may be resolved into components of amplitude $1/\sqrt{2}$ parallel and perpendicular to the marked direction. After passing through the plate the two beams have amplitudes $1/\sqrt{2}$ and $e^{i\delta}/\sqrt{2}$, and resolving back along the y-direction gives two superposed beams with amplitude $1/2$ and $e^{i\delta}/2$. These beams are both polarized parallel to the transmission direction of the polarizer, and so are transmitted to the next stage. Carrying out similar processes, individually on the two beams, shows that four beams represented by $1/4$, $e^{i\delta}/4$, $e^{i2\delta}/4$ and $e^{i3\delta}/4$ arrive at the third stage of the filter. In this way we can form the amplitude transmission of a filter with N plates as

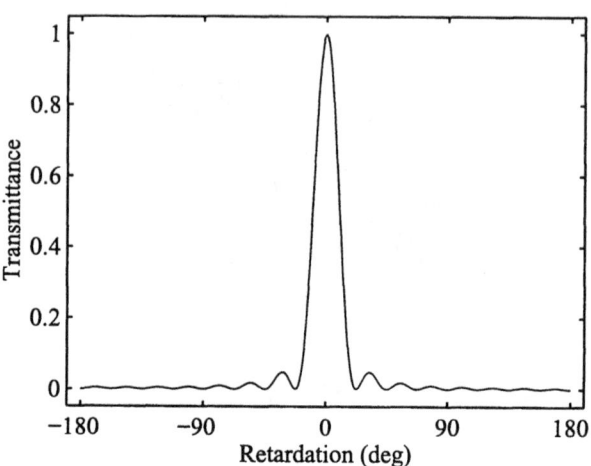

Fig. 15.7. Calculated transmittance of of a 4-stage Lyot-Ohman birefringent filter.

$$t = \frac{1}{2^N}[1 + e^{i\delta} + e^{i2\delta} + e^{i3\delta} \ldots e^{i(2^N-1)\delta}]. \tag{15.4}$$

The set of beams generated as light passes through the *longitudinally* placed plates in the Lyot-Ohman filter is the same as the set of beams generated by the *laterally* placed slits of a diffraction grating. For this reason aspects of the theory of the Lyot-Ohman filter and the theory of the diffraction grating are the same. The transmittance of the Lyot-Ohman filter, obtained by summing the geometric series and forming tt^*, can be written as

$$T = \frac{1}{2^{2N}} \frac{\sin^2 2^N \delta/2}{\sin^2 \delta/2}. \tag{15.5}$$

We have included this function in the *BTF Toolbox* as **lyot**(N, δ).

Figure 15.5, which shows the arrangement of plates in the Lyot-Ohman filter, also shows the stage-by-stage transmission of light through the filter for two cases. For a set of wavelengths defined by

$$\lambda_m = \frac{|n_e - n_o|d}{m}, \tag{15.6}$$

where m is a positive integer greater than zero called the order, the retardance of the *thinnest* plate in the filter is an integral multiple of 2π and hence every plate in the filter acts as a full-wave plate for the set of wavelengths. The *passband wavelengths*, as they are called, are transmitted through the filter without loss, and form a set of principal maxima in the transmission spectrum.

Note that the passband wavelengths of the filter are not equally spaced. The corresponding wavenumbers $\kappa_m = m/(n_e - n_o)d$ are spaced uniformly, but it is usual practice to define the working parameters in terms of wavelengths.

The free spectral range of the filter, $\Delta\lambda_{fsr}$, is the magnitude of the difference of neighbouring passband wavelengths. From the paragraph above we see that the corresponding wavenumber difference is $\Delta\kappa_{fsr} = 1/|n_e - n_o|d$. Provided that the wavenumber difference is much smaller than the wavenumber, we can relate the incremental quantities by $\Delta\lambda = \lambda^2 \Delta\kappa$. Hence

$$\Delta\lambda_{fsr} = \frac{\lambda^2}{|n_e - n_o|d}. \tag{15.7}$$

When the order for the thinnest plate is m, the order for the thickest plate is $m' = (2^{N-1})m$, and we can write $\kappa_{m'} = m'/(n_e - n_o)(2^{N-1})d$. A second set of wavenumbers, defined by $\kappa_{m'+1/2} = (m'+1/2)/(n_e - n_o)(2^{N-1})d$ sees the *thickest* plate in the filter as a half-wave plate and, following transmission through the thickest plate, these wavenumbers are lost at the last polarizer, as shown in Fig. 15.5. They form the zeros of transmittance that are closest to the principal peaks in the transmittance spectrum.

We define the *bandwidth* of the filter, also called the *minimum resolvable wavelength difference* $\Delta\lambda_{min}$, as the wavelength difference between the centre of a peak and the closest zero in the transmittance spectrum. This is equivalent to the criterion that is used for the diffraction grating. From the previous paragraph it is clear that the bandwidth of the Lyot-Ohman filter is determined entirely by the thickest plate in the filter. As a wavenumber difference the bandwidth is $\Delta\kappa = \kappa_{m'+1/2} - \kappa_{m'} = 1/2(n_e - n_o)(2^{N-1})d = 1/(n_e - n_o)2^N d$, and hence

$$\Delta\lambda_{min} = \frac{\lambda^2}{2^N|n_e - n_o|d}. \tag{15.8}$$

The *finesse* is defined as the spacing of the principal peaks in the transmittance spectrum ($\Delta\lambda_{fsr}$) divided by the fringe width ($\Delta\lambda_{min}$),

$$\mathcal{F} = \frac{\Delta\lambda_{fsr}}{\Delta\lambda_{min}}. \tag{15.9}$$

Hence for the Lyot-Ohman filter the finesse is equal to 2^N, the number of beams in the set listed in Eq.(15.4).

15.2.2 Solc Filters

Solc filters[62] overcome a disadvantage of Lyot-Ohman filters, excessive loss of energy in the large number of polarizers that are interleaved with the birefrin-

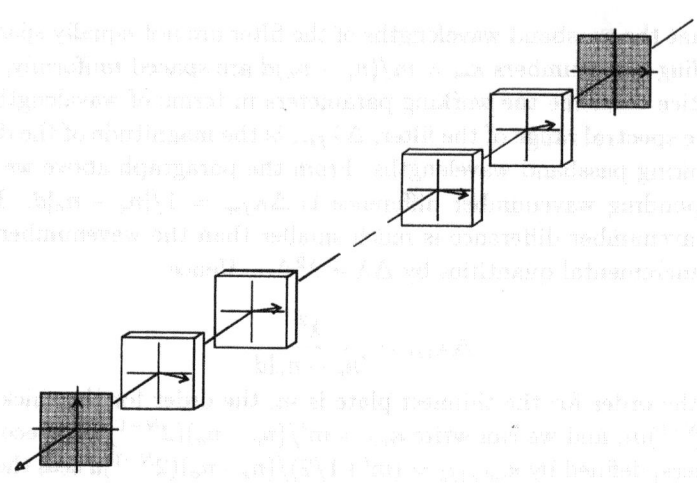

Fig. 15.8. Folded Solc filter.

gent plates. Solc filter designs use just two polarizers, one at the entrance and another at the exit, and hence the losses are smaller.

Folded Solc Filter

In the *folded Solc filter*[62] (illustrated in Fig. 15.8) the entrance and exit polarizers are crossed and the birefringent plates all have the same physical thickness d and phase thickness δ. Alternate plates are rotated by small *rocking angles* $\pm\xi_r$ defined by $\xi_r = \pi/4N$, where N is the number of plates. Wavelengths for which each plate is a half-wave plate are transmitted without loss through the filter. The transmittance curve shown in Fig. 15.9, for a folded Solc filter with 16 plates, was calculated using the function **solc**(N, δ) from the *BTF Toolbox*.

Fan Solc Filter

A typical *fan Solc filter*[62] has N identical plates, with rotation angles of ξ_r, $3\xi_r$, $5\xi_r$, $\ldots \pi - 3\xi_r$, $\pi - \xi_r$, located between a pair of parallel polarizers as shown in Fig. 15.10. The transmittance of the fan-type Solc filter is the same as for the folded Solc filter (Fig. 15.9), but shifted in phase by π.

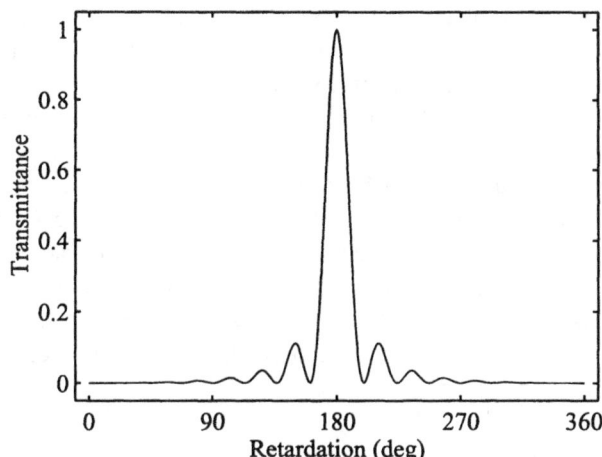

Fig. 15.9. Calculated transmittance of a 16-plate folded Solc filter.

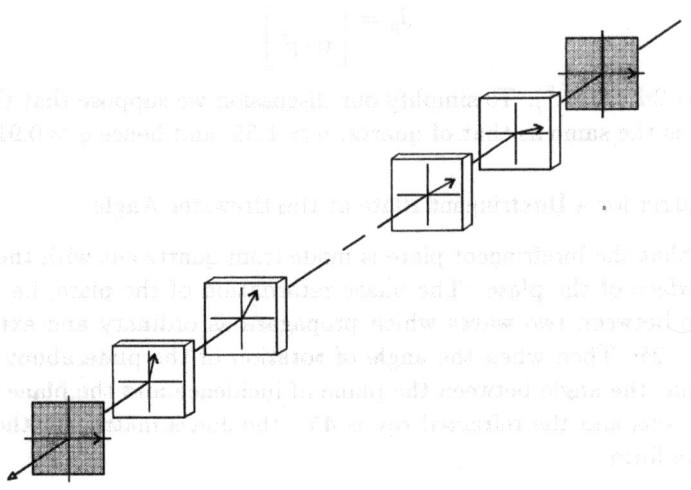

Fig. 15.10. Fan Solc filter.

15.2.3 Filters for Tuning Dye Lasers

A stack of birefringent plates can be used as an intracavity filter to tune the wavelength of a dye laser.[63] As shown in (Fig. 15.11) the plates are placed at the Brewster angle with respect to the laser beam. The p polarization is transmitted through each air-to-glass and glass-to-air interface without loss and, if the wavelength is such that the entire filter acts as a full-wave plate, the light is transmitted through the filter both without loss and without change in polarization state. The s polarization, on the other hand, suffers loss of amplitude at each Fresnel reflection and hence is unfavourable for lasing.

In some cases glass plates are added to provide more surfaces and hence increase the losses of the s state. In the remaining part of this section we use the Jones calculus to calculate the transmittance of the filter, while it is in operation in the laser cavity.[64]

Glass Plate at the Brewster Angle

The Jones matrix for a glass plate of refractive index n and placed at the Brewster angle can be determined by considering the Fresnel reflection coefficients for the p and s polarizations. The result is

$$\hat{J}_g = \begin{bmatrix} 1 & 0 \\ 0 & q^2 \end{bmatrix} \tag{15.10}$$

where $q = 2n/(1+n^2)$. To simplify our discussion we suppose that the index of the glass is the same as that of quartz, $n \approx 1.55$, and hence $q \approx 0.911$.

Jones Matrix for a Birefringent Plate at the Brewster Angle

Suppose that the birefringent plate is made from quartz cut with the optic axis in the surface of the plate. The phase retardation of the plate, i.e. the phase difference between two waves which propagate as ordinary and extraordinary waves is -2δ. Then when the angle of rotation of the plate about its normal is such that the angle between the plane of incidence and the plane containing the optic axis and the refracted ray is 45°, the Jones matrix for the plate has the simple form

$$\hat{J}_b = \begin{bmatrix} \cos\delta & iq\sin\delta \\ iq\sin\delta & q^2\cos\delta \end{bmatrix}. \tag{15.11}$$

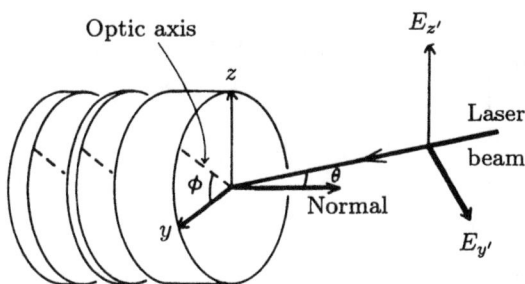

Fig. 15.11. Birefringent filter for tuning a laser. The thick lines are all in the plane of incidence of the laser beam. (Adapted from I.J. Hodgkinson and J.I. Vukusic, *Applied Optics* 17, 1944, 1978. Copyright © 1978 Optical Society of America. Reprinted with permission.)

We consider the general form of the Jones matrix for a birefringent filter, which may include any series combination of glass and birefringent plates represented by Eqs.(15.10) and (15.11). Inspection shows that the filter is represented by a Jones matrix of the form

$$\hat{J} = \begin{bmatrix} a & ib \\ ic & d \end{bmatrix},$$ (15.12)

where a, b, c, and d are real. Note that $\det \hat{J} = q^{2N}$, where N is the total number of glass and birefringent plates and q is assumed to have the same value for glass and the birefringent material.

Transmittance of Birefringent Filter

The Jones calculus is well adapted to eigenequation methods. To illustrate this, we continue with the example of the birefringent filter in a laser cavity. A condition for a mode is that the Jones vector representing the laser beam should be unchanged after a round-trip in the cavity. To simplify our discussion we assume that the birefringent plate is the only polarizing element in the cavity. Then the relevant eigenequation is

$$\hat{J}\vec{E} = t\vec{E}.$$ (15.13)

The eigenvalue t is the transmission coefficient, and the solution is

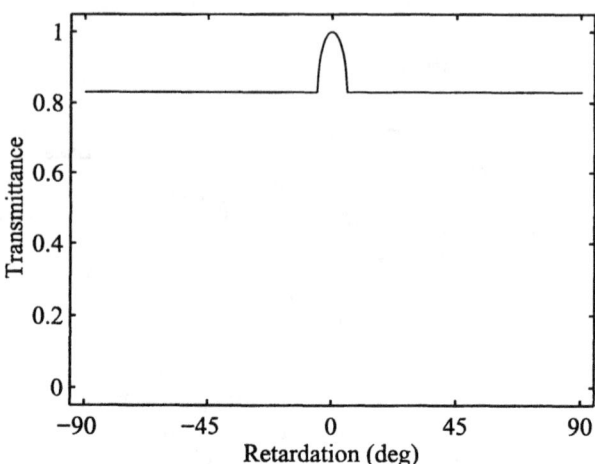

Fig. 15.12. Calculated transmittance of a single birefringent plate in a laser cavity.

$$t = \frac{1}{2}\{a + d \pm [(a+d)^2 - 4q^{2N}]^{1/2}\}. \tag{15.14}$$

Discontinuities occur in the transmittance spectrum when

$$|a + d| = 2q^N; \tag{15.15}$$

the eigenvalues are real and different and satisfy the equation $t_1 t_2 = q^{2N}$ when the argument of the square root in Eq.(15.14) is positive. Otherwise the eigenvalues are complex and given by

$$t = \frac{1}{2}\{a + d \pm i[4q^{2N} - (a+d)^2]^{1/2}\}. \tag{15.16}$$

In this case the two values of $|t|$ are equal and depend only on q and N,

$$|t| = q^N. \tag{15.17}$$

The transmittance spectrum of a single plate is illustrated in Fig. 15.12. In this case $T = |t|^2 = q^2 \approx 0.83$ at the discontinuity and the half-width value of δ is $\cos^{-1}[2q/(1+q^2)] = 5.3°$. On the same scale the free spectral range of the filter is represented by 180°.

Note that the transmittance which is plotted in Fig. 15.12 is appropriate to the device when it is in the laser cavity and is called the *active transmittance*. It is quite different to the *passive transmittance* which you would observe if you

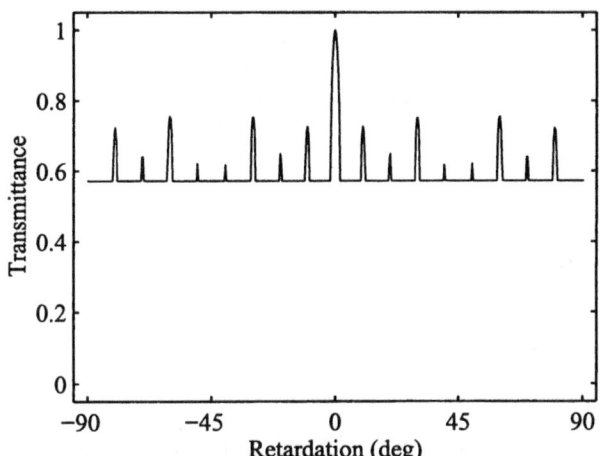

Fig. 15.13. Calculated transmittance of the birefringent filter 2 1 6. (Adapted from I.J. Hodgkinson and J.I. Vukusic, *Applied Optics* 17, 1944, 1978. Copyright © 1978 Optical Society of America. Reprinted with permission.)

removed the filter from the cavity and looked through it or made a measurement with a spectrophotometer.

The main problem with a single plate filter is the height of the flat region in the transmittance spectrum. This may be lowered by choosing an appropriate combination of birefringent plates, as shown in Fig. 15.13 for the filter with plate thicknesses in the ratio 2 1 6, and lowered even further by the addition of glass plates changing the arrangement to 2g 2 7g 1 5g 6 g (Fig. 15.14).

Filter Without Secondary Peaks

A birefringent filter can be designed so that it has no secondary peaks. The principle is to use a stack of N identical birefringent plates with optic axes aligned, as illustrated in Fig. 15.15.

An eigenvector of a single plate is now an eigenvector of the stack, and hence the transmittance spectrum of the stack is just T^N where T is the transmittance spectrum of a single plate. The transmittance spectrum has just a single peak, as illustrated in Fig. 15.16 for a stack of 10 plates of quartz. The width of the peak is the same as for a single plate, and the transmittance of the horizontal region is q^{2N}.

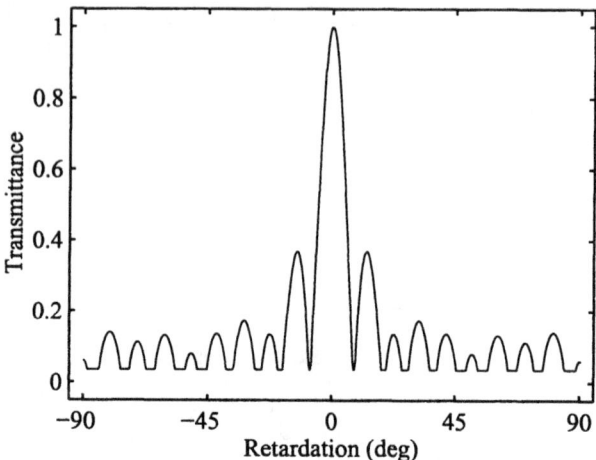

Fig. 15.14. Calculated transmittance of the filter 2g 2 7g 1 5g 6 g. (Adapted from I.J. Hodgkinson and J.I. Vukusic, *Applied Optics* 17, 1944, 1978. Copyright © 1978 Optical Society of America. Reprinted with permission.)

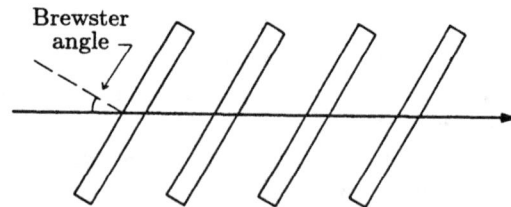

Fig. 15.15. Periodic birefringent filter. (Adapted from I.J. Hodgkinson and J.I. Vukusic, *Optics Communications* 24, 133, 1978. Copyright © 1978 North-Holland Publishing Company. Reprinted with permission.)

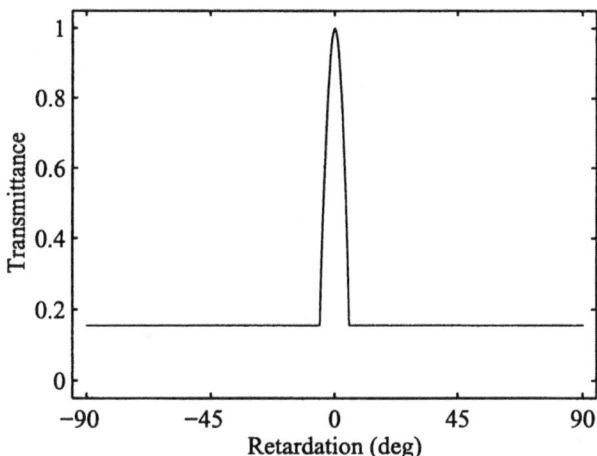

Fig. 15.16. Calculated transmittance of the filter formed by a stack of 10 identical birefringent plates with axes aligned. (Adapted from I.J. Hodgkinson and J.I. Vukusic, *Optics Communications* 24, 133, 1978. Copyright © 1978 North-Holland Publishing Company. Reprinted with permission.)

Calculations

The functions **bloom1**(*group*, δ) and **bloom2**(*group*, δ) in the *BTF Toolbox* determine the single pass transmittance (ring cavity) and the double pass transmittances (Fabry-Perot cavity) of a birefringent filter. As an example of the method of entry for *group*, the filter 2g 2 7g 1 5g 6 g is entered as [-2 2 -7 1 -5 6 1]; negative signs indicate glass and implied positive signs indicate birefringent material.

Chapter 16

Birefringent Coatings

Thin film design methods and production techniques have been developed over a long period for *isotropic* layers. Examples of coatings that are now readily available include antireflection coatings, multilayered high-reflectance stacks, narrowband filters, broadband filters, lowpass and highpass filters. In many cases these are formed by depositing alternate low-index and high-index materials, and all the films are deposited to have an optical thickness of one quarter-wave. It is relatively easy to monitor the thickness of a quarter-wave film during deposition, by waiting for a turning point in the reflectance or transmittance. One disadvantage is that only a few materials are suitable for optical coatings, and hence only a few refractive indices are available to the designer.

Birefringent thin films offer greater flexibility to the designer, but present greater challenges to the manufacturer. By choosing the deposition angle, the principal refractive indices can be selected from a continuous range of values. However, the most significant reason for using birefringent films is the unique polarizing properties that they offer for normally incident light. A birefringent coating can serve different purposes for *p* and *s* polarized light.

In this chapter, we

▨ review and add to the symbolic notation that we have introduced for birefringent stacks

▨ consider the development of birefringent coating designs

▨ conclude the book with a "sampler" of PS designs.

16.1   Isotropic Coatings

Consider an isotropic film of refractive index n and thickness d. Suppose that the optical thickness of the film is one quarter-wave at some design wavelength λ_0. This requires

$$nd = \lambda_0/4, \tag{16.1}$$

and hence the physical thickness of the film is

$$d = \lambda_0/4n. \tag{16.2}$$

For a general wavelength λ the film is specified by the index n and the ratio

$$d/\lambda = g/4n \tag{16.3}$$

where

$$g = \lambda_0/\lambda \tag{16.4}$$

is the relative wavenumber.

Many coatings are formed by alternately depositing quarter-waves with low and high indices n_L and n_H. These are represented symbolically by the letters H and L. Thus, a multilayered high reflectance stack may be represented by

$$aHLHLHLHLHg$$

where a stands for air and g stands for glass. A more compact notation is

$$aH[LH]^4g.$$

Additional insight into a design may be obtained when the refractive indices are plotted as a function of optical thickness, as shown in Fig. 16.1 for the reflecting coating with $n_L = 1.46$ and $n_H = 2.4$.

16.2   General Birefringent Coating

The concept of a quarter-wave layer is not very useful in the case of a general birefringent coating, because of cross-coupling of polarizations. A layer in a general birefringent coating requires complete specification of indices, angles, and thickness-to-wavelength ratio. We have used the format

$$[n_1 \; n_2 \; n_3 \; \eta \; \psi \; \xi \; d/\lambda]$$

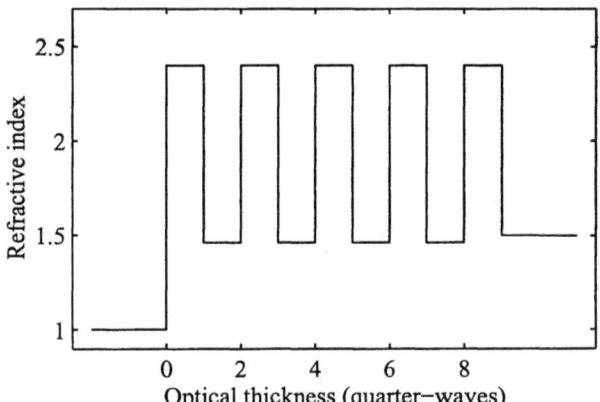

Fig. 16.1. Refractive index profile of the multilayered dielectric reflecting coating $aH[LH]^4g$.

throughout this book (and assumed that $\eta = 0$).

For the design of general birefringent coatings knowledge of n_1, n_2, n_3 and ψ as functions of deposition angle θ_v is necessary. Representative values for the materials tantalum oxide, titanium oxide and zirconium oxide are provided by the functions **tao2**, **tio2** and **zro2** in the *BTF Toolbox*, using a statement such as

$$[n_1,\ n_2,\ n_3,\ \psi] = \mathbf{tao2}(\theta_v).$$

Alternatively, the plots given in Figs. 16.2 and 16.3 can be used.

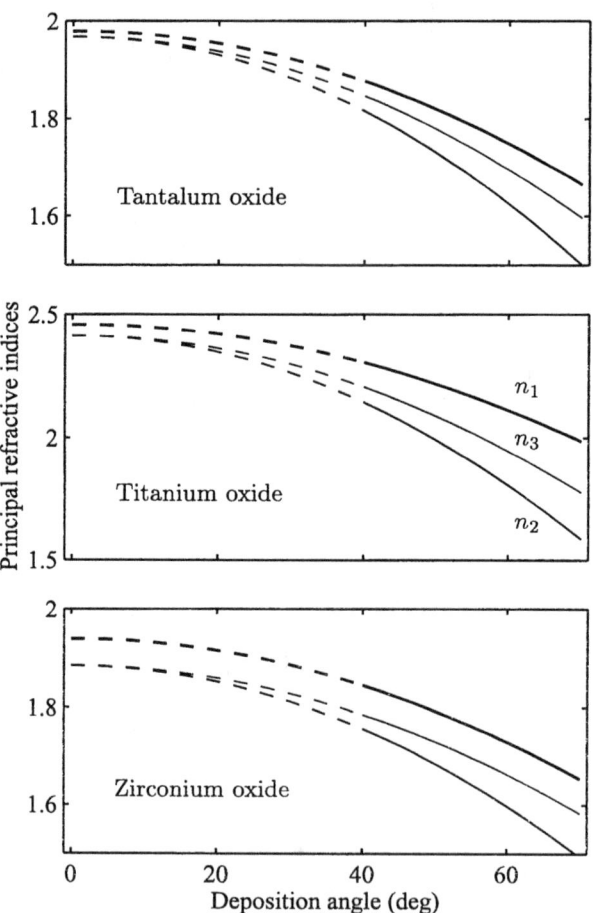

Fig. 16.2. Experimental principal refractive indices (solid lines) and extrapolated values (broken lines) for three tilted columnar thin film materials.

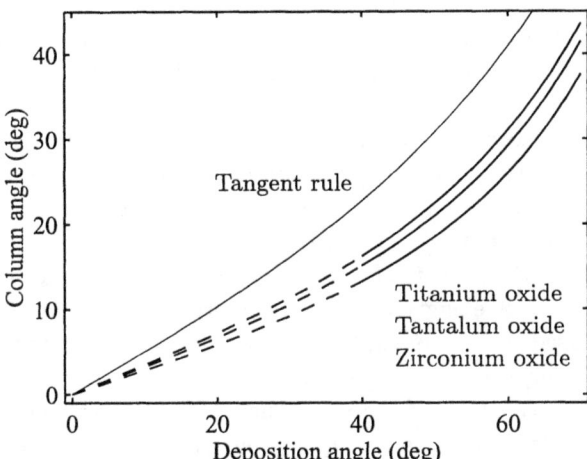

Fig. 16.3. Experimental column angles (solid lines) and extrapolated values (broken lines) for three tilted columnar thin film materials.

Table 16.1. Properties of PS coatings.

	P layer	P' layer	S layer	S' layer
β	0	0	0	0
ξ	0 or π	0 or π	$\pi/2$ or $3\pi/2$	$\pi/2$ or $3\pi/2$
n_y	n_p	n_p	n_s	n_s
n_z	n_s	n_s	n_p	n_p
$[n_y\ n_z\ d/\lambda]$	$[n_p\ n_s\ g/4n_p]$	$[n_p\ n_s\ g/4n_s]$	$[n_s\ n_p\ g/4n_s]$	$[n_s\ n_p\ g/4n_p]$

16.3   PS Coatings

We conclude our book with sections on PS coatings, formed by depositing biaxial layers in one or other of two mutually perpendicular deposition planes. PS coatings are for use with light at normal incidence, and most offer features that are unavailable in isotropic coatings. The symbolic representation used for PS coatings is defined in Sect. 4.5.3, and a summary of the properties of PS coatings is provided in this chapter in Table 16.1. The symbolic notation is referenced to external polarization p. Table 16.1 includes new features that we have added to the symbolic notation. By default, P and S are assumed to be quarter-wave layers at the design wavelength and for the external p polarization. The symbols P' and S' are used for layers that have the quarter-wave optical thickness at the design wavelength but for the external s polarization.

The functions **tao2**, **tio2**, **zro2** in the *BTF Toolbox* have a format that outputs the refractive index values n_p, n_s that are required for the design of PS coatings. For example,

$$[n_p,\ n_s] = \textbf{tao2}(\theta_v)$$

outputs the indices n_p and n_s that would be obtained by depositing tantalum oxide at angle θ_v. The indices n_p and n_s refer to light that is incident normally. For n_p the polarization is parallel to the plane of deposition and for n_s the polarization is perpendicular to the deposition plane of the layer. In Fig. 16.4 the indices n_p and n_s are plotted as functions of deposition angle for the three tilted columnar thin film materials.

16.4   Design Considerations for PS Coatings

In this section we discuss strategies for the design of PS coatings. Most of the examples that we consider involve the conversion of a *generic design*[66] that

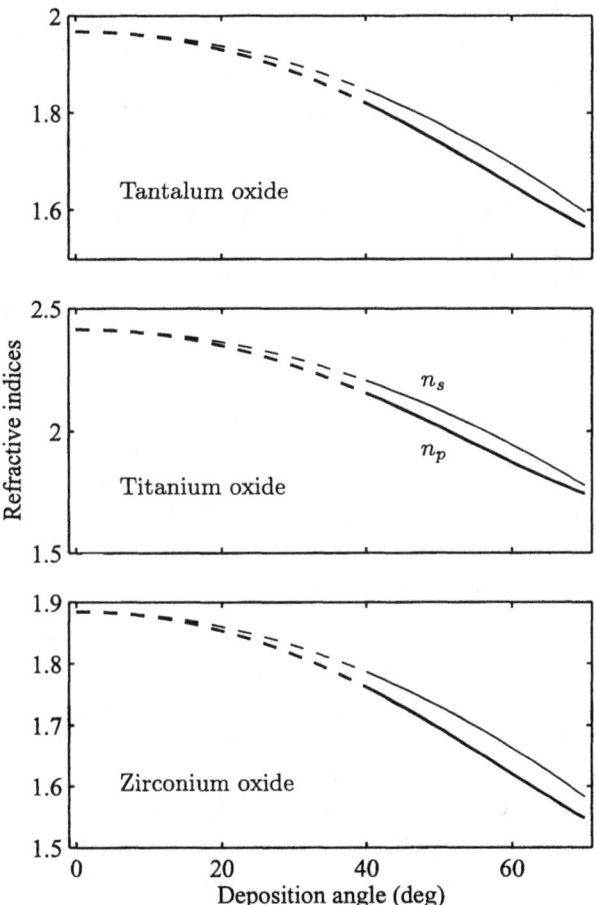

Fig. 16.4. Experimental values of the indices n_p and n_s (solid lines) and extrapolated values (broken lines) for three tilted columnar thin film materials.

already exists for isotropic materials.

Generally, a PS coating gives different responses for p and s polarizations. We assume that a particular response is required for p, often the same as the response of the generic coating, and for one or another reason a different response is required for s. Ten PS coatings are listed in Sect. 16.6 and sorted according to the difference between the p and s responses.

One possibility is that the p and s responses of the PS coating are required to be different in magnitude at the design wavelength. We refer to this as *making an anisotropic version* of the generic design and cover general principles of the design of PS coatings by describing the development of an anisotropic antireflection coating.

A potential use of a tilted-columnar film is to provide an intermediate refractive index, an index that is not available as an isotropic layer using available evaporants. We refer to this as *replacing an intermediate index*. Here the aim is to provide the intermediate index, and at the same time, minimize the difference between the p and s responses.

A third possibility is that the p and s polarizations should give *identical response profiles separated in wavelength*, as in the case of narrowband filters.

The first possibility involves compromising the s response. Finally, we show that by completely *spoiling the s response* it is possible to design thin film polarizers for use with laser light at normal incidence.

16.4.1 Making an Anisotropic Version

We consider the design of an *anisotropic antireflection coating* with the properties $R_p = 0$, $R_s > 0$ for light at normal incidence. An intracavity glass plate coated on both sides with anisotropic antireflection coatings could be used to select the polarization of a laser, in the same way as a plate at the Brewster angle, but without lateral displacement of the beam.

We start by considering the principles of designing an antireflection coating for isotropic media. The system aLg, a single quarter-wave layer of a low-index material, is the most simple antireflection coating, but requires $n_L = (n_a n_g)^{1/2} \approx 1.22$ for glass of refractive index 1.5. Such a low index is unavailable in a suitable material, and so an H layer is deposited first to make the substrate appear to have a higher index. This gives the system $aLHg$ and a condition on the ratio of the indices, $n_H/n_L = (n_g/n_a)^{1/2}$, that can be derived using Abelès matrices together with an appropriate equation for the reflection coefficient. We note that the derivation indicates that this system can be extended to multiple periods, and we have as our starting position (for the design of an anisotropic antireflection coating) the isotropic stack

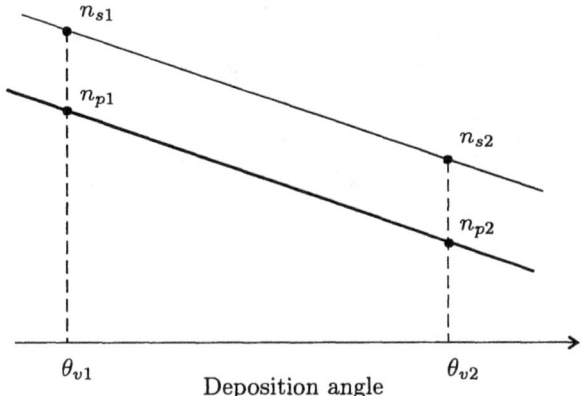

Fig. 16.5. The refractive indices of a pair of layers in a PS coating can be selected from the two pairs of values shown in the figure. The zero-reflectance condition occurs for pairs of deposition angles θ_{v1} and θ_{v2} that yield $n_{s1}/n_{p2} = (n_g/n_a)^{1/2N}$. The period P_2S_1 of the anisotropic antireflection coating comprises the layers $[n_{p2}\ n_{s2}\ g/4n_{p2}]$ and $[n_{s1}\ n_{p1}\ g/4n_{s1}]$.

$$a(LH)^N g$$

together with the condition

$$n_H/n_L = (n_g/n_a)^{1/2N} \tag{16.5}$$

on the ratio of the high and low indices.

In a PS antireflection coating the required index ratio can be achieved for a low value of the integer N by using films deposited at two angles, θ_{v1} and θ_{v2} (see Fig. 16.5). The PS design

$$a[P_2S_1]^N g,$$

is obtained from the isotropic design by translating $L \to P_2$ and $H \to S_1$. This choice means that the p polarization shown in Fig. 16.6 utilizes both the largest index n_{s1} and the smallest index n_{p2} available from two deposition angles. The condition on the indices becomes

$$n_{s1}/n_{p2} = (n_g/n_a)^{1/2N}. \tag{16.6}$$

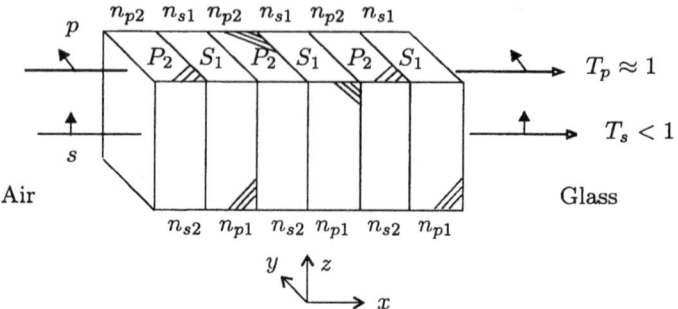

Fig. 16.6. PS antireflection coating of design $a[P_2S_1]^3g$.

Now suppose that we decide to make the coating entirely from zirconium oxide, and choose $\theta_{v2} = 60°$. Then iterative use of the *BTF Toolbox* function **zro2**, or reference to Fig. 16.4, leads to the value $\theta_{v1} = 48.5°$ for $N = 3$. The refractive index profile for the anisotropic antireflection coating with these parameters is plotted later in the book, in Fig. 16.9.

The p and s transmittances of the coating can be calculated using functions from the *BTF Toolbox* as follows. The layers P_2 and S_1 that make up the period P_2S_1 are represented by $[n_{p2}\ n_{s2}\ g/4n_{p2}]$ and $[n_{s1}\ n_{p1}\ g/4n_{s1}]$. The characteristic matrix of the stack is the matrix of the period P_2S_1 to the power of 3,

$$
\hat{M} = \mathbf{cmat}\left(\begin{bmatrix} n_{p2} & n_{s2} & g/4n_{p2} \\ n_{s1} & n_{p1} & g/4n_{s1} \end{bmatrix}\right)^3,
$$

and

$$
\hat{R} = \mathbf{reflect}(n_a,\ \hat{M},\ n_g).
$$

R_p and R_s are given by the elements $\hat{R}(1,1)$ and $\hat{R}(2,2)$.

Computed reflectance curves for the anisotropic antireflection coating are also plotted in Fig. 16.9. Further experimentation shows that the difference $R_s - R_p$ that is achieved in the initial design can be improved considerably at the expense of a small increase in R_p (from the initial value of zero) by choosing a larger value of θ_{v1} (while retaining the initial value of θ_{v2}).

A second example, the *anisotropic reflector*, is described in Sect. 16.6.2, and a third example, the *anisotropic-phase reflector*, is considered in Sect. 16.6.3. In both the anisotropic antireflection coating and the anisotropic reflector the s response is compromised by making the depth of modulation in the s refractive index profile smaller than the depth of the p profile. In Sect. 16.4.4 we show that the modulation of one index can be eliminated completely.

16.4.2 Replacing an Intermediate Index

A potential problem here is that two indices are produced, when only one is wanted. Two examples are considered in the PS sampler, an *achromatic antireflection coating* (Sect. 16.6.4) and an *achromatic fifty percent reflector* (Sect. 16.6.5). Reasonable results are obtained through choice of the refractive indices of two intermediate indices in each case.

16.4.3 Identical Response Profiles Separated in Wavelength

Some coatings, such as the *single-cavity (Fabry-Perot) narrowband filter*, have a spacer layer or "cavity" that is located between two relatively broadband reflectors and is resonant at the design wavelength. Such a cavity can be deposited as a birefringent layer, and the effect is to produce a filter with transmission peaks at two wavelengths, one for p and the other for s (Sect. 16.6.6). Similarly, a generic *two-cavity filter* design can be adapted in such a way that the cavities resonate together at a particular wavelength for p and at a different wavelength for s. Again, two peaks are produced, as shown in Sect. 16.6.7. In a third example, the *edge filter* shown in Sect. 16.6.8, both the low-pass edge and the high-pass edge occur at different wavelengths for p and s.

In each of the three examples the conversion from the generic design is achieved by making the replacement $H \rightarrow P$ in selected layers or in all layers.

16.4.4 Spoiling the s-response

Consider the *common-index thin film polarizer*

$$aP_1'[S_2'P_1']^N g$$

which is described in Sect. 16.6.9. This PS coating is formed by depositing films with the parameters $[n_{p1}\ n_{s1}\ g/4n_{s1}]$ and $[n_{s2}\ n_{p2}\ g/4n_{p2}]$. The special feature here is that the deposition angles are chosen to satisfy the index-matching condition, $n_{p1} = n_{s2}$, that is illustrated in Fig. 16.7.

Overall, the indices satisfy

$$n_{s1} > n_{p1} = n_{s2} > n_{p2}. \tag{16.7}$$

For p-polarized incident light the system behaves as a slab of index $n_{p1} = n_{s2}$ and hence $T_p \approx 1$. However, for the s polarization the system acts as a multilayered

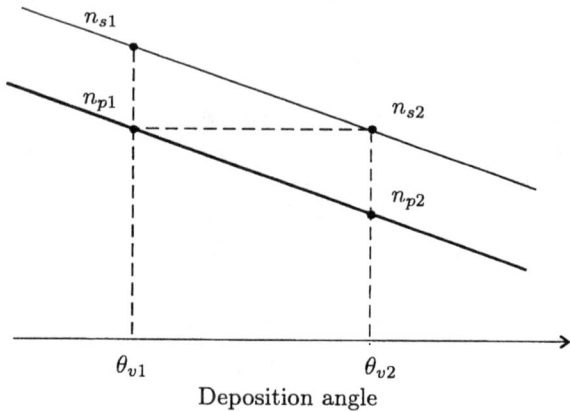

Fig. 16.7. The index-matched condition occurs for pairs of deposition angles θ_{v1} and θ_{v2} that give $n_{p1} = n_{s2}$.

reflecting stack of alternating high-index (n_{s1}) and low-index (n_{p2}) quarter-wave films. The s transmittance is $T_s = 4c/(1 + c)^2$ where $c = n_{s1}^2(n_{s1}/n_{p2})^{2N}/n_a n_g$.

We can regard the primary function of the common-index thin film polarizer (in the default orientation) as transmission of p and the function of the birefringent material as spoiling the s response. Another example in which the s response is spoiled is shown in Sect. 16.6.10. In the *double-cavity linear polarizer* the refractive indices of the two cavities are chosen so that simultaneous resonance occurs for p but not for s.

16.5   Normal and Hybrid Monitoring

For the purpose of computation we have assumed that all layers in a PS coating are monitored during deposition with light of the same external polarization, p as defined in Fig. 16.6. Thus the layer S_1 is represented by $[n_{s1}, n_{p1} \, g/4n_{s1}]$ and the layer P_2 by $[n_{p2}, n_{s2}, g/4n_{p2}]$. We refer to this situation as *normal monitoring*.

In practice, though, when we have deposited some of the coatings described in this chapter, we have found it convenient to use *hybrid monitoring*. The reason is that the substrate needs to be rotated between layers, but the monitor is most conveniently fixed in the coating chamber. In hybrid monitoring the "coating" and the "dual" are monitored alternately. Examples of hybrid monitoring are given in Fig. 16.8.

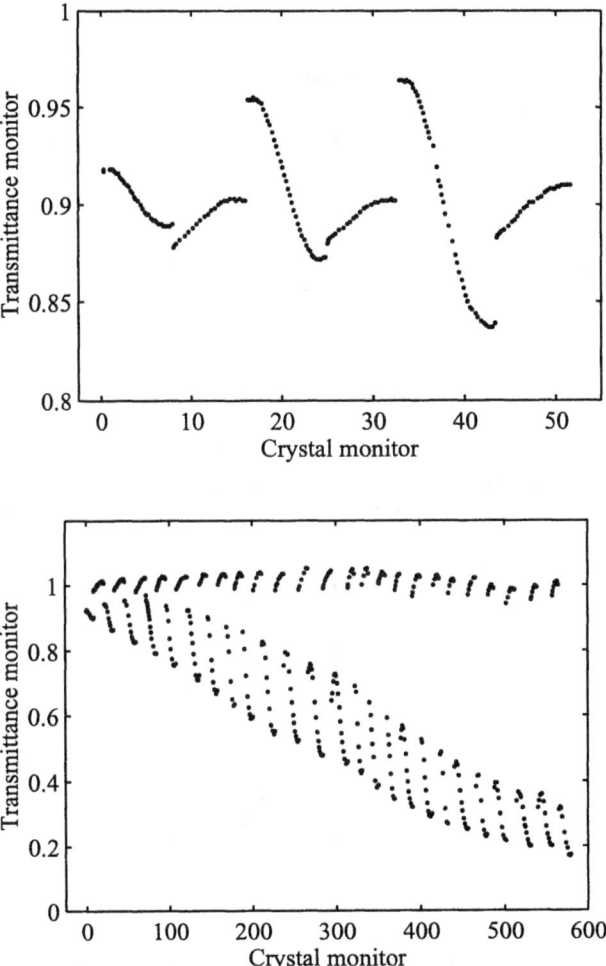

Fig. 16.8. Monitor data (not normalized) recorded during the deposition and hybrid monitoring of an anisotropic antireflection coating (upper) and a common-index thin film polarizer (lower).

16.6   PS Sampler

16.6.1   Anisotropic Antireflection Coating

Generic Design

$$a(LH)^N g$$

$$n_H/n_L = (n_g/n_a)^{1/2N}$$

PS Design

$$a[P_2 S_1]^N g$$

$$n_{s1}/n_{p2} = (n_g/n_a)^{1/2N}$$

$$n_a = 1, \; n_g = 1.5$$
$$n_{p2} = 1.621, \; n_{s2} = 1.663 \; (\text{zirconium oxide}, \; \theta_{v2} = 60°)$$
$$n_{p1} = 1.706, \; n_{s1} = 1.740 \; (\text{zirconium oxide}, \; \theta_{v1} = 48.5°)$$

$$pssystem = \begin{bmatrix} n_a & n_a & NaN \\ n_{p2} & n_{s2} & g/4n_{p2} \\ n_{s1} & n_{p1} & g/4n_{s1} \\ \vdots & \vdots & \vdots \\ n_{s1} & n_{p1} & g/4n_{s1} \\ n_g & n_g & NaN \end{bmatrix}$$

Notes

The anisotropic antireflection coating (Fig. 16.9) is designed to give zero reflectance at a specific wavelength for the p polarization and a few percent reflectance for the s polarization. The PS design mimics the generic design for the p polarization, through choice of the index ratio n_{s1}/n_{p2}. The corresponding ratio n_{p1}/n_{s2} for the s polarization doesn't satisfy the condition and hence $R_s > 0$.

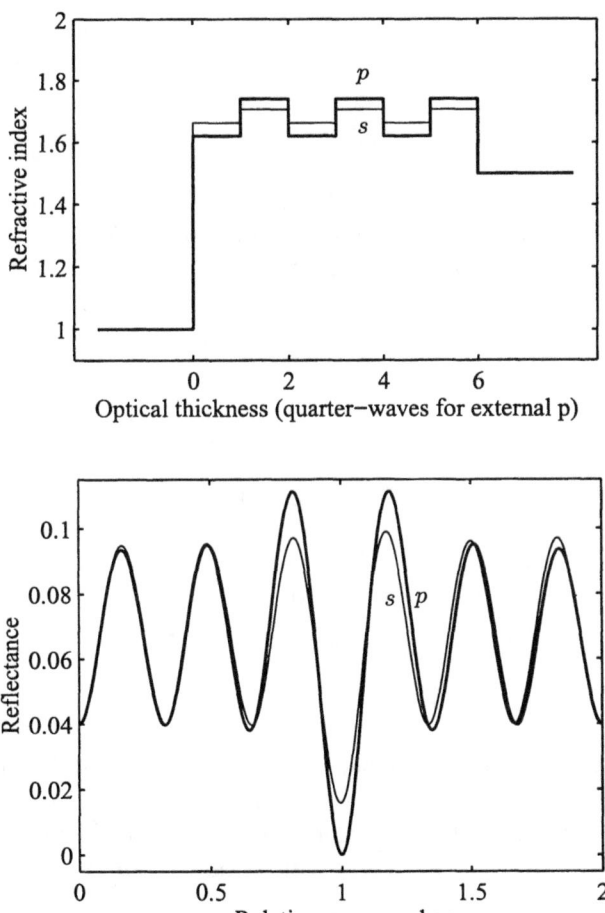

Fig. 16.9. Refractive index profile and calculated reflectance of the anisotropic antireflection coating $a[P_2S_1]^3g$.

16.6.2 Anisotropic Reflector

Generic Design

$$aH[LH]^N g$$

PS Design

$$aS_1[P_2S_1]^N g$$

$$n_a = 1, \ n_g = 1.5$$
$$n_{p1} = 2.117, \ n_{s1} = 2.173 \ (\text{titanium oxide}, \ \theta_{v1} = 43°)$$
$$n_{p2} = 1.583, \ n_{s2} = 1.624 \ (\text{zirconium oxide} \ \theta_{v2} = 65°)$$

$$pssystem = \begin{bmatrix} n_a & n_a & NaN \\ n_{s1} & n_{p1} & g/4n_{s1} \\ n_{p2} & n_{s2} & g/4n_{p2} \\ n_{s1} & n_{p1} & g/4n_{s1} \\ \vdots & \vdots & \vdots \\ n_{s1} & n_{p1} & g/4n_{s1} \\ n_g & n_g & NaN \end{bmatrix}$$

Notes

The anisotropic reflector is designed to have R_p large and R_s a few percent smaller. The refractive index profiles shown in Fig. 16.10 cause both R_p (exactly) and R_s (approximately) to mimic the reflectance profile of the generic coating. However, the high/low index ratio is smaller for the s polarization and hence $R_s < R_p$.

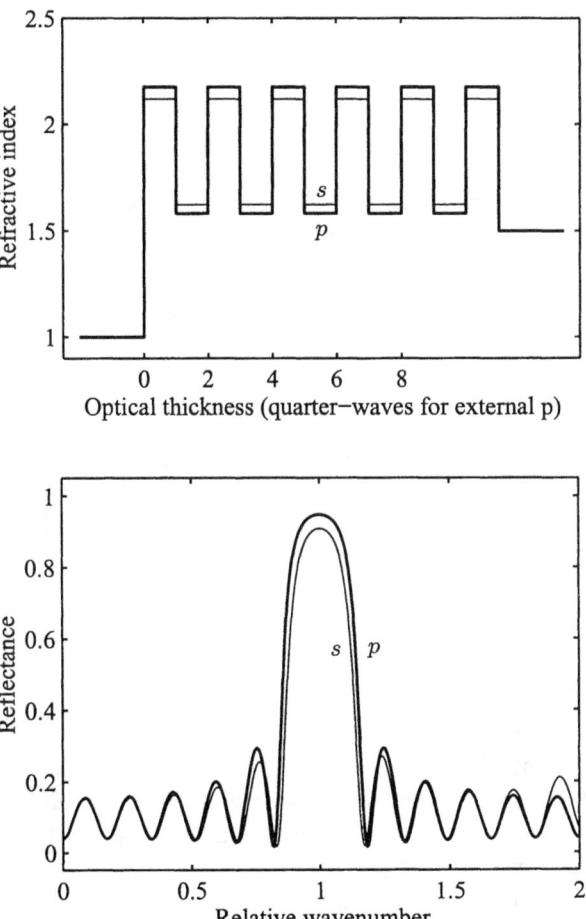

Fig. 16.10. Refractive index profile and reflectance of the anisotropic reflector $aS_1[P_2S_1]^5g$.

16.6.3   Anisotropic-Phase Reflector

Generic Design

$$aH[LH]^N g$$

PS Design

$$a(2P)H[LH]^N g$$

$$n_a = 1,\ n_H = 2.4,\ n_L = 1.46,\ n_g = 1.5$$
$$n_p = 1.917,\ n_s = 1.991\ \text{(titanium oxide, } \theta_v = 57°)$$

$$pssystem = \begin{bmatrix} n_a & n_a & NaN \\ n_p & n_s & 2g/4n_p \\ n_H & n_H & g/4n_H \\ n_L & n_L & g/4n_L \\ \vdots & \vdots & \vdots \\ n_H & n_H & g/4n_H \\ n_g & n_g & NaN \end{bmatrix}$$

Notes

The anisotropic-phase reflector (Fig. 16.11) provides the same reflectances R_p and R_s but different phase changes on reflection for the p and s polarizations. Adding a birefringent ($2P$) half-wave layer to the generic reflector design adds the polarization-dependent phase change and leaves the reflectance R_p unaltered at the design wavelength. A full-wave birefringent layer gives nearly double the phase difference, etc. In simple terms, the birefringent layer adds a polarization-dependent optical thickness in front of the reflecting coating. There is a negligible decrease in R_s. If a larger decrease can be tolerated, in both R_p and R_s, then a larger phase difference can be obtained by using odd multiples of a quarter-wave birefringent layer. With appropriate choice of birefringent material and thickness, phase differences such as 45° and 90° can be produced.

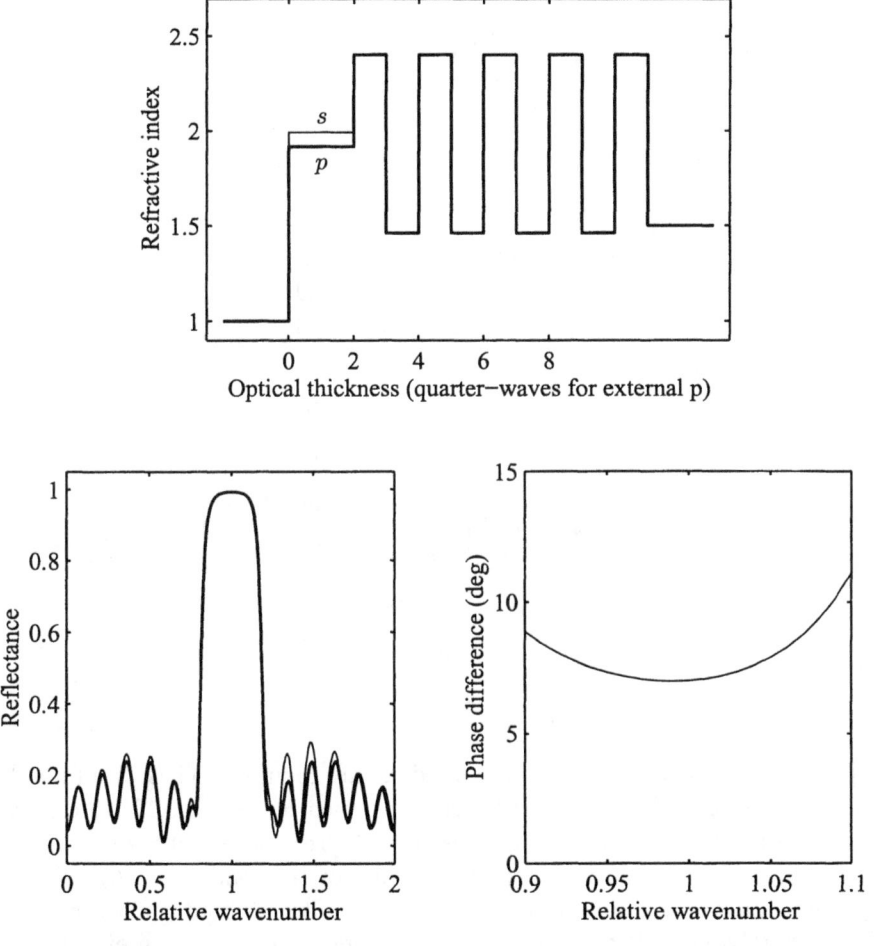

Fig. 16.11. Profiles of refractive index, reflectance, and phase difference $\delta_s - \delta_p =$ angle $\hat{r}_{22}$ − angle $\hat{r}_{11}$ for the anisotropic-phase reflector $a(2P)H[LH]^5g$.

16.6.4 Achromatic Antireflection Coating

Generic Design

$$aLI_1HI_2g$$

$$n_a = 1,\ n_L = 1.35,\ n_{I1} = 2.150,\ n_H = 2.35,\ n_{I2} = 1.780,\ n_g = 1.5$$

PS Design

$$aLP_1HS_2g$$

$$n_a = 1,\ n_L = 1.35,\ n_H = 2.35,\ n_g = 1.5$$
$$n_{p1} = 2.150,\ n_{s1} = 2.201\ (\text{titanium oxide},\ \theta_{v1} = 40.4°)$$
$$n_{p2} = 1.745,\ n_{s2} = 1.779\ (\text{titanium oxide},\ \theta_{v2} = 69.8°)$$

$$pssystem = \begin{bmatrix} n_a & n_a & NaN \\ n_L & n_L & g/4n_L \\ n_{p1} & n_{s1} & g/4n_{p1} \\ n_H & n_H & g/4n_H \\ n_{s2} & n_{p2} & g/4n_{s2} \\ n_g & n_g & NaN \end{bmatrix}$$

Notes

In the generic design two intermediate indices are required. These are provided by tilted columnar films in the PS design in which R_p (Fig. 16.12) mimics the reflectance profile R of the generic design. The refractive indices of the birefringent layers are selected to prevent spoiling of the reflectance R_s. Further improvements would result from (i) the use of materials that yield the appropriate refractive indices at small deposition angles (so that the birefringence is small), and (ii) optimization of the PS design.

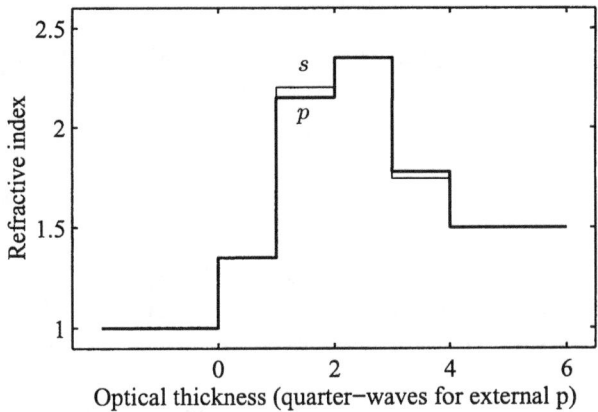

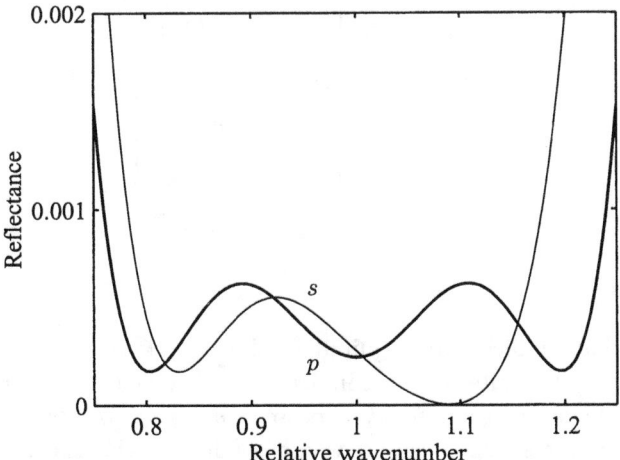

Fig. 16.12. Refractive index profile and calculated reflectance of the achromatic antireflection coating aLP_1HS_2g.

16.6.5 Achromatic Fifty Percent Reflector

Generic Design

$$aHLI_1HI_2g$$

$$n_a = 1, \; n_H = 2.35, \; n_L = 1.35, \; n_{I1} = 2.343, \; n_{I2} = 1.704, \; n_g = 1.5$$

PS Design

$$aHLP_1HS_2g$$

$$n_a = 1, \; n_H = 2.35, \; n_L = 1.35, \; n_g = 1.5$$
$$n_{p1} = 2.334, \; n_{s1} = 2.351 \; (\text{titanium oxide, } \theta_{v1} = 22°)$$
$$n_{p2} = 1.684, \; n_{s2} = 1.721 \; (\text{zirconium oxide, } \theta_{v2} = 51.5°)$$

$$pssystem = \begin{bmatrix} n_a & n_a & NaN \\ n_H & n_H & g/4n_H \\ n_L & n_L & g/4n_L \\ n_{p1} & n_{s1} & g/4n_{p1} \\ n_H & n_H & g/4n_H \\ n_{s2} & n_{p2} & g/4n_{s2} \\ n_g & n_g & NaN \end{bmatrix}$$

Notes

Two intermediate indices are required in the generic design. In the PS design (Fig. 16.13) the intermediate indices are provided by birefringent films. Anisotropic effects are reduced by choosing $(n_{p1} + n_{s1})/2 \approx n_{I1}$ and $(n_{p2} + n_{s2})/2 \approx n_{I2}$. Further improvements would result from the use of materials with small birefringence and through optimization of the PS design.

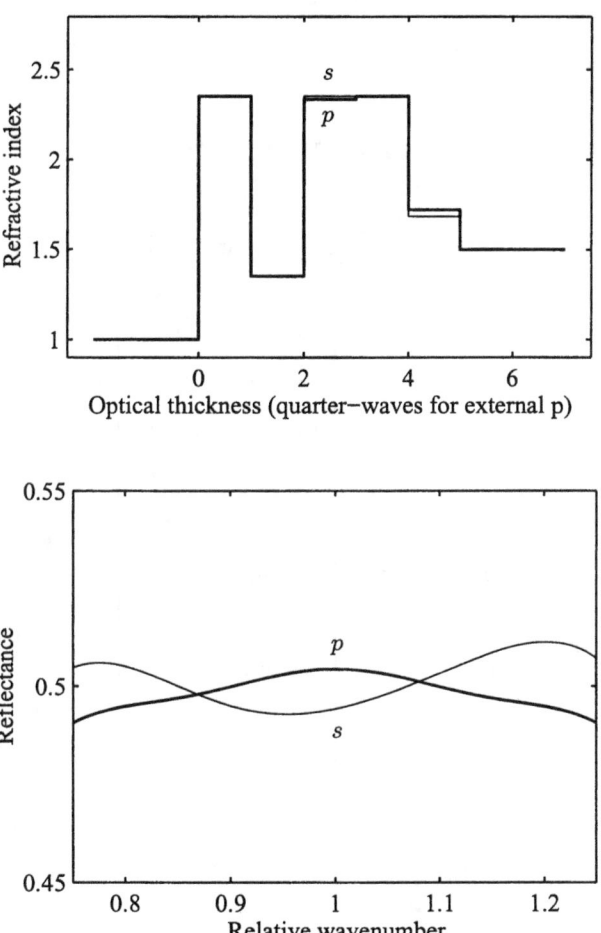

Fig. 16.13. Refractive index profile and calculated reflectance of the achromatic 50% reflector $aHLP_1HS_2g$.

16.6.6   Single-Cavity Narrowband Filter

Generic Design

$$a[HL]^N(2H)[LH]^Ng$$

PS Design

$$a[HL]^2(10P)[LH]^2g$$

$$n_a = 1,\ n_H = 2.4,\ n_L = 1.46,\ n_g = 1.5$$
$$n_p = 1.873,\ n_s = 1.945 \text{ (titanium oxide, } \theta_v = 60°)$$

$$pssystem = \begin{bmatrix} n_a & n_a & NaN \\ n_H & n_H & g/4n_H \\ n_L & n_L & g/4n_L \\ n_H & n_H & g/4n_H \\ n_L & n_L & g/4n_L \\ n_p & n_s & 10g/4n_p \\ n_L & n_L & g/4n_L \\ n_H & n_H & g/4n_H \\ n_L & n_L & g/4n_L \\ n_H & n_H & g/4n_H \\ n_g & n_g & NaN \end{bmatrix}$$

Notes

In this example we transform the single-cavity narrowband filter design to an anisotropic design. The generic design consists of a spacer layer (the cavity) bounded by two reflecting stacks. The spacer layer is most sensitive to change of optical thickness, and for this reason it is converted to a birefringent layer in the PS design. The refractive index profile and the resulting p and s transmittance peaks are plotted in Fig. 16.14.

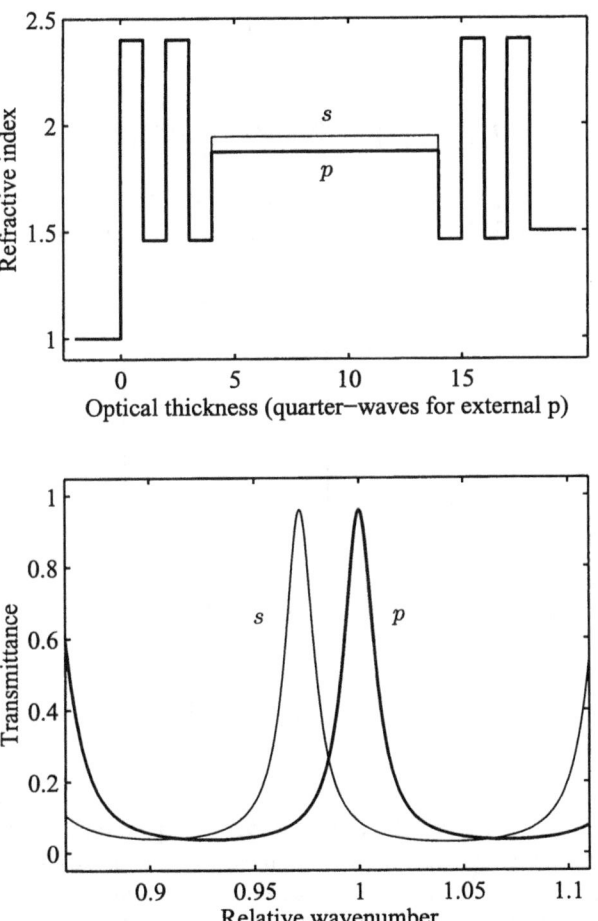

Fig. 16.14. Refractive index profile and calculated transmittance of the single-cavity filter $a[HL]^2(10P)[LH]^2g$.

16.6.7 Multi-Cavity Narrowband Filter

Generic Design

$$a[HL]^N(2H)[LH]^N C[HL]^N(2H)[LH]^N g$$

PS Design

$$a[HL]^2(10P)[LH]^2 C[HL]^2(10P)[LH]^2 g$$

$$n_a = 1, \ n_H = 2.4, \ n_L = 1.46, \ n_c = 1.68, \ n_g = 1.5$$
$$n_p = 1.643, \ n_s = 1.684 \ (\text{zirconium oxide}, \ \theta_v = 57°)$$

$$pssystem = \begin{bmatrix} n_a & n_a & \text{NaN} \\ \vdots & \vdots & \vdots \\ n_p & n_s & 10g/4n_p \\ \vdots & \vdots & \vdots \\ n_c & n_c & g/4n_c \\ \vdots & \vdots & \vdots \\ n_p & n_s & 10g/4n_p \\ \vdots & \vdots & \vdots \\ n_g & n_g & \text{NaN} \end{bmatrix}$$

Notes

In this example we transform the double-cavity narrow-band filter design to an anisotropic double-cavity design. The generic design consists of two identical single-cavity (Fabry-Perot) filters coupled by a middle layer C. An advantage of the design is a fairly flat top on the transmittance peak. In the PS coating the spacer layers are replaced by thicker birefringent layers, in order to separate the p and s transmittance peaks. The refractive index and transmittance profiles are shown in Fig. 16.15. The index of the coupling layer has an influence on the ripple at the top of the passband – decreasing the index decreases the ripple but decreases the bandwidth, and hence a compromise needs to be made. We have specified C as an isotropic layer of index $n_c = 1.68$, but a birefringent layer can be used without compromising the design.

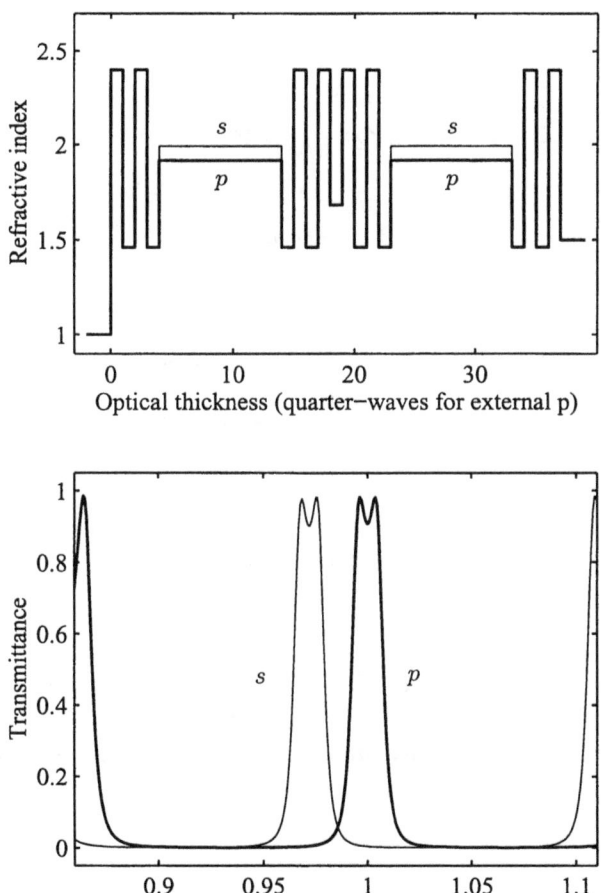

Fig. 16.15. Refractive index profile and calculated transmittance of the double-cavity filter $a[HL]^2(10P)[LH]^2C[HL]^2(10P)[LH]^2g$.

16.6.8   Edge Filter

Generic Design

$$a[(uH)(vL)(uH)]^N g$$

PS Design

$$a[(2P)L(2P)]^N g$$

$$n_a = 1, \ n_L = 1.46, \ n_g = 1.5$$
$$n_p = 2.034, \ n_s = 2.101 \ (\text{titanium oxide}, \ \theta_v = 49°)$$

$$pssystem = \begin{bmatrix} n_a & n_a & NaN \\ n_p & n_s & 2g/4n_p \\ n_L & n_L & g/4n_L \\ n_p & n_s & 2g/4n_p \\ \vdots & \vdots & \vdots \\ n_g & n_g & NaN \end{bmatrix}$$

Notes

The anisotropic edge filter has displaced low-pass and high-pass edges as shown in Fig. 16.16. The refractive index profiles ensure that both T_p (exactly) and T_s (approximately) mimic the transmittance profiles of the generic design.

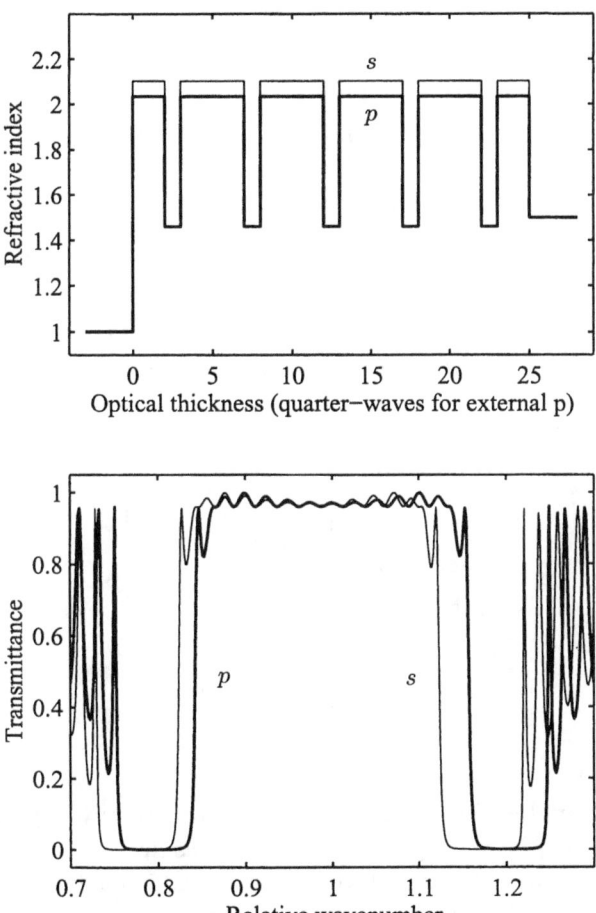

Fig. 16.16. Refractive index profile and calculated transmittance of the edge filter $a[(2P)L(2P)]^5g$.

16.6.9   Common-Index Thin Film Polarizer

Generic Design

$$aH[LH]^N g$$

PS Design

$$aP'_1[S'_2 P'_1]^N g$$

$$n_{s1} > n_{p1} = n_{s2} > n_{p2}$$

$$n_a = 1,\ n_g = 1.5$$
$$n_{p1} = 1.624,\ n_{s1} = 1.667\ \text{(zirconium oxide, } \theta_{v1} = 59.5°)$$
$$n_{p2} = 1.583,\ n_{s2} = 1.624\ \text{(zirconium oxide, } \theta_{v2} = 65°)$$

$$pssystem = \begin{bmatrix} n_a & n_a & NaN \\ n_{p1} & n_{s1} & g/4n_{s1} \\ n_{s2} & n_{p2} & g/4n_{p2} \\ \vdots & \vdots & \vdots \\ n_{p1} & n_{s1} & g/4n_{s1} \\ n_g & n_g & NaN \end{bmatrix}$$

Notes

The deposition angles of the zirconium oxide layers satisfy the index-matching condition $n_{p1} = n_{s2}$ for the s polarization. Thus, as illustrated in Fig. 16.17, the stack behaves as a slab of material with the matched index for the s polarization and as a high/low reflecting stack for the p polarization.

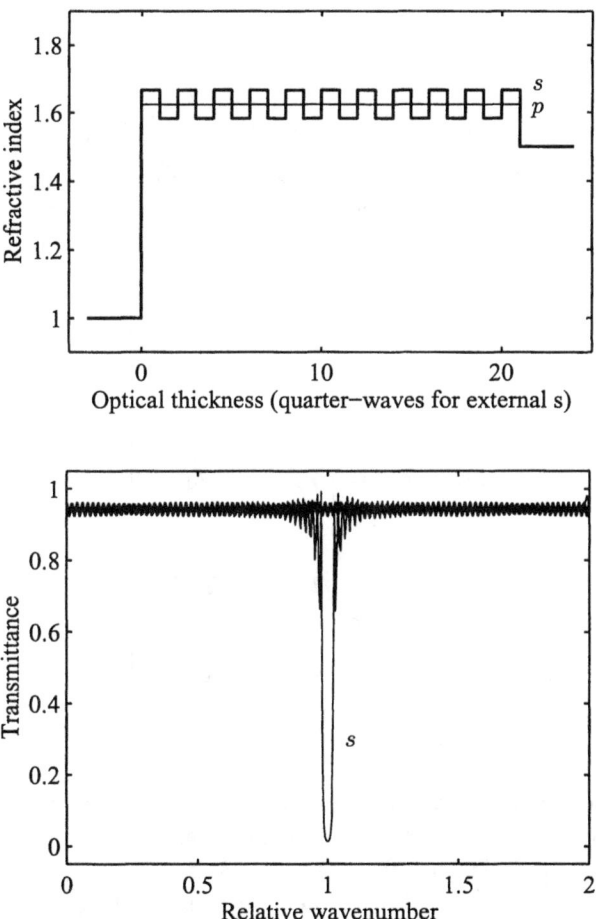

Fig. 16.17. Refractive index profile (illustrated for $N = 10$) and transmittance of the common-index thin film polarizer $aP_1'[S_2'P_1']^N g$ (calculated for $N = 50$).

16.6.10   Multi-Cavity Linear Polarizer

Generic Design

$$a[HL]^N(2H)[LH]^N C[HL]^N(2H)[LH]^N g$$

PS Design

$$a[HL]^2(10P)[LH]^2 C[HL]^2(10S)[LH]^2 g$$

$$n_a = 1, \ n_H = 2.4, \ n_L = 1.46, \ n_c = 1.68, \ n_g = 1.5$$
$$n_p = 1.643, \ n_s = 1.684 \text{ (zirconium oxide, } \theta_v = 57°)$$

$$pssystem = \begin{bmatrix} n_a & n_a & NaN \\ \vdots & \vdots & \vdots \\ n_p & n_s & 10g/4n_p \\ \vdots & \vdots & \vdots \\ n_c & n_c & g/4n_c \\ \vdots & \vdots & \vdots \\ n_s & n_p & 10g/4n_s \\ \vdots & \vdots & \vdots \\ n_g & n_g & NaN \end{bmatrix}$$

Notes

In this example we transform the generic two-cavity filter design, in which T is insensitive to polarization, to a polarizing filter in which T_p mimics T and $T_s \approx 0$. The generic design has two identical cavities coupled by a middle layer C. The PS design achieves the planned outcomes through the P, S cavity arrangement (see Fig. 16.18). P and S are quarter-waves at $g = 1$ for the p polarization ($n_p d_p = n_s d_s = \lambda_0/4$), but have different optical thicknesses for the s polarization ($n_s d_p \neq n_p d_s$). The index of the coupling layer has an influence on the ripple at the top of the passband – decreasing the index decreases the ripple, but decreases the bandwidth; and hence a compromise needs to be made. We have specified $n_c = 1.68$, but a birefringent layer can be used.

The PS two-cavity filter has the properties of a linear polarizer at the design wavelength. The extinction ratio for the polarizer is shown in Fig. 16.19.

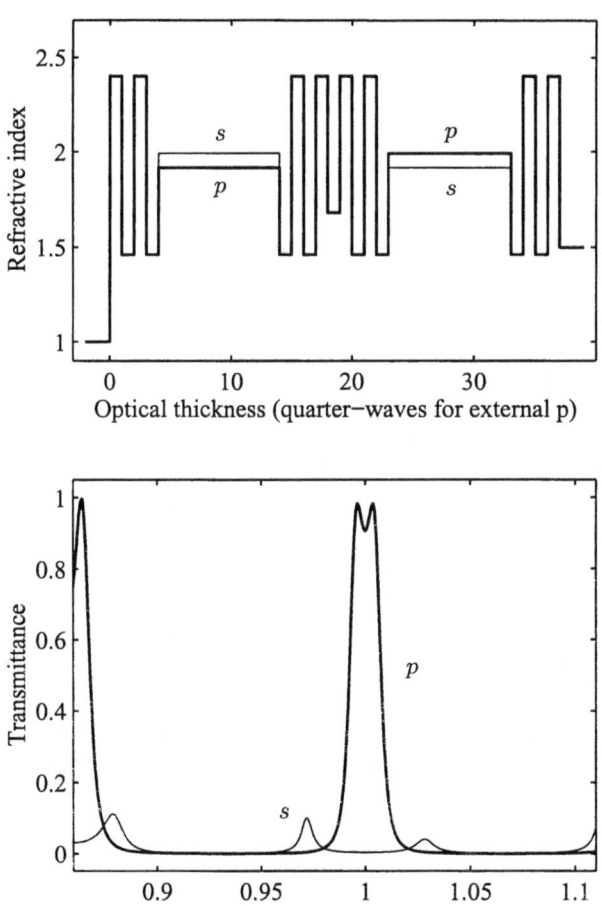

Fig. 16.18. Refractive index profiles and calculated transmittance of the double-cavity linear polarizer $a[HL]^2(10P)[LH]^2C[HL]^2(10S)[LH]^2g$.

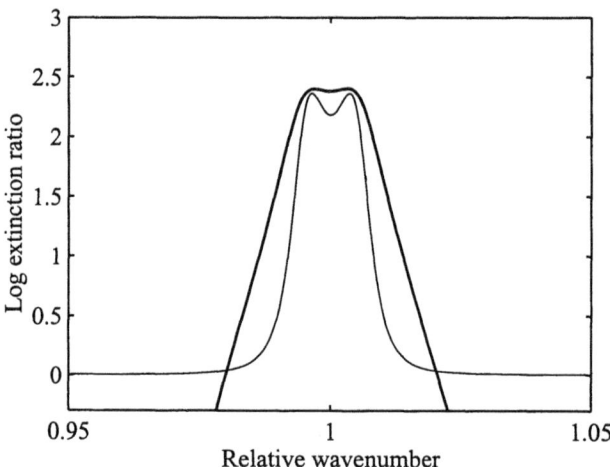

Fig. 16.19.   Calculated extinction ratio of the PS double-cavity linear polarizer $a[HL]^2(10P)[LH]^2C[HL]^2(10S)[LH]^2g$, shown superposed on the profile of the transmitted peak.

Appendix A

Birefringent Thin Films Toolbox

A.1   Quick Reference

Constants

clight	c	Speed of light in vacuum
epsilon0	ε_0	Permittivity of vacuum
mu0	μ_0	Permeability of vacuum
z0	z_0	Impedance of vacuum

Materials

alumin	Optical constants of aluminium
bpcd	Bragg-Pippard cd indices
bpcdi	Inverse of Bragg-Pippard cd equations
bpvd	Bragg-Pippard vd indices
bpvdi	Inverse of Bragg-Pippard vd equations
gold	Optical constants of gold
mgf2	Dispersion of magnesium fluoride
quartz	Dispersion of quartz
silver	Optical constants of silver
tao2	Refractive indices of tantalum oxide films
tio2	Refractive indices of titanium oxide films
zro2	Refractive indices of zirconium oxide films

Mueller calculus

mmat	$\hat{M}$	Mueller matrix
rmmat	R_m	Rotation matrix

Jones calculus

bloom1		Single pass birefringent filter
bloom2		Double pass birefringent filter
solc		Folded Solc filter
jmat	$\hat{J}$	Jones matrix
lyot		Lyot-Ohman filter
rjmat	R_j	Rotation matrix

Berreman calculus

cmat	$\hat{M}$	Characteristic matrix
epsilon	$\hat{\varepsilon}$	Relative permittivity matrix
fmat	$\hat{F}$	Field matrix
pmat	$\hat{A}_d$	Phase matrix
poynting	$\hat{S}$	Poynting vector
reflect	$\hat{R}$	Reflectance and transmittance
rxmat, rymat, rzmat	$\hat{R}_x, \hat{R}_y, \hat{R}_z$	Rotation matrices
smat	$\hat{A}$	System matrix

Script files

fresnel	Computation of fields
herpin	Effective birefringent media
reverse	Film parameters from $\hat{M}$
vretard	View retardation maps
vscatter	View scatter maps

Input arguments

General birefringent coating		
indices	=	[n1 n2 n3]
angles	=	[eta psi xi]
material	=	[indices angles]
layer	=	[material dw]
cover, substrate	=	[material NaN]
stack	=	[layer1
		layer2
		......
		layerN]
system	=	[cover
		stack
		substrate]

PS coating		
psmaterial	=	[ny nz]
pslayer	=	[psmaterial dw]
pscover	=	[nC nC NaN]
pssubstrate	=	[nS nS NaN]
psstack	=	[pslayer1
		pslayer2
		......
		pslayerN]
pssystem	=	[pscover
		psstack
		pssubstrate]

A.2 Commands and Functions

alumin

Purpose

Complex refractive index of aluminium.

Synopsis

[w, n] = alumin

Description

Table of wavelength and refractive index values (experimental values from L.G. Schulz and L.G. Schulz and F.G. Tangherlini[55]).

Examples

The values in Table 12.1 can be generated using alumin.

See Also

gold, mgf2, quartz, silver, tao2, tio2, zro2

References

See Sect. 12.3.

bloom1, bloom2

Purpose

Transmittance of birefringent filter.

Synopsis

T = bloom1 (group, delta)
T = bloom2 (group, delta)

Description

The functions bloom1 and bloom2 use Jones matrices to compute the transmittance of a birefringent filter constructed as a stack of birefringent plates positioned at the Brewster angle. The argument group is a row vector in which -3 (for example) indicates 3 glass plates and 3 indicates a birefringent plate of thickness three units. The second argument delta is the phase retardation of a plate of thickness one unit. bloom1 corresponds to a single pass (ring cavity) and bloom2 is for a double pass (Fabry-Perot cavity).

Example

```
for j = 1:501
delta = -pi/2+(j-1)*pi/500;
T1(j) = bloom1 (1,delta);
T2(j) = bloom2 ([2 1 6],delta);
T3(j) = bloom2 ([-2 2 -7 1 -5 6 -1],delta);
T4(j) = bloom1 ([1 1 1 1 1 1 1 1 1 1 1],delta);
deltap(j) = delta*180/pi;
end
figure plot (deltap,T1)
figure plot (deltap,sqrt(T2))
figure plot (deltap,sqrt(T3))
figure plot (deltap,T4)
```

produces Figs. 15.12, 15.13, 15.14, 15.16.

Algorithm

Equations from Sect. 15.2.3.

See Also

lyot, solc

References

See Fig. 15.11 for the optical configuration of the birefringent filter.

bpcd, bpcdi, bpvd, bpvdi

Purpose

Modelling form birefringence.

Synopsis

$[n1, n2, n3]$ = bpcd $(nc, nv, L1, L2, L3, p)$
$[n1, n2, n3]$ = bpvd $(nc, nv, L1, L2, L3, p)$
$[nc, L1, L2, L3, p]$ = bpcdi $(n1, n2, n3, nv)$
$[nc, L1, L2, L3, p]$ = bpvdi $(n1, n2, n3, nv)$

Description

Bragg-Pippard equations giving the three principal refractive indices of crystallite-defined and void-defined birefringent media. The input arguments for the forward equations are the refractive indices of the crystallites and the voids, the depolarizing factors (which should add to unity), and the packing density. For the inverse equations, the input parameters are the principal refractive indices and the void index.

Examples

See the example at the end of Sect. 8.3.2.

Algorithm

Implements Eqs.(8.19), (8.20), (8.24)–(8.29).

clight

Purpose

Constant used in electromagnetism.

Synopsis

y = clight

Description

Speed of light in vacuum, c.

Examples

clight produces 29792458.

See Also

epsilon0, mu0, z0

References

See Eq.(2.32) and Table 2.2.

cmat

Purpose

Characteristic matrix used in Berreman calculus.

Synopsis

General birefringent coating
M = cmat (layer, beta)
M = cmat (stack, beta)
M = cmat (system, beta)

PS coating
M = cmat (pslayer)
M = cmat (psstack)
M = cmat (pssystem)

Description

The function cmat calculates the 4×4 characteristic matrix M of a layer or a stack of layers.

The general form is M = cmat (arg1, beta)
(i) When arg1 has the layer or pslayer format the function returns the characteristic matrix of the layer.
(ii) When arg1 has the stack or psstack format the function returns the characteristic matrix of the stack of layers.
(iii) When arg1 has the system or pssystem format the function neglects the cover and substrate rows and returns the characteristic matrix of the remaining layer or stack of layers.

Algorithm

General birefringent coating
m = size (arg1, 1);
m1 = 1; m2 = m;
if arg1 (1, 7) == NaN
m1 = 2;

```
end
if arg1 (m, 7) == NaN
m2 = m-1;
end
M = eye (4);
for j = m1:m2
[F, alpha] = fmat (arg1(j,:),beta);
M = M*F*pmat (alpha, arg1(j,7))*inv (F);
end
```

PS coating

Implements Eqs.(4.63)–(4.65), with $\beta = 0$.

See Also

fmat, pmat

References

See Sect. 4.4.5.

epsilon

Purpose

Relative permittivity matrix used in Berreman calculus.

Synopsis

e = epsilon (material)
e = epsilon (layer)

Description

The function epsilon uses the rotation matrices rxmat and rzmat to transform the relative permittivity matrix from the material frame to the propagation frame.

Examples

See the example at the end of Sect. 2.6.3.

Algorithm

(arg1 is the input argument.)
e123 = [arg1(1)^2 0 0; 0 arg1(2)^2 0; 0 0 arg1(3)^2];
e = rxmat (arg1(4))*e123*rxmat (-arg1(4));
e = rzmat (arg1(5))*e*rzmat (-arg1(5));
e = rxmat (arg1(6))*e*rxmat (-arg1(6));

See Also

rxmat, rzmat

References

The algorithm implements Eq.(2.78).

epsilon0

Purpose

Constant used in electromagnetism.

Synopsis

y = epsilon0

Description

Permittivity of vacuum, ϵ_0.

Examples

epsilon0 produces 8.8542e-012.

Algorithm

y = 1/mu0/clight/clight;

See Also

clight, mu0, z0

References

See Sect. 2.2.1 and Table 2.2.

fmat

Purpose

Field matrix used in Berreman calculus.

Synopsis

F = fmat (material, beta)
F = fmat (layer, beta)

[F, alpha] = fmat (material, beta)
[F, alpha] = fmat (layer, beta)

[F, alpha, E, H] = fmat (material, beta)
[F, alpha, E, H] = fmat (layer, beta)

Description

The function fmat determines the 4×4 field matrix F for a birefringent material or layer. The columns of F are the four basis vectors. The second form returns alpha, a row vector with elements equal to the four values of $n \cos \theta$. The third form returns E and H, 3×4 matrices of electric and magnetic field components.

Algorithm

A check is made for an isotropic layer, in which case F is computed using Eqs.(4.25) and (4.26). Otherwise the matrix Mbeta is computed, using the definition in Eq.(3.49) and the function epsilon to obtain the relative permittivity matrix for the propagation plane. The eigenequation (Eq.(3.52))

[F,Alpha]=eig (Mbeta);

determines F and a diagonal matrix Alpha. Subsequent processing, which includes use of the function poynting to determine senses of energy flow, sorts the columns of F and the related row vector alpha. Sorting is discussed in Sect. 5.2.

F contains the y and z components of E and H; the x components are determined using Eqs.(3.47) and (3.48).

See Also

epsilon, poynting

gold

Purpose

Complex refractive index of gold.

Synopsis

[n] = gold (w)
[w, n] = gold

Description

Optical constants of gold. The first option uses a fitted polynomial (supplied by H.A. Macleod[67]) to estimate the complex index of refraction for a given wavelength in nm units. The second option returns a table of wavelength and refractive index values, (experimental values from L.G. Schulz and L.G. Schulz and F.G. Tangherlini[55]).

Examples

gold (550) produces 0.4015 + 2.3412i.

The values in Table 12.1 can be generated using **gold**.

See Also

alumin, mgf2, quartz, silver, tao2, tio2, zro2

References

See Sect. 12.3.

jmat

Purpose

Matrices used in Jones calculus.

Synopsis

J = jmat (xi)
J = jmat (Delta, xi)

Description

The function jmat calculates the Jones matrix for a linear polarizer at angle
xi to the y-axis in the first option, and the Jones matrix for a wave plate of
retardation Delta aligned at angle xi to the y-axis in the second option.

See Also

rjmat

lyot

Purpose

Transmittance of Lyot-Ohman filter.

Synopsis

T = lyot (N, delta)

Description

The function lyot computes the transmittance of a Lyot-Ohman birefringent filter with N birefringent plates. delta is the phase retardation of the thinnest plate.

Examples

```
for j = 1:501
delta = -pi+(j-1)/500*2*pi;
T(j) = lyot (4,delta);
deltap(j) = delta*180/pi;
end
plot (deltap,T)
```

produces Fig. 15.7.

Algorithm

The function **lyot** implements Eq.(15.5).

See Also

bloom1, bloom2, solc

References

See Fig. 15.5 for the optical configuration of the Lyot-Ohman birefringent filter.

mgf2

Purpose

Dispersion equation for crystalline magnesium fluoride.

Synopsis

$y = \mathsf{mgf2}\,(x)$

Description

The function mgf2 uses a Sellmeier dispersion equation for computation of the refractive indices of crystalline magnesium. The input argument x may be a single wavelength in micrometre units $(\lambda/\mu\,\text{m})$, or a row vector of wavelengths. The output is a matrix of values with columns representing $n_e\ n_o\ dn_e/d\lambda\ dn_e/d\lambda$.

Examples

Algorithm

The functions implement Eqs.(14.3) and (14.4).

See Also

alumin, gold, quartz, silver, tao2, tio2, zro2

References

Table 14.1. The constants used in the dispersion equation are listed in the CVI catalogue.[60]

mmat

Purpose

Matrices used in Mueller calculus.

Synopsis

M = mmat (xi)
M = mmat (Delta, xi)

Description

The function mmat calculates the Mueller matrix for a linear polarizer at angle xi to the y-axis in the first option, and the Mueller matrix for a wave plate of retardation Delta aligned at angle xi to the y-axis in the second option.

See Also

rmmat

mu0

Purpose

Constant used in electromagnetism.

Synopsis

y = mu0

Description

Permeability of vacuum, μ_0.

Examples

mu0 produces 1.2566e-006.

Algorithm

y = 4*pi*1.0e-007;

See Also

clight, epsilon0, z0

References

See Sect. 2.2.1 and Table 2.2.

pmat

Purpose

Phase matrix used in Berreman calculus.

Synopsis

Ad = pmat (alpha, dw)

Description

The function **pmat** computes the phase matrix **Ad** for a birefringent layer. The input row vector **alpha** is the set of four values of $n \cos \theta$ for the basis vectors, and **dw** is the thickness of the layer divided by the wavelength of the incident light.

Algorithm

Implements Eq.(4.33).

See Also

fmat, cmat

References

See Sect. (4.4.4).

poynting

Purpose

Poynting flux vectors used in Berreman calculus.

Synopsis

p = poynting (E, H)
p = poynting (F)

Description

The function poynting (E, H) takes as input the 3×4 matrices E and H for the electric and magnetic fields of the four basis vectors. The output p is a 3×4 matrix with rows representing Poynting fluxes for the x, y, and z directions.

The alternative form poynting (F) accepts the field matrix F as input and yields a row vector of Poynting fluxes for the x direction.

Algorithm

p = [real (E(2,:).*conj (H(3,:))-E(3,:).*conj (H(2,:)))
real (E(3,:).*conj (H(1,:))-E(1,:).*conj (H(3,:)))
real (E(1,:).*conj (H(2,:))-E(2,:).*conj (H(1,:)))]/2;

p = real ([F(1,:).*conj (F(2,:))-F(3,:).*conj (F(4,:))])/2;

See Also

reflect, fmat

References

See Sect. 3.3.6.

quartz

Purpose

Dispersion equation for crystalline quartz.

Synopsis

$y = \text{quartz}(x)$

Description

The function quartz uses a Laurent dispersion equation for computation of the refractive indices of crystalline quartz. The input argument x may be a single wavelength in λ/μm units, or a row vector of wavelengths. The output is a matrix of values with columns representing n_e n_o $dn_e/d\lambda$ $dn_e/d\lambda$.

Examples

Algorithm

The functions implement Eqs.(14.1) and (14.2).

See Also

alumin, gold, mgf2, silver, tao2, tio2, zro2

References

Table 14.1. The constants used in the dispersion equation are listed in the CVI catalogue.[60]

reflect

Purpose

Reflection and transmission coefficients in Berreman calculus.

Synopsis

General birefringent coating
[R, r] = reflect (system, beta)
[R, r] = reflect (Fc, M, Fs)

PS coating
[R, r] = reflect (pssystem)
[R, r] = reflect (nC, M, nS)

Description

The function **reflect** returns the 4×4 matrix **R** that contains 8 reflection coefficients and 8 transmission coefficients for a birefringent layer or stack of layers surrounded by cover and substrate media.

Algorithm

General birefringent coating
When just two input parameters are provided the function **cmat** is used to calculate the characteristic matrix of the layer, or stack of layers. Next **fmat** is used to determine the field matrices **Fc**, **Fs** for the cover and the substrate. In both cases the poynting fluxes **pc**, **ps** are found using the function **poynting**, and then the system matrix **A** is computed using **smat**. Finally, r and R are computed using

r = inv ([I(:,2) I(:,4) -A(:,1) -A(:,3)])*[-I(:,1) -I(:,3) A(:,2) A(:,4)];

R = abs ([pc(2) pc(4) ps(1) ps(3)]'*[1/pc(1) 1/pc(3) 1/ps(2) 1/ps(4)]).*abs (r).^2;

To allow evanescent inputs to be handled, input power fluxes of small magnitude, pc(1), pc(3), ps(2), and ps(4), are replaced by NaN's.

PS coating
The algorithm is based on direct computation for the special case of PS coatings.

See Also

cmat, fmat, poynting, smat

References

See Sect. 5.1. The meanings of r and R are defined by Eqs. (5.6) and (5.8). The algorithms for computing r and R are based on Eqs.(5.6) and (5.9) for the general birefringent coating, and on Eqs.(5.23), (5.24), (5.26) and (5.27) for the PS coating.

rjmat

Purpose

Rotation matrix used in Jones calculus.

Synopsis

Rj = rjmat (angle)

Description

The function rjmat provides a rotation matrix for transformation between a standard coordinate system and a new system rotated about the x-axis. The argument angle is input in radian units.

Algorithm

The algorithm implements Eq.(4.14).

rmmat

Purpose

Rotation matrix used in Mueller calculus.

Synopsis

Rm = rmmat (angle)

Description

The function rmmat provides a rotation matrix for transformation between a standard coordinate system and a new system rotated about the x-axis. The argument angle is input in radian units.

Algorithm

The algorithm implements Eq.(4.4).

rxmat, rymat, rzmat

Purpose

Rotation matrices used in Berreman calculus.

Synopsis

Rx = rxmat (angle)
Ry = rymat (angle)
Rz = rzmat (angle)

Description

The functions rxmat, rymat and rzmat provide rotation matrices for transformations between material frame coordinates and propagation frame coordinates. The argument angle is input in radian units.

Algorithm

The algorithms implement Eqs.(2.71) and (2.72).

See Also

epsilon

silver

Purpose

Complex refractive index of silver.

Synopsis

[n] = silver (w)
[w, n] = silver

Description

Optical constants of silver. The first option uses a fitted polynomial (supplied by H.A. Macleod[67]) to estimate the complex index of refraction for a given wavelength in nm units. The second option returns a table of wavelength and refractive index values (experimental values from L.G. Schulz and L.G. Schulz and F.G. Tangherlini[55]).

Examples

silver (550) produces 0.0540 + 3.3164i.

The values in Table 12.1 can be generated using silver.

See Also

alumin, gold, mgf2, quartz, tao2, tio2, zro2

References

See Sect. 12.3.

smat

Purpose

System matrix used in Berreman calculus.

Synopsis

A = smat (system, beta)
A = smat (Fc, M, Fs)

Description

The function **smat** computes the system matrix A for a system of cover, birefringent layers, and substrate. In the second form the input arguments Fc and
Fs are the field matrices for the cover and substrate, and M is the characteristic
matrix of the layer or stack of layers.

Algorithm

In the first option, the number of rows m of system is determined, then

Fc = fmat (system (1,:), beta)
M = cmat (system, beta)
Fs = fmat (system (m,:), beta)

Finally

A = inv (Fc)*M*Fs;

See Also

fmat, cmat

References

The algorithm implements Eq.(4.41) for the system matrix.

solc

Purpose

Transmittance of folded Solc filter.

Synopsis

T = solc (N, delta)

Description

The function solc uses Jones matrices to compute the transmittance of a folded Solc birefringent filter with N plates. delta is the phase retardation of each plate, and N is assumed to be an even integer.

Examples

```
for j = 1:501
delta = (j-1)/500*2*pi;
T(j) = solc (16,delta);
deltap(j)=delta*180/pi;
end
plot (deltap,T)
```

produces Fig. 15.9.

Algorithm

epsilonr = pi/4/N;

determines the rocking angle, and the Jones matrix is computed with

```
J = jmat (delta,-epsilonr)*jmat (delta,epsilonr);
J = J^(N/2);
```

The crossed polarizers transmit amplitude $J(2,1)$, hence
T = abs (J(2,1))^2;

See Also

bloom1, bloom2, jmat, lyot

References

See Fig. 15.8 for the optical configuration of the folded Solc filter.

tao2

Purpose

Typical refractive indices and column angle of tantalum oxide films.

Synopsis

y = tao2 (thetav)
[np, ns] = tao2 (thetav)
[n1, n2, n3, psi] = tao2 (thetav)

Description

Uses constants derived experimentally for deposition angles **thetav** in the range 40° to 70°, and by extrapolation in the range 0° to 40°, for calculation of the refractive indices and column angle **psi** of tantalum oxide films. When the number of output arguments is zero or one, the data is returned in the order n1, n2, n3(=ns), psi, np. Here n1, n2, n3 are the principal refractive indices and ns, np are refractive indices appropriate to propagation at normal incidence; p and s are referenced to the deposition plane. When two (or four) output arguments are specified, the function returns data as required for 2×2 (or 4×4) matrix calculations.

Examples

tao2 (60*pi/180) produces 1.7483 1.6248 1.6948 0.5088 1.6517 for a deposition angle $\theta_v = 60°$; converting the column angle to degree gives $\theta_v = 29°$.

Algorithm

A quadratic approximation is used to relate the principal refractive indices to deposition angle, and a modified version of the tangent rule relates column angle to deposition angle. Eq.(3.34) is used for computation of np from n1, n2, n3, psi.

See Also

alumin, gold, mgf2, quartz, silver, tio2, zro2

References

See Sect. 8.2.

tio2

Purpose

Typical refractive indices and column angle of titanium oxide films.

Synopsis

y = tio2 (thetav)
[np, ns] = tio2 (thetav)
[n1, n2, n3, psi] = tio2 (thetav)

Description

Uses constants derived experimentally for deposition angles thetav in the range
$40°$ to $70°$, and by extrapolation in the range $0°$ to $40°$, for calculation of the
refractive indices and column angle psi of titanium oxide films. When the number of output arguments is zero or one, the data is returned in the order n1, n2,
n3(=ns), psi, np. Here n1, n2, n3 are the principal refractive indices and ns, np
are refractive indices appropriate to propagation at normal incidence; p and s
are referenced to the deposition plane. When two (or four) output arguments
are specified, the function returns data as required for 2×2 (or 4×4) matrix
calculations.

Examples

tio2 (60*pi/180) produces 2.1112 1.8053 1.9450 0.5412 1.8733 for a deposition
angle $\theta_v = 60°$; converting the column angle to degree gives $\theta_v = 31°$.

Algorithm

A quadratic approximation is used to relate the principal refractive indices to
deposition angle, and a modified version of the tangent rule relates column angle
to deposition angle. Eq.(3.34) is used for computation of np from n1, n2, n3,
psi.

See Also

alumin, gold, mgf2, quartz, silver, tao2, zro2

References

See Sect. 8.2.

zro2

Purpose

Typical refractive indices and column angle of zirconium oxide films.

Synopsis

y = zro2 (thetav)
[np, ns] = zro2 (thetav)
[n1, n2, n3, psi] = zro2 (thetav)

Description

Uses constants derived experimentally for deposition angles **thetav** in the range
40° to 70°, and by extrapolation in the range 0° to 40°, for calculation of the
refractive indices and column angle **psi** of zirconium oxide films. When the
number of output arguments is zero or one, the data is returned in the order n1,
n2, n3(=ns), psi, np. Here n1, n2, n3 are the principal refractive indices and ns,
np are refractive indices appropriate to propagation at normal incidence; p and
s are referenced to the deposition plane. When two (or four) output arguments
are specified, the function returns data as required for 2 × 2 (or 4 × 4) matrix
calculations.

Examples

zro2 (60*pi/180) produces 1.7294 1.5975 1.6628 0.4530 1.6205 for a deposition
angle $\theta_v = 60°$; converting the column angle to degree gives $\theta_v = 26°$.

Algorithm

A quadratic approximation is used to relate the principal refractive indices to
deposition angle, and a modified version of the tangent rule relates column angle
to deposition angle. Eq.(3.34) is used for computation of np from n1, n2, n3,
psi.

See Also

alumin, gold, mgf2, quartz, silver, tao2, tio2

References

See Sect. 8.2.

z0

Purpose

Constant used in electromagnetism.

Synopsis

y = z0

Description

Impedance of free space, z_0.

Examples

z0 produces 376.7303.

Algorithm

z0 = sqrt (mu0/epsilon0);

See Also

clight, epsilon0, mu0

References

See Eq.(2.31) and Table 2.2.

Notes and References

1. C. Kittel, *Introduction to Solid State Physics* (John Wiley, New York, 1986)

2. M. Born and E. Wolf, *Principles of Optics* (Pergamon Press, London, 1959)

3. For an early report of anisotropy in thin films deposited obliquely see A. Kundt, *Ann. Phys. Chem.* **27** (1886) 59. Double refraction in bulk crystalline material (calcite) was reported two centuries earlier, by E. Bartholinus in 1669. See the account of anisotropy in R.D. Guenther, *Modern Optics* (John Wiley, New York, 1990).

4. P. Bousquet, *Ann. Phys.* **2** (1957) 163; P. Bousquet and Y. Delcourt, *J. Phys. Rad.* **18** (1957) 447; K.D. Sinel'nikov, I.N. Shklarevskii and N.A. Vlasenko, *Opt. Spectrosc. (USSR)* **2** (1957) 651.

5. F. Horowitz and H.A. Macleod, *Proc. Soc. Photo-Opt. Instrum. Eng.* **380** (1983) 83.

6. MATLAB is a trademark of The MathWorks, Inc., 24 Pine Park Way, Natick, MA 01760, USA.

7. G.R. Fowles, *Introduction to Modern Optics* (Holt, Rinehart and Winston, New York, 1975)

8. E. Collett, *Polarized Light* (Marcel Dekker, New York, 1993)

9. For a recent review article see Chapt. 27, *Ellipsometry* by R.M.A. Azzam in *Handbook of Optics*, Vol. 2, ed. M. Bass, (McGraw-Hill, New York, 1995).

10. For a recent article on the Mueller matrix and its application see Chapt. 22, *Polarimetry* by R.A. Chipman in *Handbook of Optics*, Vol. 2, ed. M. Bass, (McGraw-Hill, New York, 1995).

11. We refer to the 4×4 matrix framework for biaxial layers as the Berreman calculus, acknowledging the pioneering work of D.W. Berreman, *J. Opt. Soc. Am.* **62** (1972) 502. Another foundation paper is S. Teitler and B.W. Henvis, *J. Opt. Soc. Am.* **60** (1970) 830. For a selection of more

recent articles on the 4 × 4 matrix method see:- P.J. Lin-Chung and S. Teitler, *J. Opt. Soc. Am. A* **1** (1984) 703; R.S. Weis and T.K. Gaylord, *J. Opt. Soc. Am. A* **4** (1987) 1720; K. Eidner, *J. Opt. Soc. Am. A* **6** (1989) 1657; E. Georgieva, *J. Opt. Soc. Am. A* **12** (1995) 2203.

12. O.S. Heavens, *Optical Properties of Thin Solid Films*, (Dover, New York, 1965)

13. F. Abelès, *Progress in Optics*, ed. E. Wolf, **2**, 251 (North Holland Publishing, Amsterdam, 1963)

14. I.J. Hodgkinson and D. Endelema, *Appl. Opt.* **29** (1990) 4424.

15. M.O. Vassel, *J. Opt. Soc. Am.* **64** (1974) 176.

16. See also E.A. Kolosovski, D.V. Petrov, A.B. Tsarev and I.B. Yakovkin, *Opt. Commun.* **43** (1982) 21.

17. H. König and G. Helwig, *Optik* **6** (1950) 111.

18. J.M. Nieuwenhuizen and H.B. Haanstra measured column angles in aluminium films deposited at a range of oblique angles, and the tangent rule was proposed by G.W. van Oosterhout. See J.M. Nieuwenhuizen and H.B. Haanstra, *Philips Tech. Rev.* **27** (1966) 87.

19. The microstructural asymmetry described as elongation of column thickness or bunching of columns perpendicular to the deposition plane is referred to as the *first anisotropy*; the columnar structure is called the *second anisotropy*. For early observations of elongated and bunched particles in permalloy, iron and cobalt see M.S. Cohen, *J. Appl. Phys.* **32** (1961) 87S; W.J. Schuele, *J. Appl. Phys.* **35** (1964) 2558; D.E. Speliotis et al., *J. Appl. Phys.* **36** (1965) 972. See also the work of K. Okamoto et al., much of which is published in *Journal of the Physical Society of Japan* and *Thin Solid Films*.

20. K. Robbie, L.J. Friedrich, S.K. Dew, T. Smy and M.J. Brett, *J. Vac. Sci. Technol. A* **13** (1995) 1032.

21. We are grateful to K. Robbie, Department of Electrical Engineering, University of Alberta, Edmonton, Alberta, Canada T6G 2G7, for providing the photographs in Figs. 7.7 and 7.9.

22. N.O. Young and J. Kowal, *Nature* **183** (1959) 104.

23. K. Robbie, M.J. Brett and A. Lakhtakia, *J. Vac. Sci. Technol. A* **13** (1995) 2291.

24. For a review of the early literature on computer simulation of the growth of thin films, including both 2–D and 3–D serial deposition, the reader may consult the reference I.J. Hodgkinson and P.W. Wilson, *CRC Crit.*

Rev. Solid State Mater. Sci. **15** (1988) 27. In this book we restrict our discussion to 3–D serial deposition. For specific examples of 2–D simulation, see K.H. Guenther, *Appl. Opt.* **23** (1984) 3806; M.R. Jacobson, F. Horowitz and B. Liao, *Proc. Soc. Photo-Opt. Instrum. Eng.* **505** (1984) 228; M. Sikkens, I.J. Hodgkinson, F. Horowitz, H.A. Macleod and J.J. Wharton, *Proc. Soc. Photo-Opt. Instrum. Eng.* **505** (1984) 236; K.-H. Müller, *J. Appl. Phys.* **58** (1985) 2573; P. Meakin, *Phys. Rev. A* **38** (1988) 994; M.J. Brett, *J. Vac. Sci. Technol. A* **6** (1988) 1749; J.S. Gau and B. Liao, *Thin Solid Films* **176** (1989) 309; R.N. Tait, T. Smy and M.J. Brett, *Thin Solid Films* **187** (1990) 375; S. Müller-Pfeiffer and H.J. Anklam, *Vacuum* **42** (1991) 113.

25. D. Henderson, M.H. Brodsky and P. Chaudhari, *Appl. Phys. Lett.* **25** (1974) 641.

26. J.R. Gee, Ph.D. thesis, University of Otago, Dunedin, New Zealand (1992).

27. J.L. Finney, *Proc. R. Soc. London A* **319** (1970) 479.

28. C.H. Bennett, *J. Appl. Phys.* **43** (1972) 2727.

29. J.L. Finney, *Nature* **266** (1977) 309.

30. The acronym PIE (perpendicular incidence ellipsometry) was coined by R.M.A. Azzam, see R.M.A. Azzam, Opt. Comm. **19** (1976) 122.

31. T. Motohiro and Y. Taga, *Appl. Opt.* **28** (1989) 2466.

32. A. Zuber, H. Jänchen and N. Kaiser, *Appl. Opt.* **35** (1996) 5553.

33. H. Wang, *J. Opt. Soc. Am. A* **11** (1994) 2331.

34. See also E. Pelletier, F. Flory and Y. Hu, *Appl. Opt.* **28** (1989) 2918. For a comparison of envelope (photometric) and waveguide methods see F. Horowitz and S.B. Mendes, *Appl. Opt.* **33** (1994) 2659.

35. Technical information from Patinal Newsletter No. 3, June 1994, E. Merc, Patinal-Centre, 64578 Gernsheim, Germany.

36. The equations that we refer to as the Bragg-Pippard (BP) equations were reported first by O. Wiener and then independently by W.L. Bragg and A.B. Pippard. See O. Wiener, *Abh. Math.-Phys. Kl. Saechs.* **32** (1912) 575, and W.L. Bragg and A.B. Pippard, *Acta Crystallogr.* **6** (1953) 865.

37. B. Abeles and J.I. Gittleman, *Appl. Opt.* **15** (1976) 2328.

38. H.K. Pulker and E. Jung, *Thin Solid Films* **9** (1971) 57.

39. S. Ogura, Ph.D. thesis, Newcastle upon Tyne Polytechnic, Newcastle upon Tyne, England (1975).

40. M. Harris, M. Bowden and H.A. Macleod, *Opt. Commun.* **51** (1984) 29.

41. H.A. Macleod, *J. Vac. Sci. Technol. A* **4** (1985) 418.

42. I.J. Hodgkinson and P.W. Wilson, *CRC Crit. Rev. Solid State Mater. Sci.* **15** (1988) 27.

43. L.I. Epstein, *J. Opt. Soc. Am.* **42** (1952) 806.

44. H.K. Pulker, *Optik* **32** (1971) 496.

45. H.A. Macleod and D.A. Richmond, *Thin Solid Films* **37** (1976) 163.

46. D.R. Gibson and P.H. Lissberger, *Appl. Opt.* **22** (1983) 269.

47. S. Tolansky, *Surface Microtopography*, (Longmans Green, London, 1960)

48. P. Giacomo, P.W. Baumeister and F.A. Jenkins, *Proc. Phys. Soc. London* **73** (1959) 480.

49. G. Koppelmann, *Optik* **17** (1960) 416.

50. For a review of the early literature on nucleation, growth, and post-deposition processes in thin metal films the reader may consult the reference I.J. Hodgkinson and P.W. Wilson, *CRC Crit. Rev. Solid State Mater. Sci.* **15** (1988) 27.

51. G.B. Smith, *Appl. Phys. Lett.* **46** (1985) 716; P.J. Martin, W.G. Sainty and R.P. Netterfield, *Appl. Opt.* **23** (1984) 2668; P.J. Martin, W.G. Sainty, R.P. Netterfield and A.N. Buckley, *Vacuum* **35** (1985) 621; R.P. Netterfield and P.J. Martin, *Appl. Surf. Sci.* **25** (1986) 265.

52. J.G.W. van de Waterbeemd and G.W. van Oosterhout, *Philips Res. Rep.* **22** (1967) 375.

53. For a comprehensive and cross-referenced list of about 700 articles on aggregated metal films see the review paper G.C. Papavassiliou, *Prog. Solid State Chem.* **12** (1979) 185.

54. F.L. Pedrotti and L.S. Pedrotti, *Introduction to Optics* (Prentice-Hall, New Jersey, 1987)

55. L.G. Schulz, *J. Opt. Soc. Am.* **44** (1954) 357, and L.G. Schulz and F.G. Tangherlini, *J. Opt. Soc. Am.* **44** (1954) 362.

56. S. Norrman, T. Andersson, C.G. Granqvist and O. Hunderi, *Phys. Rev. B* **18** (1978) 674.

57. For a review of polarizers and retardation plates see Chapt. 3 on *Polarizers* by J.M. Bennett in *Handbook of Optics*, Vol. 2, ed. M. Bass, (McGraw-Hill, New York, 1995).

58. A. Dinca, M.E. Trifan, V. Lupei and M.P. Dinca, *J. Mod Opt.* **43** (1996) 1615.

59. E. Hecht, *Optics* (Addison-Wesley, Reading, Massachusetts, 1987)

60. Product information, *Optics and Coatings*, CVI Laser Corporation, Box 11308, Albuquerque, New Mexico, 87192, USA.

61. Z. Knittl, *Optics of Thin Films* (Wiley and Sons, New York, 1976)

62. P. Yeh, *Optical Waves in Layered Media* (John Wiley, New York, 1988)

63. J.M. Yarborough, and J. Hobart, at IEEE/OSA Conference on Laser Engineering and Applications (Washington, D.C. 1973); A.L. Bloom, *J. Opt. Soc. Am.* **64** (1974) 447; G. Holtom and O. Teschke, *IEEE J. Quantum Electron.* **QE-10** (1974) 557.

64. In a more advanced analysis the 4×4 matrix method is applied to take into account interference from multiple-reflections. See T. Mavrudis, J. Mentel and M. Schumann, *Appl. Opt.* **34** (1995) 4217.

65. Dispersion equations for a large range of crystals and glasses have been listed by W.J. Tropf, M.E. Thomas and T.J. Harris in *Properties of Crystals and Glasses*, Chapt. 33 of *Handbook of Optics*, Vol. 2, ed. M. Bass, (McGraw-Hill, New York, 1995).

66. Most of the generic thin film designs for isotropic materials that we have used are described by J.A. Dobrowolski in *Optical Properties of Films and Coatings*, Chapt. 42 of *Handbook of Optics*, Vol.1, ed. M. Bass, (McGraw-Hill, New York, 1995).

67. H.A. Macleod, Thin Film Center Inc., 2745 E Via Rotonda, Tucson, Arizona 85716-5227, USA.

Index

A

Abelès calculus, 65
Abelès matrices, 66
  deposition plane, 69
  isotropic layers, 67
Achromatic:
  antireflection coating, 299, 308
  half-wave plate, 261
  quarter-wave plate, 261
  reflector, 299, 310
  wave plate, 259
Active transmittance, 284
Airy function, 249
alumin, 221, 328
Aluminium:
  anisotropy during deposition, 209
  computer modelling, 231
  refractive index, 221, 328
  telescope mirrors, 212
angles, 26, 71, 326
Angular frequency, 14
Angularly-selective coatings, 209
Anisotropic:
  antireflection coating, 296, 302
  reflector, 298, 304
  scatter, 167
    dependence on deposition angle, 177
    from fluid patches, 188
    from herring-bone stacks, 177
    from stress-related cracks, 169
    in situ measurement, 175
    into the air, 168
    into the substrate, 175
    patterns on film, 174
    simple theory, 180
    stress, in birefringent films, 169
Anisotropic-phase reflector, 298, 306
Anisotropy:
  aging of metal films, 216
  during etching, 217
  fluid transport, 185
  modelling for metals, 217
  scatter, 188
Auxiliary matrix, 47

B

Babinet compensator, 265
Basis vectors, 36, 40
  in Berreman calculus, 46
  in Jones calculus, 40
  in Mueller calculus, 36
Berek compensator, 267
Berreman calculus, 57
  characteristic matrix, 60
  computations, 69
  field matrix, 57
  phase matrix, 60
  system matrix, 62
Berreman matrix, 49
Berreman vector, 32
Biaxial media, 119
Birefringence:
  computer modelling, 131
  values, 144

Birefringent Fabry-Perot polarizer, 247
Birefringent films, 4
  columnar microstructure, 5
  form birefringence, 5
  model, 5
  optical classification, 5
  structure, 4
Birefringent filters, 276
  fan Solc, 280
  folded Solc, 280
  for tuning dye laser, 282
  Lyot-Ohman, 276
  without secondary peaks, 285
Birefringent media:
  crystallite defined, 150
  void defined, 150
bloom1, 287, 329
bloom2, 287, 329
bpcd, 150, 202, 331
bpcdi, 152, 201, 331
bpvd, 150, 331
bpvdi, 152, 331
Bragg-Pippard equations, 149, 223
  inversion, 151
Brewster angle, 282
BTF toolbox, 7, 69, 90

C

Characteristic matrix, 60
  computation of film properties, 63
  properties, 62
  stack of layers, 61
  transformation property, 60
Charge density, 12
Circular birefringence, 275
Circular polarizer, 274
clight, 17, 332
Closed transmittance, 241
cmat, 71, 73, 298, 333
Coated-plate polarizer, 244

Coatings:
  birefringent, 289
  isotropic, 290
  PS, 294
Coherence, 32
Coherent states, 38
Column angle, 131
Common-index polarizer, 299, 318
Computer modelling:
  anisotropic resonance in metal, 225
  anisotropy in metals, 217
  birefringence, 131
  column angle, 131
  deposition of aluminium, 231
  deposition of dielectrics, 127
  deposition of gold, 228
  deposition of metal, 227
  deposition of silver, 231
  deposition/etch paths, 231
  etching of metal, 227
  form birefringence, 149
  HBC model, 127
  isotropic resonance in metal, 223
  periodic boundary conditions, 127
  radial distribution function, 128
  structural hysteresis loops, 231
  two-dimensional angular distribution, 131
  visual analysis, 127
cover, 71, 326
Conductivity, 12
Crystallite-defined media, 150
Crystals, 2
  axes, 2
  optical classification, 2
  structural classification, 2
  unit cell, 2
  principal dielectric axes, 4
  principal refractive indices, 4
Current density, 12

D

Degree of polarization, 35, 241
Depolarization factors:
  dielectrics, 150
  metals, 222
Depolarization field, 149
Depolarizer, 275
Deposition, 117
  biaxial media, 119
  columnar structures, 119
  computer modelling, 127
  helical microstructures, 125
  optically active media, 125
  parameters, 119
  uniaxial media, 119
  vacuum, 118
  wavy anisotropic media, 123
  zig-zag anisotropic media, 123
Dichroic:
  glass polarizer, 243
  polarizer, 241
  sheet polarizer, 242
Dual-wavelength wave plate, 257

E

Edge filter, 299, 316
Effective column angle, 153
  biaxial layers:
    common deposition plane, 155
    different deposition planes, 164
  isotropic layers, 154
Effective media, 153
Effective refractive indices, 153
  biaxial layers:
    common deposition plane, 155
    different deposition planes, 164
  isotropic layers, 154
Electric displacement, 12
Electric field, 12
Electric susceptibility, 12

Electron beam evaporator, 118
Ellipsometer, 32
Ellipsometric parameters, 41
Ellipsometry, 41
Elliptical polarization, 38
Embedded thin film polarizer, 246
epsilon, 26, 335
epsilon0, 17, 336
Evaporation rate and thickness controller, 118
Extinction ratio, 241

F

Fabry-Perot interferometer, 247
  free spectral range, 250
  fringe finesse, 250
  spectral range without overlap, 250
FECO, 201
Field coefficients, 58
Field matrix, 47, 57
  isotropic medium, 58
  reciprocal in isotropic medium, 58
  sorting columns, 81
  transformation property, 59
Fluid patches, 184
  anisotropic scatter, 188
  change of birefringence, 197
  in MDM filters, 185
  recording, 185
  scatter, 188
  theory of scatter, 191
Fluid transport, 183
  anisotropy, 185
  influence on birefringence, 197
fmat, 48, 57, 337
Form birefringence, 135
  Bragg-Pippard equations, 149
  computer modelling, 149
  depolarizing factors, 150
  depolarizing field, 149

Form birefringence (*Continued*)
  parallel isotropic layers, 155
  structure fraction, 151
  void defined media, 150
Free spectral range, 250
fresnel, 29, 325
Fresnel's equation, 20, 43
  deposition plane, 44
  isotropic media, 45
  uniaxial media, 45
Fringe finesse, 250

G

Glan-air polarizer, 252
Glan-Foucault polarizer, 252
Glan-laser polarizer, 252
Glan-Thompson polarizer, 252
Glass plate polarizer, 243
gold, 221, 339
Gold:
  anisotropy during deposition, 209
  computer modelling, 228
  refractive index, 221, 339
group 287, 329

H

Harmonic wave plate, 259
HBC model, 127
Heavens calculus, 59
Heavens matrices, 66
  deposition plane, 69
  isotropic layers, 67
herpin, 164–5, 325
Herpin indices, 153
  biaxial layers:
    common deposition plane, 155
    different deposition planes, 164
  isotropic layers, 154

I

Impedance of vacuum, 16, 17

indices, 26, 71, 326
Ion gun, 118
Isotropic coatings, 290

J

jmat, 56, 340
Jones calculus, 52
  basis vectors, 40
  periodic arrangements, 56
Jones matrix, 49, 52
  birefringent plate, 282
  elements in series, 55
  glass plate at Brewster angle, 282
  linear polarizer, 54
  quarter-wave plate, 54
  real polarizer, 240
  retardation plate, 54
  rotated elements, 55
  with interference, 74
  with major reflections, 75
Jones vector, 32, 38, 41, 49

L

Laurent equation, 256
layer, 71, 326
Linear polarizer, 274
  birefringent Fabry-Perot, 247
  closed transmittance, 241
  coated-plate, 244
  degree of polarization, 241
  dichroic, 241
  dichroic glass, 243
  dichroic sheet, 242
  embedded thin film, 246
  extinction ratio, 241
  Glan-air, 252
  Glan-Foucault, 252
  Glan-laser, 252
  Glan-Thompson, 252
  glass plate, 243

open transmittance, 241
Rochon prism, 253
tilted glass plate, 243
total transmittance, 241
transmission efficiency, 241
wire grid, 242
Wollaston prism, 253
lyot, 278, 341
Lyot-Ohman filter, 276

M

Magnesium fluoride:
  dispersion, 256
  refractive indices, 256
Magnetic field, 12
Magnetic induction, 12
Magnetic susceptibility, 12
Magnetization, 12
material, 26, 71, 326
Material axes, 22
MATLAB mathematical software, 7
Maxwell's equations, 12, 26
  for anisotropic media, 18
  for isotropic media, 17
  for vacuum, 12
Maxwell-Garnet equation, 151
Metal films, 203
  aging, 216
  aluminium, 209
  anisotropy, 204
  etching, 204
  gold, 209
  growth, 204
  morphology, 206
  recording optical anisotropy, 206
  silver, 209
Metals, anisotropic resonance, 225
  computer modelling, 225
  depolarization factors, 222
  isotropic resonance, 223

refractive index of bulk metal, 221
structural hysteresis loops, 231
Metre, 16
mgf2, 256-7, 342
mmat, 52, 343
Modal:
  modal condition, 94
    general case, 94
    isotropic bounding media, 95
    poles of reflection coefficient, 98
    uncoupled modes, 97
  modal contours, 100
  modal cutoffs, 100
  modal field structure, 104
  modal order, 111
  modal overlap, 108
  modal polarization, 106
Moisture patches, 184
  anisotropic scatter, 188
  change of birefringence, 197
  in MDM filters, 185
  recording, 185
  scatter, 188
  theory of scatter, 191
Monitoring:
  hybrid, 300
  normal, 300
Mueller calculus, 50
  basis vectors, 36
Mueller matrix, 49, 50, 52
  elements in series, 51
  rotated elements, 50
Multi-cavity:
  filter, 299, 314
  polarizer, 300, 320
Multiple reflection in a prism, 253
Multiple-order wave plate, 257
mu0, 17, 344

O

Open transmittance, 241
Order of interference, 249

P

Partial polarization, 35
Passive transmittance, 284
Permeability, 12
  of vacuum, 17
Permittivity, 12
  of vacuum, 17
Perpendicular incidence ellipsometry,
    136
  absorbing anisotropic film, 141
  characteristic curves, 138
  computation of parameters, 136
  homogeneous anisotropic film, 142
  ideal anisotropic film, 138
Phase change:
  on reflection, 90
  on transmission, 90
Phase matrix, 60
  transformation property, 60
Phase velocity, 15
  surface, 20
Photons, 41
  spin, 41
Planar waveguides, 93
  modal condition, 94
  modal contours, 100
  modal cutoffs, 100
  modal field structure, 104
  modal order, 111
  modal overlap, 108
  modal polarization, 106
  power flow, 111
  prism coupler, 112
Planck's constant, 41
pmat, 60, 345
Polarization, 12

degree of, 35
  partial, 35
  unpolarized light, 35
Polarization state filter, 274
  circular polarizer, 274
  depolarizer, 275
  linear polarizer, 274
  scrambler, 275
Polarized light, 36
  circular, 39
  elliptical, 38
  linear, 40
Power, 48
  flow, 111
poynting, 48, 346
Poynting vector, 22
Principal refractive indices:
  biaxial layers:
    common deposition plane, 155
    different deposition planes, 164
  from photometric measurements, 147
  in situ measurement, 145
  isotropic layers, 154
  parallel isotropic layers, 155
  using narrowband filters, 146
  values, 144
  waveguide method, 147
Prism coupler, 112
Propagation, 12
  axes, 24
  in a common direction, 26
  in anisotropic media, 18
  in isotropic media, 17
  in vacuum, 12
PS coatings, 72, 91, 294
  achromatic antireflection coating,
    299, 308
  achromatic reflector, 299, 310
  anisotropic antireflection coating,
    296, 302

anisotropic reflector, 298, 304
anisotropic-phase reflector, 298, 306
common-index polarizer, 299, 318
edge filter, 299, 316
monitoring, 300
multi-cavity filter, 299, 314
multi-cavity polarizer, 300, 320
single-cavity filter, 299, 312
pscover, 73, 326
pslayer, 73, 326
psmaterial, 73, 326
psstack, 73, 326
pssubstrate, 73, 326
pssystem, 73, 326

Q

quartz, 256–7, 347
Quartz:
 dispersion, 256
 refractive indices, 256
Quartz crystal sensor, 118

R

Radial distribution function, 128
Ray velocity surface, 20
reflect, 90–1, 298, 348
Reflectance coefficients, 80
 all media isotropic, 89
 crystal-crystal interface, 81
 isotropic bounding media, 85
Reflection, 77
 anisotropy, 209
 phase change, 90
Refractive index surface, 20
Refractive indices:
 aluminium, 21, 328
 gold, 221, 339
 magnesium fluoride, 256, 342
 quartz, 256, 347
 silver, 221, 353

 tantalum oxide, 146, 357
 titanium oxide, 146, 359
 zirconium oxide, 146, 361
Residual gas pressure controller, 118
reverse, 65, 325
rjmat, 56, 350
rmmat, 52, 351
Rochon prism polarizer, 253
Rotation matrix, 24
 for Jones calculus, 55
 for Mueller calculus, 51
rxmat, 26, 352
rymat, 352
rzmat, 26, 352

S

Scatter, 167
 anisotropy, 188
 dependence on deposition angle, 177
 from fluid patches, 188
 from herring-bone stacks, 177
 from stress-related cracks, 169
 in situ measurement, 175
 into the air, 168
 into the substrate, 175
 patterns on film, 174
 simple theory, 180
Scrambler, 275
Second, 16
Sellmeier equation, 256
SI units, 16
silver, 221, 353
Silver:
 anisotropy during deposition, 209
 computer modelling, 231
 refractive index, 221, 353
Single-cavity filter, 299, 312
Snell's law quantity, 43
smat, 62, 71, 354
solc, 280, 355

Solc filter, 279
  fan, 280
  folded, 280
Soleil-Babinet compensator, 266
Spectral range without overlap, 250
stack, 71, 326
Stokes parameters, 33
Stokes vector, 32, 34, 41, 49, 50
Stress:
  in birefringent films, 169
Structural hysteresis loops, 231
Structure fraction, 151
substrate, 71, 326
Substrate temperature controller, 118
system, 71, 326
System matrix, 62
  transformation property, 62

T

tao2, 146, 291, 294, 357
Telescope mirrors, 212
Thermal evaporator, 118
tio2, 146, 291, 294, 359
Total field, 59
Total transmittance, 241
Tourmaline, 241
Transmission, 77
  anisotropy, 209
  efficiency, 241
  phase change, 90
Transmittance coefficients, 80
  all media isotropic, 89
  crystal-crystal interface, 81
  isotropic bounding media, 85
Two-dimensional angular distribution,
    131

U

Uniaxial media:
  deposition, 119

Unpolarized light, 35
  representation of, 37

V

Variable phase compensator, 265
  Babinet, 265
  Berek, 267
  Soleil-Babinet, 266
Velocity of light in vacuum, 17
Void-defined media, 150
vretard, 120, 325
vscatter, 169, 325

W

Wave plate, 256
  achromatic, 259
  achromatic half-wave, 261
  achromatic quarter-wave, 261
  dual-wavelength, 257
  for tuning dye laser, 282
  harmonic, 259
  multiple order, 257
  uncompensated retardance, 261
  zero-order, 259
Wave vector, 14
  surface, 20
Wavelength, 14
Wavelength filter, 276
  fan Solc, 280
  folded Solc, 280
  for tuning dye laser, 282
  Lyot-Ohman, 276
  without secondary peaks, 285
Wavy anisotropic media, 123
Wide-field element, 264
Wire grid polarizer, 242
Wollaston prism polarizer, 253

Z

Zero reflectance condition, 297

Zero-order wave plate, 259
Zig-zag anisotropic media, 123
z0, 17, 363
zro2, 146, 291, 294, 298, 361